PERGAMON BIOLOGICAL SCIENCES SERIES Vol. 1

Editor: Peter Gray
University of Pittsburgh

Physiology of Plants and Their Cells

About the Author

James A. Goss (Ph.D., University of California at Los Angeles) is currently Associate Professor of Biology at Kansas State University and Treasurer of the Kansas Academy of Sciences. In the interest of his work, Dr. Goss has travelled extensively throughout the world and has written numerous articles, some of which have appeared internationally. His research in plant physiology has been sponsored by the U.S. Atomic Energy Commission, the U.S. National Institute of Mental Health, and the Kansas Agricultural Station. He has been awarded travel grants by the Ford Foundation, the U.S. Agency for International Development, and Sigma Xi. Dr. Goss has also held the positions of President and Secretary of the Kansas State University Chapter of Sigma Xi.

Physiology of Plants and Their Cells

James A. Goss

Division of Biology
Kansas State University

PERGAMON PRESS INC.

New York · Toronto · Oxford · Sydney · Braunschweig

PERGAMON PRESS INC.
Maxwell House, Fairview Park, Elmsford, N.Y. 10523

PERGAMON OF CANADA LTD.
207 Queen's Quay West, Toronto 117, Ontario

PERGAMON PRESS LTD.
Headington Hill Hall, Oxford

PERGAMON PRESS (AUST.) PTY. LTD.
Rushcutters Bay, Sydney, N.S.W.

VIEWEG & SOHN GmbH
Burgplatz 1, Braunschweig

Contribution No. 1047

Division of Biology and Kansas Agricultural Experiment Station

Printed in the United States of America
0–08 017036 6

Contents

Preface

THIS BOOK is only an introduction to a vast and interesting field of science, plant physiology. It is also a very dynamic subject. Change is the keyword and active research describes the activity of its adherents. One cannot read all of the research reports published weekly on this subject throughout the world. They are too numerous. In fact, it is not even likely that one can keep up with the review articles. This change is exciting and challenging. It is typical of our contemporary social and political life, but it also means that our knowledge of plant physiology is so vast that no text or set of books can entirely cover all of our knowledge of this subject.

Although this book has been written as a text for a beginning course in plant physiology, it will also be useful as a general reference for teachers and scientists who have not studied plant physiology before but now find themselves interested in certain aspects of the field, and for students of biology and agriculture who would like to know more about how the plant functions.

As a text, this book has not been written for plant physiologists, but rather for students who plan to use the knowledge of plant physiology to help them in their life's endeavors in other fields. It is a known fact that very few of the students enrolled in a beginning course in plant physiology ever intend to become plant physiologists. Still, many introductory courses are taught ignoring this fact. As a result, the student ends up with a great deal of confusion but with little that will be of use to him in his future.

Rarely, if ever, is there an introductory course in plant physiology which is not composed of at least some undergraduate students. Even when graduate students are present, it is the obligation of the instructor to teach the course on a level that is understandable for the undergraduates.

Most writers of such books as this ignore the fact, and are more concerned about the potential plant physiologist, who comprises less than 10% of the class, and ignore the majority. This book does not.

It has often been said that truth does not change, but only our interpretation of it. This is logical if truth is defined as that which is infallible and unchanging. Man is continually striving to find truth, to know what is true and what is not. However, he is plagued by the many inherent limitations. These limitations affect his ability to know the truth even when it exists.

The only world a man can really know is the world created for him by his senses. The things he sees, feels, smells, tastes, and hears, are the things he really knows. Everything else is merely symbolic, although useful in helping to explain the phenomena revealed by his senses. As an example, a seed can be kept under favorable environmental conditions and soon it will germinate. We can observe it through various stages of development including the seedling, maturity, reproduction, senescence, and, finally, death. Our senses reveal the plant in these various stages, so we are certain of this knowledge concerning it. However, a plant physiologist is not satisfied with this knowledge. He desires to explain the causes of these phenomena, and as a result must revert to symbolism. The chemical elements are merely symbols which allow chemists to explain phenomena. We shall never know if they represent reality. Perhaps other symbols would serve just as well.

And so the goal of the plant physiologist is to symbolically explain the cause of the phenomenon of life in the plant world. We might say that he is interested in knowing what makes the plant tick. His ultimate goal is to be able to interpret plant behavior. Since he will probably never truly know whether or not these symbols, which he uses to explain a cause, represent reality, the best he can hope to achieve is to find an arrangement of symbols (a theory or hypothesis) which infallibly explains these phenomena. This is the source of disagreement over our theories. This is why we often have several theories to explain the mechanism of a phenomenon and why we cannot always agree on the right one.

Since this book is written to give the student a firm foundation of knowledge on which to base his future studies, weight is not given equally to all theories. In fact, usually only the one which is most widely accepted is given. This gives the student some confidence in what he is learning, rather than leaving him in a state of confusion. This also gives the instructor an excellent opportunity to supplement the text, rather than to repeat it, in his lecture or recitation sessions. Using the text as a basis, he can build on the student's knowledge either by using the historical approach to the subject to show how various theories have developed, or he can use the experimental approach to show contemporary efforts that are being made in the field, or he can compare alternate theories. The student can, and should support the facts found in this book with his own observations, with supplemental reading, with laboratory experience where possible, and with lectures given by qualified teachers and scientists. However, be cautious to always keep the facts and hypotheses separated when learning about contemporary research achievements. Facts seldom change but hypotheses often do.

Plant physiology is a study of the living activity of the plant. It is the study of what makes a plant function. It is a study of the nutrition of a plant, of its metabolic activities, and of how this metabolism is regulated to cause the plant to grow, flower, and to produce seeds and fruit. It is the study of life itself. What could be more exciting? It can be a study that will help the student understand how to grow better and larger plants, to grow better flowers in his or her backyard, to get better yields from crops even under adverse conditions, to help alleviate pollution, and to understand why plants are so important in helping to solve many of the international problems with which we are faced at the present time. It will help us to lead this world in the future on a path of happiness and prosperity for all rather than to a state of impoverishment and starvation, from which we can never recover.

I've enjoyed writing this book, and I hope that you, the reader, will enjoy reading it.

I wish to offer my thanks to the many people who have helped make this book a reality. First, to the many past and present researchers who have discovered the facts presented in this book. Second, to my wife for typing part of this manuscript and to her and my children Larry, Lynn, Linda, Gerald, Liana, and Lori, for getting by without a husband and father for so long while the manuscript was being prepared. I wish, too, to thank Lillian Woolley for whose interest in writing stimulated my work on this book, and to Ted Barkley for the encouragement needed from the administration at Kansas State University.

I wish to extend my appreciation to the many authors and publishers who allowed me to use materials previously published. These will be specifically acknowledged with the presentation of their material in this book.

And last, but not least, I want to thank the staff at Pergamon Press for the help and encouragement they extended me. Perhaps this list of people to whom I owe a great deal appears extensive, but success in any endeavor can only be accomplished by the help of many associates.

Division of Biology   JAMES A. GOSS
Kansas State University

Chapter
One | # The Plant Cell
and Its Nutrition

VALUES OF PLANTS TO MAN

The assumption can be safely made that as long as man is on this earth he will be interested in, and in need of, plants. They currently serve many functions, such as food, drugs, raw materials for industry, shelter, as ornamentals whereby they satisfy some aesthetic value, and as an aid in the conservation of our natural resources.

Plants as Food

Within the next 50 years, plants will continue to serve as man's principal source of food. Synthetic diets are currently being produced that are changing and will continue to change our eating habits so that we may become less dependent upon natural foods, but these dietary components are themselves derived from plant materials. Of course, carbon dioxide can be taken from the atmosphere and used by man to produce organic chemicals that can be used for food, but the amount of energy required and the expense involved do not make this a practical undertaking, nor is there any hope of it becoming so in the future. Plants can produce foods much more cheaply and more efficiently, so they will continue to be our chief source of food and a great deal of research effort will be devoted to improving the plant and its environment. Plant breeding will develop better plants; a more favorable use of fertilizers will be practiced; pesticides will be improved and their use extended; supplemental irrigation will be used more extensively and plant growth regulators of numerous types will be developed and used.

Plants as Drugs

The use of drugs has increased tremendously within the past century and will continue to increase as rapidly, or more so, in the future, as the world population increases and as the level of civilization continues to rise in all countries. In the past, most of these drugs have come from plants. This source has certain disadvantages, such as problems of extraction and purification, so the future will find more synthetic drugs produced. However, the basic chemicals from which these are synthesized will largely come from plants.

Plants as Raw Materials

The uses of plant products as raw materials for industry certainly have not declined. In spite of the fact that less coal is used today than in the past, and less wood is being burned as fuel in the United States, the vast interest in and use of plastics and similar polymers, which may be made from plant products, more than compensate for the reduced use of plants as fuel, and there are good reasons to anticipate even greater demands for these polymers in the future. Also, the use of oil, a plant product, has increased greatly.

Plants as Wood

The use of wood for construction has been extensive in the past and will continue into the future. It is easy to work with, inexpensive, and a good building material especially for areas where the building may be subjected to stresses and strains, such as those due to earthquakes, strong winds, etc. Perhaps the limiting factors in the future will be the lack of available trees for lumber rather than the presence of more desirable building materials. Even today, the diameter of the trunks of trees being harvested for timber is smaller than in the past.

Oxygen Source

Both plants and animals, including man, must have oxygen to breathe or they will soon perish. In nature, many chemical reactions are occurring continuously that remove oxygen from the atmosphere. If no method existed by which the oxygen could be restored to the atmosphere, man would soon die. Plants provide that method of oxygen renewal. Through the process of photosynthesis, they are continually freeing oxygen and returning it to the atmosphere, where we can obtain it for breathing. If for no other reason than this plants are essential.

Plants as Energy Sources

We have considered so far only how plants are valuable because of the chemicals they produce, chemicals that we use. But plants are also very important for the energy they make available. It is difficult for man to grasp the importance of energy in his life, perhaps because energy cannot usually be seen and because it is usually so abundant that it becomes common and therefore uninteresting. Life itself is an energy-requiring and energy-using process. All animals, including man, must use energy continuously to survive. Any time this energy is not available, we die. The energy needed is supplied by plants, either directly or indirectly, through the foods we eat. Food is necessary not only for the chemicals it furnishes, chemicals that serve as building blocks for our bodies, but also for the energy the foods contain. Remember the calories?

Not only are we dependent upon plants for the energy needed for survival, but also for the energy needed to maintain our high standard of living, such as energy to drive our automobiles, energy required by industry, energy to drive our airplanes, trains, ships, etc., and energy to heat or cool our buildings. The statement has been made, and is certainly justified, that there is not enough available energy on earth today to allow all nations to have the standard of living that we enjoy. How will this problem be solved in the future? By the available energy going to the strongest nation?

Plants as Ground Cover

The importance of plants as ground cover is often ignored or un-recognized. At times, flowers are planted to cover up an ugly bare spot in our yards, but plants as cover crops provide even more important services, namely, those of soil and water conservation. How much soil would exist in much of the United States if it were not for plants? Certainly in the Eastern and Southern states there would be little left, since it would be eroded away by rain nearly as soon as the soil was formed. Only in desert regions, where rainfall is light, would soil cover the rock of the earth. This loss of soil would not only be due to water carrying the soil away as it runs off, but also to increased runoff. More of the water coming down in the form of precipitation would run off, with less sinking into the soil. Plants are very helpful in reducing the amount of runoff, aiding a greater percentage of the precipitation to enter the soil and remain as reserve water. So it is evident that even the weeds are helpful, since they serve as ground cover necessary to soil and water conservation.

There are a number of other important uses of plants, such as agents for the study of life, avocational subjects, and their aesthetic values. However, if one needed ample evidence to support the contention that plants are of practical value to us, such evidence has now been given.

Now, what has this to do with plant physiology? Of course, the answer is that plant physiology is the study of the functions of plants, so if you are going to study plants, knowing something of their value adds motivation to such a study.

WHAT IS PLANT PHYSIOLOGY?

Plant physiology is very much related to agriculture, and should be. If agriculture is defined as the science and art of the production of plants and animals that are useful to man, and since man is interested in all plants, then agriculture embraces any study of plants. However, this book is not just limited to those plants that are normally considered to be crop plants, but considers any plant that has served to reveal knowledge about plant physiology. Nevertheless, it is evident to one who has been interested in plant physiology for several years that relatively little is known about the physiology of most plants that are not crop plants.

Plant physiology is defined as the study of the function of plants, or, in other words, what makes plants tick. The study of plant physiology is a study of life itself. It is the study of the living plant in all of its living activities. It is an attempt to interpret this living phenomenon in such a way that it may be used to control the plant. We can then cause it to develop into the type of plant desired to better serve the needs of mankind. The process of life is not fully known today, but we are approaching this goal and are doing quite well.

What a plant is and what it may become depends upon two factors: the genetic makeup of the plant and the plant's environment. This principle is true of all living entities, including ourselves. The potential of each plant is set by the genes and cannot be surpassed. However, no plant, and no student, reaches his potential because of the environment which determines how closely the individual will approach its potential. However, genetic makeup and environment can act only as they alter the physiology of the plant, as illustrated in Fig. 1.1. For example, diseases, low temperatures, heat, etc. can only affect a plant by altering its physiology.

To be of value, plants must grow. We define growth as an increase in size, which is often accompanied by an increase in dry matter as well as an increase in water. This growth requires water, a source of energy,

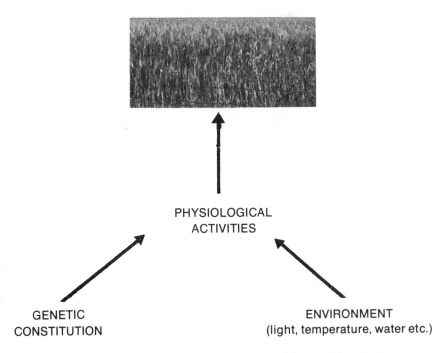

Fig. 1.1   An illustration of the relationship between physiology and plant development.

and a variety of nutrients. The source of all these is the environment. And yet, we cannot mix water, energy, and other nutrients from the environment and get growth. Such growth is possible only when controlled by the genes present in every cell of that plant. Although genes seem to be small and simple, we cannot do the things they can do, otherwise it would be possible to change the potential of a plant to make it grow to our specifications. Someday, as our knowledge increases, we shall be able to do this. Maybe you will show us how. This book will explain how these genes are able to control the physiology of the plant to help it towards reaching its potential.

THE LIVING CELL

This book assumes that the cell is the basic unit of life. This means that all living entities either consist of one cell or of a number of cells. If it does not contain one or more cells, it is not living.

To the question, what is a living cell?, the answer can be given that it is the smallest living part of any living entity. However, this does not tell much about the living cell.

Cell Similarities

In what respects are all living cells similar? By all living cells we mean those of both plants and animals, with no exceptions. Now, if you recall the properties of all the different cells studied in your previous biology courses and consider their diversity, you will probably throw up your hands and say that it is impossible to show any similarities. Cells are so different from one another. In fact, no two living cells are the same. Nevertheless, cells have similarities, too.

Some of their chemical constituents are similar, at least as far as the groups of chemicals found are concerned. For instance, all cells contain a great deal of water, and they contain carbohydrates, amino acids, proteins, lipids, nucleotides, and nucleic acids of one type or another. They do differ though as to the species of each of these chemicals they contain.

All cells have a definite physical structure, too, however this structure differs greatly from cell to cell. In fact, the only such structure that is common to all cells is the *plasma membrane* (plasmalemma), a thin, outer membrane made up largely of proteins and lipids. Structurally, the basic living cell consists of a colloidal system, with the dispersion medium of water in which are dissolved many chemicals, limited by a plasmalemma. Such a cell is shown in Fig. 1.2. I do not wish to insinuate that this basic cell represents the ancestor of all cells nor that this is the most primitive cell. However, it is the simplest cell structurally and can perform all of the activities necessary for life that the more complex cells perform. In view of this fact, one might wish to ask why cells of more complexity exist. There is no necessity of this complexity as far as the cell is concerned. However, it must be noted that all cells of similar simplicity are very small cells, much smaller than the average cell size. They include the unicellular organisms known as bacteria and cyanophyta, cells known as prokaryotic cells. *Prokaryotic cells* are cells almost devoid of internal physical structure, and although they are intriguing, this text will not be further concerned with them. If you wish to learn more about them, consult a book on microbial or bacterial metabolism.

If, as indicated above, a cell is nothing more than a colloidal system surrounded by a plasma membrane, it may be possible to construct such a structure. However, would it be alive? This brings up the interesting question, what activities must be associated with a living cell?

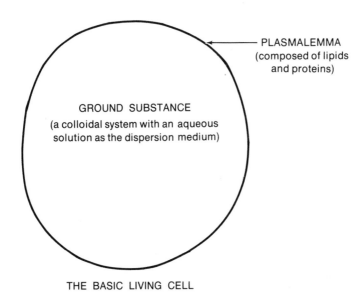

THE BASIC LIVING CELL

Fig. 1.2

All living cells perform a number of activities. Which of these activities are common to all? Only the following are: First, cells must be able to exchange certain chemicals with their environment without losing excessive amounts of the chemicals they contain. This is in part the function of the plasma membrane. Second, they must take up and use energy. Third, they must be able to synthesize the chemicals they need and to replace those that they lose, either to their environment or through destruction. Fourth, they must be adaptable so that they can change as their environment changes. Finally, every living cell must eventually either divide or fuse or it will die. Any time it cannot do the activities listed above, the cell dies. Our basic living cell diagrammed in Fig. 1.2 is capable of performing these activities.

Cell Differences

Barring mutation, all cells of an individual should have the same genetic makeup. However, no two cells would have the same environment. Thus, one would expect that differences among cells of the same individual would be those imposed upon it by the environment, whereas differences between cells of two individuals would be due to differences caused both by differences in genetic makeup and environment. In any respect, each cell is different from every other cell.

That each cell is different can be demonstrated in various ways. If one germinates pollen grains in a nutrient medium and observes them as they germinate, not all pollen grains in a culture germinate at the same time nor do the tubes formed grow at the same rate or to the same extent. One might also consider the population effect. If one pollen grain is placed in the germination medium, it usually will not germinate. If a number of pollen grains is added, even from the same anther, the original pollen grain germinates. The percentage of germination increases with the increase in the size of the population in a medium. This illustrates the *population effect*. It can best be explained by assuming that certain chemicals are necessary for germination and that these chemicals are produced by the individual pollen grains. However, each produces different amounts of a given, needed chemical. Some produce amounts inadequate for germination, such as our original pollen grain did. Others produce this in excess. When these pollen grains are placed in the same medium, some of the excess produced is lost to the medium and moves to the surface of the deficient pollen grain and is subsequently taken up by it, supplying the deficient quantities and thereby allowing the original pollen grain to germinate. Each cell accumulates different amounts of a given metabolite than any other cell.

Another source of differences in cells arises from the differences in the number of organelles within each cell. Individual cells differ in the number of mitochondria per cell, or number of ribosomes per cell, etc. These differences should make the cells different. Plant cells also differ in the size, shape, and pattern of the cell wall. The above evidence does suffice to demonstrate that cells do differ, and is ample to show that no two cells are exactly alike.

The origin of these differences can be explained by the theory of evolution, a theory which no good biologist will deny. In order to form a new species, two requirements must be met. First, mutations are necessary, and second, there must be an isolation of gene pools.

The fundamental activities given above are required by all living organisms, but there is no one mechanism by which all plant cells perform these. It is irrelevant how they are carried out. The important factor is that the activity be performed. Let us assume that the first living cell was able to carry out these activities in a specific manner. When the next species was evolved from it, its gene pool was isolated, but mutations continued. Therefore, it was possible that other mechanisms evolved to perform the same activities, and such variations increased as more species formed. As more and more species were formed, each was required to carry out the same activities but new ways of doing this evolved in some and not in others. This has led to the differences that exist among plants, differences that may be gross or minute. On the other hand, certain similarities are widespread, since mutations have not been instrumental in eliminating these mechanisms. Some mechanisms are not eliminated as alternates evolve. Therefore, a plant may have the ability to use any one of several pathways to accomplish a specific activity. In fact, a specific function may be carried out by more than one pathway in the same individual at the same time. For instance, respiration may be occurring via two different pathways at the same time even in the same cell. Life is wonderfully diverse.

Now, what does this have to do with a book such as this? The above discussion explains why this book is not a compendium of mechanisms that occur within the plant or plant cell. Rather, I have tried to find the mechanism or theory that is most common or about which we know most, and presented this as the way in which the plant entity performs a given function. If the previous contentions are true, we are wasting our time searching for one mechanism of photosynthesis or of respiration, or of permeability or of any other physiological activity. The history of research achievements begins with the discovery of one "universal" mechanism and is later altered as additional mechanisms, for the same function, are discovered. If you wish to learn other mechanisms, you can do so by supplemental reading, taking more advanced courses or, by experimentation.

CELL CLASSIFICATION

Perhaps, in view of the fact that no two cells are alike, the reader may conclude that any system of classification of cells is useless. If cells are not similar they cannot be classified, since systems of classification depend upon similar characteristics among the cells. Of course, any classification system is artificial, and none can be devised that has no exceptions, but such systems are useful in aiding our understanding of and in allowing us to converse about cells. Due to the many differences among cells, there are many systems of classification that could be devised and most would be equally valid. One such system for plant cells is given in Table 1.1.

The cell diagrammed in Fig. 1.2 is a typical prokaryotic cell, but shall

Table 1.1   A system of plant cell classification.

Cell Types	Where found	Specialties
	Unicells	
Prokaryotic cells	bacteria, cyanophyta	general
Eukaryotic cells	algae, fungi	general
	Eukaryotic Tissue Cells	
Meristematic cells	plant meristems	cell production
Parenchyma cells	root cortex, leaves, etc.	general, photosynthesis
Collenchyma cells	leaves, stems	support, general
Fiber cells	vascular bundles, stems	support
Vessel cells	xylem	conducting
Tracheid cells	xylem, stems	conducting
Schlereid cells	bark, seeds	protection
Sieve tube and companion		
cells	phloem	conducting
Gamete cells	pollen, spores, ovules	reproduction
Guard cells	leaves, fruits	control of water loss

not be further studied in this text, other than to occasionally refer to physiological studies that have used it. This text shall be mostly concerned with the eukaryotic cell, both eukaryotic unicells and the eukaryotic tissue cells. The structure of a eukaryotic tissue cell is diagrammed in Fig. 1.3. Note how it differs from Fig. 1.2.

PLANT CELL STRUCTURE

The eukaryotic tissue cell is a very complex unit composed of numerous structures, some of which are visible under the light microscope. These structures are not physiologically independent but interact remittently. However, for convenience in discussing the various parts of the cell, it is possible and customary to assign names to various cell parts, as outlined in Table 1.2. For instance, the cell can be said to be composed of cell wall and a protoplast.

Cell Wall

The cell wall is the rigid layer of polysaccharides and related chemicals which surround the protoplast. Some authors do not consider the cell wall to be part of the living cell but rather a non-living, external secretion of the cell. However, if one considers only those structures within the cell that are essential to the life of the cell, one could justify excluding mitochon-

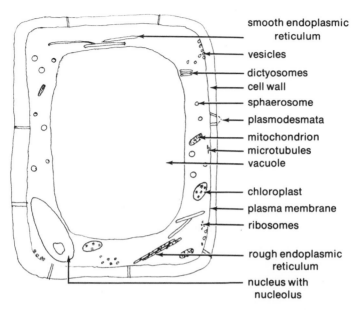

smooth endoplasmic
reticulum

vesicles

dictyosomes

cell wall

sphaerosome

plasmodesmata

mitochondrion

microtubules

vacuole

chloroplast

plasma membrane

ribosomes

rough endoplasmic
reticulum

nucleus with
nucleolus

Fig. 1.3   A drawing of a thin, longitudinal sectional view of a plant eukaryotic tissue cell.

dria, nuclei, etc. on the basis that these are not necessary for prokaryotic cells. Since essentially all plant cells have a cell wall, let us consider the wall as an integral part of the cell. Therefore, this book shall consider the cell as being composed of the cell wall and protoplast, but shall also recognize that some cells do not have a cell wall, in which case the protoplast is the cell.

The cell wall is porous and offers little resistance to chemicals that move through it. In addition, in tissue cells, the wall is penetrated by strands of protoplasm that connect adjacent protoplasts. These are called *plasmodesmata* and appear in all cells of multicellular plants, but do not occur in either prokaryotic or eukaryotic unicells. They are formed at the time the cell divides when the cell wall begins to form. Each plasmodesma is a membranous tube probably filled with an aqueous-colloidal solution,

Table 1.2   An outline of cell structure.

cell = cell wall + protoplast
protoplast = cytoplasm + nucleus + vacuome
cytoplasm = plasma membrane + mesoplasm + vacuolar membrane
mesoplasm = ground substance + inclusions
ground substance = ectoplasm + endoplasm

but it is doubtful that plasmodesmata contain other membranous cell structures. Perhaps they function as tubes through which adjacent proto-plasts exchange materials.

Basically, the physical properties of the cell wall are due to the cellulose molecules it contains. These cellulose molecules form the wall skeleton and a network on which other substances, such as lignin, are deposited. Walls of adjacent cells are cemented together by polyuronides, especially protopectins and pectates, which form the *middle lamella* of the cell wall. The cellulose framework is deposited on the surfaces of the middle lamella.

The cell wall differs greatly in thickness and morphology from cell to cell. Such differences form the basis for much of the classification scheme in Table 1.1. Some cells have a thin wall, cells such as meristematic and parenchyma cells. Their walls consist of the middle lamella and the *primary wall*. Other cells have thicker walls with the walls being relatively thick in vessel elements, and extremely thick in fiber, tracheid, and stone cells. In these there is an additional layer or more of cell wall material deposited on the primary wall, and this is called the *secondary wall*. However, even in a given cell the wall is not uniformly thick around the cell surface, but often has areas known as *pit fields*, where the wall is thinner than normal. These pit fields often have a number of plasmo-desmata penetrating them. The cell wall will be studied in more detail in subsequent chapters.

Cell Protoplast

The protoplast consists of the cytoplasm, a nucleus, and the vacuome. The *vacuome* consists of one or more vacuoles. In mature plant cells there is often just one vacuole present and it often comprises most of the volume of the cell, but in immature cells numerous small vacuoles may be found. The vacuole is an aqueous solution composed of various inorganic and organic chemicals dissolved in water. Sometimes it is colored due to the presence of pigments such as the anthocyanins, but often it is colorless and transparent. It is doubtful if active enzymes exist in the vacuole so the organic chemicals found there must be synthesized in the cytoplasm and secreted into the vacuole. Such organic chemicals include metabolites and toxic materials. The inorganic chemicals are stored in the vacuole after entering the cell from the external environment. The vacuole can be therefore considered as a metabolic pool. The pH of the vacuole is often quite low, being about 2–3. This is in contrast to the pH of the cytoplasm which is believed to be about 7. Since the vacuole is quite acid, when the cell is ruptured, as is often done to extract enzymes and chemicals for study, the acidity often destroys cytoplasmic enzymes and causes the chemicals of the cytoplasm to change into other chemicals. Such changes are often prevented by rupturing the cells in the presence of a buffer, such as calcium carbonate.

Cell Nuclei

The number of nuclei in a plant cell will vary from none to many. Prokaryotic cells and the sieve tube elements have no nuclei, whereas the cells of certain fungi and certain algae have many nuclei per cell. However, most tissue cells have one nucleus. This nucleus is large and often spherical. It is limited on the outside by a *nuclear membrane*, which differs from most other cell membranes in that it is single layered and is permeated with large pores. The nuclear membrane separates the nucleoplasm (nuclear sap) from the cytoplasm, and is present in the cell except during certain stages of nuclear division. This *nucleoplasm* is somewhat granular but is perhaps mostly a colloidal system. Within the nucleus can be found certain structures suspended within the nucleoplasm. *Chromatin* is present during interphase. This is stringy material that stains darkly and appears to be transformed into the chromosomes. In fact, it has been isolated from the cell and shown to act as chromosomes do in the synthesis of messenger RNA (*m*-RNA) that controls protein synthesis. The production of this *m*-RNA is the prime function of the chromatin and more will be said about this in Chapter 10. During the prophase of cell division, the chromatin contracts to form the chromosomes. After cell division, the chromosomes again change into the chromatin material.

Cell Nucleoli

Also present in the nucleoplasm is one or more *nucleoli*, with the actual number per cell being specific for each plant species. These structures are often dense, spherical, and often contain one or two vacuoles. The nucleoli are formed on a specific area of a specific chromosome during telophase, and part of a chromosome or chromatin remains attached to and is part of the nucleolus. They disappear during late prophase of nuclear division and are absent until telophase when they reform. Their size is similar in any one type of cell, but is larger in young cells and in leaf cells in the light, perhaps due to the influence of nutrition on nucleolar size. Size also increases with the degree of ploidy, or with the number of sets of chromosomes. The function of the nucleolus appears to be the synthesis of ribosomal RNA and ribosomes. These *ribosomes* will be described in more detail later, but consist of about equal amounts of ribosomal RNA and protein and are found in the cytoplasm, nucleoli, mitochondria, and chloroplasts. They function in protein synthesis by representing the site of polypeptide synthesis.

Whereas the vacuole is acid and the cytoplasm is neutral in pH, the nucleus often has a pH above 7. This is due to the presence of histones; proteins that have more basic groups than acid groups.

Cell Cytoplasm

The cytoplasm is limited on the outside by the plasma membrane, on the inside by the vacuolar membrane, and between these two is the *mesoplasm*. The plasma and vacuolar membranes are typical cell membranes in that they are composed of proteins and phospholipids and are double layered. The organization of the protein and lipid molecules in the membrane is not known but the leading theory postulates a structure as shown in Fig. 1.4. These membranes are differentially-permeable mem-

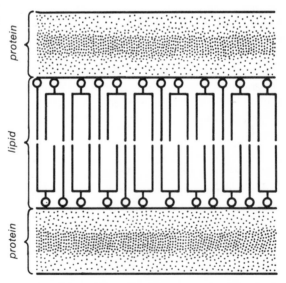

Fig. 1.4   Diagrammatic representation of membrane structure. (From Thompson, G. A. Jr. in *Plant Biochemistry* by Bonner and Varner. Academic Press, 1965. Reproduced with permission.)

branes which means that they allow certain chemicals to pass through them and restrict the passage of other chemicals. Generally small molecules and those that are lipid-soluble pass through these membranes more rapidly than do others. More will be said about these membranes in subsequent chapters. They seem to be formed from vesicles which move to the membrane surface and become incorporated into the existing membranes.

Cell Mesoplasm

The mesoplasm consists of the *ground substance* and its inclusions. The ground substance is an aqueous solution with a pH value near 7. It has a variable viscosity that can change from a viscosity near that of water to that near a solid gel. Generally it is more viscous near the periphery and this area of the ground substance is referred to as the *ectoplasm*. The less viscous, inner area is called *endoplasm*. The ground substance demonstrates the remarkable phenomenon of protoplasmic streaming and represents the site of the mechanism of this streaming.

Suspended within the ground substance are numerous cell organelles whose structure consists largely of membranes. These organelles vary considerably in size, shape, and numbers, both within a given cell and from cell to cell, adding to the cell's individuality. Some, including the plastids, mitochondria, and sphaerosomes, are visible with the light microscope. Others, including the microtubules, vesicles, dictyosomes, endoplasmic reticulum, and the ribosomes require the electron microscope to be seen.

Cell Plastids

The *plastids* are the largest of these suspended organelles, being 4–6 μ in diameter and lens-shaped. They are found in all plant cells but are absent from animal cells, and thereby help to characterize plant cells. Plastids do not arise *de novo*, but are formed either by division of pre-existing plastids or by division and differentiation of proplastids. They are classified according to their most obvious contents into chloroplasts, amyloplasts, or chromoplasts depending upon whether they contain large amounts of chlorophyll, starch, or carotenoids respectively. All plastids are interchangeable and therefore can change from one type of plastid to another, but all of the plastids of a given cell are of one type. Basically, the plastid consists of a plastid membrane which is a double membrane, like the plasma membrane, limiting the plastid. Inside this membrane is the *stroma* or transparent colloidal solution. Embedded in the stroma may be starch grains, globules, or lamellae.

Amyloplasts have one or more starch grains embedded in the stroma. They are especially abundant in storage tissues such as roots, cotyledons, endosperm, and tubers, and function as sites of starch storage. Here starch is formed when sugars are present in excess and starch is broken down when sugars become deficient. They may also function in root tips in geotropism which causes the roots to grow down instead of up.

Chromoplasts are plastids which contain large amounts of carotenoid pigments. These pigments give the yellow, orange, or red colors to fruits, to certain roots, such as carrot and sweet potato, and rarely to flowers. The carotenoids are associated with the globular lipoidal and the fiber structures embedded in the stroma of the chromoplast. They function by coloring the fruits and some flowers as an aid to reproduction. Chromoplasts contain all of the carotenoids found in chloroplasts as well as many not so found. The red color of tomato fruits is due to chromoplasts which contain 87% lycopene, 7% β-carotene and 3% phytofluene. The chromoplasts of red pepper fruit contain mostly capsanthin, but a large variety of other carotenoids are also present.

Cell Chloroplasts

Of all the plastids, the *chloroplasts* have been studied most intensively because it is here that photosynthesis occurs. The chloroplasts may number one or many per cell. They are not found in all plant cells, but are located in most algal cells, and in the mesophyll cells and guard cells of leaves, and in some of the outer parenchyma cells of herbaceous and woody plant stems. Most of the cells of green plants do not contain chloroplasts, and in those that do, the number of chloroplasts per cell varies from cell to cell, even in the same multicellular plant.

The chloroplast too is limited by a double membrane and contains stroma as do other plastids. However, embedded in the stroma of the chloroplast is a dense network of membranous structures called the *lamellae (see* Fig. 1.6). These membranous structures are composed of flattened membranous sacks that look much like pennies and are called disks or *thylakoids*. These thylakoids pile up to form dense structures known as *grana*, with 40–60 grana being present in each chloroplast. Such grana can be seen with the light microscope, but electron microscopy is needed to see the thylakoids. Some of the thylakoids extend between the grana joining adjacent grana to interconnect all grana in the chloroplast. These interconnecting lamellae are called stroma lamellae.

The chloroplasts are capable of self-reproduction by division. They are also formed from proplastids. In their development from proplastids, the chloroplast membrane develops first and is a double membrane similar to that of the plasma membrane. Then membranes develop within the chloroplast, after the plastid is exposed to light, and these become oriented to form dense grana, connected by less dense membranes and bathed in the stroma of the chloroplast. It is on the surface of the lamellae that the chloroplast pigments, including chlorophyll, are found and where the light reaction of photosynthesis occurs, while the dark reaction of photosynthesis occurs in the surrounding stroma.

Ellipsoidal starch grains, one to four per chloroplast, also exist in the stroma of the chloroplast, if the cell has recently been kept in bright light, but not if kept in the dark. Apparently this starch arises from photosynthesis.

Also embedded in the stroma of the chloroplast are ribosomes which function as sites for protein synthesis. The chloroplast does contain a great deal of protein since about 70% of the dry weight of the chloroplast is protein. Another 21% is lipoidal with about 0.02–0.1% DNA and 1.0–7.5% RNA. Details of the chemical composition of chloroplasts will be considered with the discussion of photosynthesis in Chapters 6 and 7.

It is interesting to consider that one of our theories of chloroplast origin in the first cell states that chloroplasts were originally unicellular organisms and that they invaded non-green cells to form a successful symbiotic relationship that has existed for much of geological time. Support for this theory comes from the self-duplicating nature of the chloroplasts. Perhaps someday we shall be able to culture chloroplasts free of the cell.

Cell Mitochondria

When the author attended high school, he was told by his biology instructor that mitochondria were present in all cells but that they had no function to perform within the cell. Now we know differently. They have a very important function to perform, namely, they are the site of cell respiration. They represent the area of energy release within the cell, energy that is essential at all times if the cell is to remain alive, just as power plants serve as sources of energy for civilization. Mitochondria are sometimes referred to as the power plants of the cell.

If one looks at an electron micrograph of mitochondria, (see Fig. 1.5), the external features of these structures can be seen to be of uniform

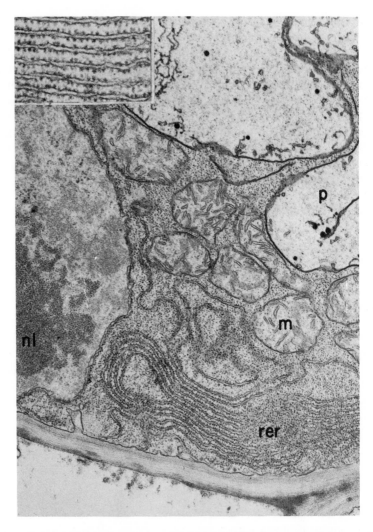

Fig. 1.5 Rough endoplasmic reticulum and mitochrondria of pea root cells as revealed by electron microscopy. (After Bouck, G. B. Stratification and Subsequent Behavior of Plant Cell Organelles, *Journal of Cell Biology*, 1963, **18**: 441. Reproduced with permission.)

shape and size within a given cell and remarkably uniform among plant cells generally. In size they are smaller than the plastids but larger than sphaerosomes. They are somewhat ellipsoidal with a double membrane. The outer membrane is pitted with pits of irregular size and distribution, and surrounds the mitochondrion. The inner membrane is much invaginated to form both a wall around the inside of the mitochondria and projec-tions or *cristae* that extend out into the stroma and at times even bridge the diameter of the mitochondrion, or join ends to form rings. These cristae vary in number per mitochondrion but are more numerous in cells with high respiration rates, and do appear less uniform in size and orienta-tion in plant cells as compared with those of animal cells. Their surface is covered with numerous, minute knob-like projections of unknown function.

Now if one compares this structure with that of a mitochondrion in a living cell, and especially one in which protoplasmic streaming is active, one is in for a surprise. In the living cell no two mitochondria are the same size or shape. Not only that, but their size and shape vary continuously as they interact among themselves and with other cell organelles. They may be thread-like, dumbbell-shaped, or round, and often branched. Also they can be seen to divide often, as if pulled apart by mechanical forces.

Mitochondria are found in all cells, whether they be plant or animal. They are located in the cytoplasm, suspended in the ground substance, and their number per cell varies from a few in the generative cell of pollen to several thousand in some plant cells. No doubt their number varies from one cell to another even within the same multicellular plant, giving individuality to each cell. They are more numerous in cells that have high respiration rates.

Mitochondria are formed by division of pre-existing mitochondria. They synthesize much of their own DNA and RNA, so they are nearly autonomous. They divide by fission and such division can be observed under the light microscope. Perhaps someday we shall be able to culture them *in vitro*. We cannot at the present time, but they can be rather easily removed from the cell and will continue functioning as sites of respiration for several hours after isolation. As was observed with chloroplasts, there is also the possibility that mitochondria originated as a cell that became parasitic to another cell and has since existed by this continued symbiotic relationship.

The substrate for the mitochondrion is pyruvic acid, which it oxidizes, giving off carbon dioxide, water, and ATP, an energy carrier. This is also where most of the carbon dioxide comes from, which we expel from our bodies when we breathe.

Cell Sphaerosomes

Sphaerosomes are the most numerous organelles that can be seen in a cell with the light microscope. They are smaller than the mitochondria but vary considerably in size even within the same cell. Their shape is spherical and they are usually in motion due either to Brownian movement or protoplasmic streaming. They correspond to the *lysosomes* of animal cells.

Structurally, the sphaerosomes are limited by a membrane, but this membrane differs from the membranes of most cell organelles since it is only one layer in thickness. Most of the volume of the sphaerosome is filled with lipid. In fact, over 98% of the sphaerosome is made up of lipid materials. Perhaps they represent sites of lipid storage although they apparently are not able to synthesize the lipids they contain.

Cell Ground Substance

If one looks at the ground substance through the light microscope it appears that it is transparent and consists only of an aqueous solution. However, by use of the electron microscope to give greater magnification and better resolution, it soon becomes evident that such is not the case. The ground substance does have a structure, and this structure is composed of a network of minute but long and interconnected membranous tubes called the *endoplasmic reticulum*, abbreviated ER. This is found in all cells, and can be seen in Fig. 1.5. Some of the ER are smooth and some are rough. When rough, it is because the surface is covered with numerous spherical ribosomes or polyribosomes arranged in definite patterns. The ER gives rise to the vacuolar membrane by swelling and pinching off and possibly also to the nuclear membrane and plasmodesmata. The walls of the ER consist of the double membrane similar to that found in most other cell organelles.

Cell Ribosomes

The *ribosomes* are composed of about equal amounts of protein and ribosomal RNA. They are often spherical and are found either bound to the surface of the ER or free in the ground substance. Some of them are joined together to form larger structures called *polyribosomes* or polysomes. Polyribosomes are believed to be a number of ribosomes held together by an RNA molecule and undergoing polypeptide synthesis. The ribosomes are synthesized by the nucleoli of the nucleus and probably move out into the cytoplasm through the nuclear pores. They function as sites of polypeptide and consequently protein synthesis. They are also found, and believed to be synthesized, in chloroplasts and mitochondria.

Cell Dictyosomes

The Golgi apparatus or *dictyosomes* are tubes that are usually shorter than the ER and often of larger diameter. They do not have ribosomes attached to their surface, and are fewer in number than the ER. They function by producing vesicles, which form at the ends of the dictyosomes and pinch off from them thereby being released into the cytoplasm. These vesicles are of various sizes and spherical, much like the spaerosomes but much smaller. They are limited by a double membrane and are transparent inside. They function in plasma membrane and cell wall synthesis. It is believed that they contain chemicals that contribute to cell wall synthesis and they move to the site of wall synthesis where they coalesce with the plasma membrane extruding their contents to the outside where these materials are used in wall synthesis. The vesicle membrane becomes an integral part of the plasma membrane thereby adding to this membrane and perhaps this is how the plasma membrane is formed.

Cell Microtubules

Also present within the cytoplasm and in the nucleus are numerous *microtubules*. In the nucleus these tubules function in bundles to form spindle fibers which cause chromatid migration during nuclear division as will be discussed in a later chapter. Their function in the cytoplasm is unknown although they seem to become oriented just within the plasma membrane and one theory assigns to them the role of orienting the cellulose molecules in the cell wall during cell wall synthesis. Microtubules can be seen in Fig. 1.6. They are composed largely of protein and possibly originate from microfilaments.

In the last few pages, an attempt has been made to either introduce the student to the structure of the plant cell or to review this structure with him. In future discussions, mention will be made of such structures and more details will be given, especially in regards to their functions. However, it is well to keep in mind the location in the book of this information as you may wish to refer to it periodically for a review.

INTRACELLULAR TRANSPORT

Before embarking on the subject of cell nutrition, since nutrition involves not only the uptake of nutrients by the cell but also the distribution of these to the cell organelles where they will be used, it is well to consider

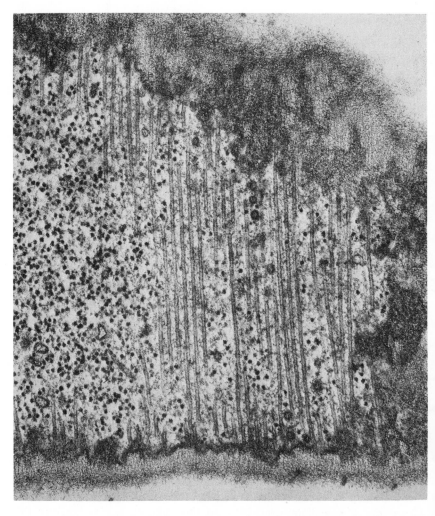

Fig. 1.6 Electron micrograph of microtubules running parallel beneath the primary wall. (Courtesy of Newcomb, E. H. in *Ann. Rev. Plant Physiol.*, Ann. Rev. Inc., 1969.)

how these materials move within the cell or a study of intracellular translocation.

After a nutrient molecule moves through the plasma membrane it can get to a specific cell organelle either by its movement to the organelle or by the organelle moving to the nutrient. No doubt both activities occur within the living plant cell.

Diffusion

The simplest mechanism by which molecules move within the cell is by *diffusion*. Diffusion is a passive mechanism, meaning the required energy is not derived from metabolic activity of the cell but rather comes from heat in the environment. The higher the temperature, the greater the kinetic energy of each molecule and therefore the more rapid the rate of diffusion. Movement by diffusion is from an area of high concentration to one of lower concentration of the chemical concerned. It is nevertheless a slow process and applies only to small molecules. The mechanism of diffusion is explained in Chapter 2.

Diffusion is caused by the kinetic energy of the diffusing molecule. Each molecule contains kinetic energy. Because of this, each small molecule is moving. As it moves, it strikes other molecules and larger suspended materials. If one looks at organelles such as sphaerosomes, one observes that they are moving about first in one direction and then in another, changing direction very often. This is due to their bombardment by smaller moving molecules which knock them first in one direction and then in another. *Brownian movement*, named after Robert Brown who first reported it, is the result of this bombardment. It does not result in a net movement in any one direction, but does keep a suspended organelle in constant motion allowing it to come in contact with a greater nutrient supply. The extent of motion decreases as the size of the organelle increases, having little effect on plastids or nuclei. Brownian movement occurs in both living and dead cells but is more evident in dead cells.

Protoplasmic Streaming

Perhaps the most interesting and extensive movement within the cell is *protoplasmic streaming*. By protoplasmic streaming not only do nutrients move great distances within the cell and thereby become widely distributed, but also all of the cell organelles move constantly throughout the cell when protoplasmic streaming is active. This is no doubt one of the most important living phenomena. It is occurring at all times in all living cells so if protoplasmic streaming cannot be observed, one cannot tell whether the cell is alive or dead, but it does not occur in dead cells and therefore if it is occurring, the cell must be alive. Diffusion and Brownian movement occur in dead and in living cells but protoplasmic streaming only occurs in living cells.

Protoplasmic streaming refers to the flow of cytoplasm inside the cell. It can be observed in several different patterns, one or more of which may be occurring in a cell at a given time. These patterns are illustrated in Fig. 1.7. One pattern is called agitation or turbulent motion. Agitation is the most common type and the least organized. In fact sometimes it is difficult to separate this type from Brownian movement. The difference is that with agitation, movement is statistically directional whereas in Brownian movement, movement is non-directional. It can often be observed in pollen tubes and in many other cells.

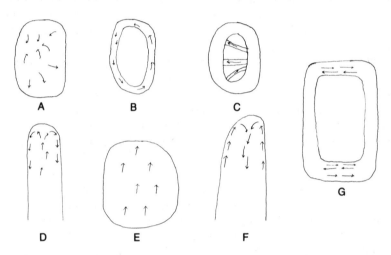

Fig. 1.7   A drawing to illustrate the types of protoplasmic streaming in plant cells. A = agitation, B = rotation, C = circulation, D = fountain, E = tidal, F = counter-fountain, and G = multistriate.

The second type is rotation. This occurs primarily in cells where the protoplasm is limited to the outside of the cell, a condition that is usually the case in mature cells. No strands of cytoplasm are evident and the cytoplasm is found only just inside the cell wall. It too is common in pollen tubes as well as in certain other cells.

The third type of streaming is called circulation. In this case, movement is along strands of cytoplasm that extend through a large vacuole. This pattern is common in pollen grains during tube development and in many other cells that have a large vacuole.

The most interesting pattern to observe is the fountain and counter-fountain types of streaming. These are particularly evident in growing pollen tubes. In the fountain type, the cytoplasm moves up through the center of the cell toward one end, and then flares out and moves down the sides of the cell just inside the cell wall. The counter-fountain pattern occurs when the cytoplasm moves along the cell just inside the cell wall, and near one end flows to the center of the cell and down the center.

Often one observes a multistriate type of streaming. In this case, one can see cytoplasm flowing along definite and adjacent tracks through the cytoplasm. Multiple channels exist and the direction of flow will vary from one channel to the next within the same cell. Tobacco hair cells and others illustrate this.

In fungi, one can often observe the type of streaming known as tidal streaming wherein the mass of cytoplasm within the cell will flow first in one direction and then reverse and flow in the opposite direction. This is not only a means by which nutrients move to organelles, and organelles move to nutrients, but is often associated with movement of the entire cell.

The rate of streaming is extremely variable, varying from near zero to in excess of one millimeter per second. It sometimes appears to be very rapid as one observes this under the microscope. However, since the rate increases as the size of particle observed decreases, and since this rate cannot be determined with electron microscopy as this kills the cell, we actually do not know what the maximum rate may be.

The rate of streaming has been shown to be influenced by many environmental factors. Also, the cell must be alive and carrying out respiration so the energy for streaming must come from that released during respiration. In most cases, the mechanism seems to be located at the boundary between the sol and gel of the ground substance. If you have not observed this streaming in cells you should certainly do so. It is definitely a good means of intracellular translocation.

CELL NUTRITION

The subject of plant nutrition will be the concern of the next several chapters of this book. It can be defined as the uptake and utilization of chemicals by the plant. There are a number of chemicals that the plant requires for its normal growth and development, just as we require certain nutrients for ours.

For its nutrition, the plant cell needs a source of energy, a source of carbon, and a source of certain inorganic elements, known as mineral nutrients. Those cells which contain chlorophyll can capture the energy of sunlight as their source of energy and use inorganic carbon dioxide from the atmosphere as the source of carbon. Those cells which do not contain such chlorophyll must obtain, from their environment, certain organic compounds which can be used both as a source of energy and as a source of carbon. Sugars usually function in this capacity, but in certain cells, fats, amino acids, and other molecules may suffice.

It should be noted that the green plant can do very well with a few mineral nutrients, water, and oxygen and carbon dioxide as nutrients. However, this does not mean that each cell of that plant has the same nutrient requirements. Indeed, most of the cells of that plant, if it is one of the *Tracheophyta*, do not contain chlorophyll so they must have sugars also, and some even require other materials, especially hormones. All cells of the multicellular plant are interdependent. The green cells form excess sugars, which are transported to the colorless cells for their use or for storage, and the latter cells may also function as part of the transport system of the plant and in this way support the green cells.

Chapters 2, 3, 4, 5, and 6 will emphasize nutrition of the multicellular plant, and subsequent chapters will be concerned more with the utilization of nutrients.

Chapter Two | Water Enters the Plant

Plants are composed largely of water. If we obtain plants from the field or greenhouse their water content can be determined in the following manner. First weigh the plants, then place them in an oven overnight at a temperature of about 100°C. Next morning again weigh the plants. If the initial weight was 100 grams and the final weight 15 grams then this indicates a water content of 85% of the fresh weight, as illustrated in Table 2.1. This is the water content often found in plant leaves or in herbaceous plants, and indicates that the plant is composed largely of water. That this high water content is necessary is evident from the fact that if the water content is reduced, the plant will wilt and eventually die.

Submerged aquatic plants, such as *Elodea* and some algae, obtain water directly through their epidermal cells. Some land-plants obtain much water by absorbing it through their foliage or stems, such as in the case of the coast redwood and some desert and tropical plants that obtain much of their water from the fog or dew. However, most of the water found in plants enters the plant through its roots. Since these roots are in the soil, it is reasonable to assume that the water required by a plant is obtained from the soil in which the plant is growing. The soil also serves as a source of mineral nutrients, so the soil is the source of most of the foods

Table 2.1   Determination of the water content of plant leaves.

Fresh weight of leaves	100 grams
Dry weight of leaves	15 grams
Weight of water	85 grams
Water content = 85% of the fresh weight	

needed by plants. Since what a plant is and what it may become depends upon its genetic makeup and upon its environment, and since most of the foods required by the plant come from the soil, it is necessary to learn something about the soil if we are to understand the physiology of the plant. Actually much of the plant has the soil as its environment.

SOIL STRUCTURE

Soil is a complex mixture of many substances. Pick up a handful of soil and study it closely, and you will observe that the soil is made up of *soil particles*. These particles vary in size, shape, and composition. They are composed either of *mineral matter* or of *organic matter*. The organic matter is the complex chemicals of leaves, roots, and other plant parts that have resisted decay after falling to the ground. As these plant organs decompose, those chemicals that decompose easily disappear first and eventually all that remain are the chemicals that are very resistant to decay. We call this organic matter of the soil, the *humus*. If observed under the electron microscope the humus can be seen to consist of very small particles about 0.002 millimeters (mm) or less in diameter. These particles are so small that they have the properties of colloids, which will be discussed later.

Soil Particles

The soil mineral matter is derived from the decomposition of rock, rock called *parent material*. This rock makes up much of the lithosphere of the earth or the earth's crust. Through the processes of weathering, the rock is broken into smaller and smaller particles. First it becomes boulders, then gravel, sand, silt, and finally clay with these categories determined by the size of the particle, as seen in Fig. 2.1. The rock, boulders, sand, and clay all comprise the mineral matter of the soil but

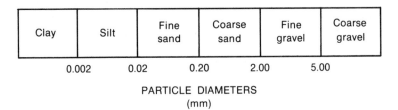

Clay	Silt	Fine sand	Coarse sand	Fine gravel	Coarse gravel

0.002   0.02   0.20   2.00   5.00

PARTICLE DIAMETERS
(mm)

Fig. 2.1 Soil particle size classification.

perhaps the clay is the most important of these to the plant. *Clay* particles are very small, being about 0.002 mm or less in diameter. Due to the small size, clay is colloidal and therefore has properties that are very important in soil-plant relations, properties such as the ability to retain water and mineral nutrients.

Colloids are important because of their small size which gives them a tremendously large surface area in relation to their volume. To get some idea of the magnitude of the surface of these soil particles, consider a soil particle in the form of a cube of 1.0 cm length. This would be the size of coarse gravel, and would have a surface of 6 cm^2 and a volume of 1 cm^3. Now if this cube is broken up into smaller particles, the total volume of all the particles would not change but the total surface would. The surface to volume ratio can be determined by

$$S/V = K\frac{1}{x}$$

where S = the surface area
V = the volume
K = a constant which is equal to 6, for a cube
x = the length of the cube

Since the total volume remains constant, $S = K(1/x) = 6/x$. From this equation, we can determine the increase in surface area as the size of the soil particle is decreased, and representative values from such a calculation are given in Table 2.2. Therefore, by cutting a 10 mm cube into pieces the size of colloids, one increases the surface area ten thousand times, with no change in volume. It is this large surface area that gives soil colloids their important properties.

Since colloids are surrounded by water, they form what is known to physical chemists as a two-phase system, with a boundary between these

Table 2.2 The surface area of some soil particles.

Particle diameter (mm)	Classification	Surface area (cm^2)
10.0	coarse gravel	6
2.5	fine gravel	24
1.0	coarse sand	60
0.1	fine sand	600
0.01	silt	6000
0.001	clay	60,000

two phases. This boundary, at the surface of the colloid, has an electrical charge, usually negative in sign. Although this electrical charge is small, it is important, because it is able to attract and hold both water and mineral nutrients to the surface of the colloid, as seen in Fig. 2.2, or as we say the colloid *adsorbs* these substances. Notice in Fig. 2.2 how the water molecules are oriented and packed around the colloid when they are adsorbed. The adsorption of water to the colloidal surface is called *hydration* of the colloids. Both humus and clay are soil colloids. Colloids are also found in plant cells, such as the cell walls and the proteins of the cell, and these too bind water.

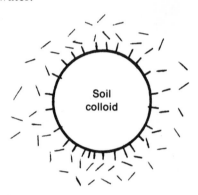

Fig. 2.2   The adsorption of water molecules by soil colloids.

Soil Pore Space

The humus and the mineral matter comprise the soil particles and, in a good soil, should make up about 50% of the soil volume. The remaining 50% of the soil volume is composed largely of *pore space*. This is the space among the soil particles, (*see* Fig. 2.3), and is filled with water or air. As the water content of the soil increases, the water occupies more and more of the pore space, displacing the air. As the water content decreases, more and more air enters the soil.

Variations in Soil Structure

In addition to the soil particles and pore space, a soil contains varying amounts of living organisms: worms, insects, algae, fungi, bacteria, and others. These too are important components of the soil as we shall see later. Soil is composed then of soil particles, pore space, water, and living

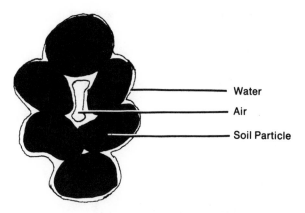

Fig. 2.3   The distribution of water among the soil particles.

organisms, and all of these combine to make up the soil with which we are familiar and which is so important for plants. Since these components can vary in so many ways, particularly in amounts of each present in the soil and in the size of the soil particles, there are an infinite number of types of soil. If you have four jars, one containing mineral matter, one humus, one water, and the other living soil organisms, and you mix these four constituents together, you have a soil. Since you can mix various amounts of each of these and the size and nature of the mineral matter and organic matter particles can vary, the types of soil you can make will be numerous. That is why soil varies so much from one location to another, because of the variety of mixtures possible. Some of these soils will support plants better than others.

SOIL PROFILE

Having considered the components which make up a soil, let us consider the distribution of these components in a soil or what is called the *soil profile*. If a deep hole is dug into the soil, and the soil on the sides of this hole is studied, we find that the soil differs from one depth to another, as illustrated in Fig. 2.4. This difference is due to differences in the soil components with depth. Near the top of the soil, we can see a dark layer which is referred to as the *A horizon* of the soil, or *topsoil*. This is dark because it contains more humus than the lower layers. Below this is a deeper layer of lighter-colored soil called the *B horizon* or *subsoil*.

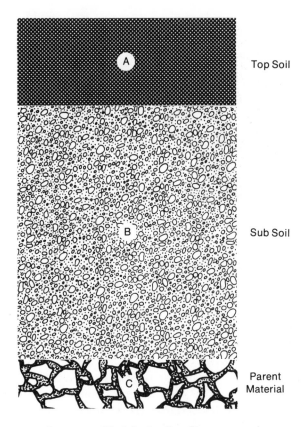

Top Soil

Sub Soil

Parent
Material

Fig. 2.4   A soil profile.

At lower depths, we find the parent material from which the soil was originally formed. Plant roots grow very well in the topsoil, which is rich in organic matter and in mineral nutrients and oxygen. They grow less well in the subsoil and perhaps not at all in the parent material. The depth of the topsoil and subsoil layers varies from soil to soil so the distribution of roots in the soil would likewise vary. Sometimes the subsoil has a high clay content with reduced pore space so oxygen is deficient and this reduces root growth. Sometimes the subsoil has a hard soil layer in it, which we call *hardpan*, and this reduces root growth. Sometimes excess water is present in the subsoil and occupies all of the pore space reducing root growth.

Table 2.3 Types of soil water.

A. Free water, sometimes called gravitational water.
B. Bound water.
1. Capillary water.
2. Hygroscopic water.

SOIL WATER

Water within the soil, as illustrated in Table 2.3, is found either as *free* water, located in the pore space, or as *bound* water that is either adsorbed to the soil colloids or is held by *capillarity* within the soil by the particles. As water is added to the soil surface, either by irrigation or precipitation, the water enters the soil. The rate of entrance depends on the size of the soil particles. It enters rapidly and moves rapidly downward through the soil if the soil is sandy or gravelly. This movement is greatly reduced if the soil is largely clay since the clay particles are so small that they do not leave large pore spaces through which the water can rapidly move. Nevertheless, due to the large number of small pore spaces, the clay soils do hold a great deal of water once the water gets into the soil. For instance, a sandy soil may hold only 1 cm or less of water per 12 cm of soil depth, whereas a clay soil may hold 2 cm or more per 12 cm of soil.

Soil Water Movement

As water is added to a soil, the top layer of the soil becomes saturated with water, or in other words, the pore space becomes filled with water to the exclusion of air. As more water is added to the soil, deeper layers become saturated, as the free water moves downward under the influence of gravity. One cannot partially saturate a soil by adding water to it. The more water added, the deeper the soil is saturated. The thickness of the soil layer saturated when water a cm deep is placed on the surface varies with the soil but is often about 6 cm. The free water enters the larger pores and moves downward, primarily through channels left by decayed roots and soil animals, such as worms and insects, as well as through cracks and crevices. Sometimes the free water is impeded from downward movement and remains in the soil, giving a poorly drained soil. Poorly drained soil has a deficiency in oxygen, and it also warms up slowly in the Spring and cools slowly in Winter since water has about five times the heat capacity of soil particles. This heat capacity allows the soil ability to adsorb heat with little temperature increase, as discussed in Chapter 4. About 2–3 days following irrigation, all the free water has moved to deeper layers of soil leaving only the bound water behind. At this point, we say that the water content of the soil is at *field capacity*.

Water-holding Capacity

The water supply in the soil is determined by the amount of water added to the soil by precipitation and irrigation as well as by the *water-holding capacity* of the soil or its ability to retain the water added to it. This ability of the soil to retain the water added to it is determined by the properties of the soil and varies from one soil to another.

Types of Soil Water

The free water represents only a small portion of the water that enters the plant because it is not present during most of the growing period of the plant and also because when it is present, little oxygen is present and therefore the water cannot get into the plant. So it is the bound water which is most available to plants throughout their growing season. This bound water is held in the soil by capillarity and by adsorption to the colloids. As the water is removed from the soil by plants or by evaporation, the water held by capillarity is lost first and eventually only the water adsorbed to the colloids remains. The latter we call *hygroscopic water*, and as we shall explain later, this water also is not available to the plants. Therefore, most of the water in the soil that plants use is the bound water held by capillarity and sometimes called *capillary water*.

The water supply in the soil is determined by the amount of water added to the soil by precipitation and irrigation as well as by the *water-holding capacity* of the soil or its ability to retain the water added to it.

THE SOIL SOLUTION

Water in the soil is never pure water, but always contains gases and salts dissolved in it. The water plus these dissolved gases and salts comprise the *soil solution*. A *solution* can be defined as a homogeneous mixture of two or more substances dispersed in the molecular or ionic form. The substance present in greatest amount is known as the *solvent*, and that present in lesser amount is known as the *solute*. In the soil, as well as in most solutions within the plant, the solvent is water, so we refer to these solutions as *aqueous solutions*. Water is an excellent solvent. In fact, it dissolves so many chemicals that it has been referred to as the universal solvent. The reasons for its solvent action are listed in Chapter 4. The solute of the soil solution consists of gases and of many molecules and ions of inorganic and organic chemicals that are present in the soil. The ratio of the amount of the solute to solvent is known as the *concentration* of the solution, which can be expressed by various terms such as *percent*, *molar* or *molal* concentration, or *parts per million* (ppm).

Diffusion

The concentration of a solution may also be expressed in terms of its *osmotic pressure* (OP). The osmotic pressure refers to the concentration of particles in the solution without reference to the chemicals composing these particles. To understand how osmotic pressure is determined and its importance to soil-plant relationships, it is necessary to consider the phenomenon known as osmosis, one you learned in general botany, biology, or chemistry. We can define osmosis as the movement of water through a differentially-permeable membrane. All molecules have *kinetic energy*, and due to the presence of this energy they are always in motion moving in a straight line until they hit another molecule and bounce off in a straight line but in a different direction. As a result of this movement, there is a net movement from an area of high concentration to one of low concentration, a movement called *diffusion*. Chemicals do tend, through diffusion, to maintain an equal concentration of their molecules wherever they are not inhibited from doing so by a barrier. Adding molecules or ions of another chemical dilutes their concentration. Pure water has the highest concentration of water molecules possible. Adding salt to this water dilutes the water molecules, and water molecules will then move from pure water into the solution to restore a uniform water concentration. Such movement will increase the volume of the solution or if the volume is confined by the walls of the container, a pressure will be produced as a result of the tendency of the water molecules to move in, increasing the total number of particles. Note that OP depends upon the concentration of solute particles present irrespective of the size or nature of these particles.

Osmotic Pressure Determination

The OP value of a solution, such as the soil solution, can be determined by several means. First, it can be calculated if the concentration of the solution is known. For instance, if the solutes are non-dissociated molecules, such as sucrose, the OP, in bars, is equal to the following:

$$OP = \frac{22.7\,mt}{T}$$

where m is the molal concentration of the solution, t is the prevailing absolute temperature, T is the absolute temperature at 0°C, or 273°K, and the value of 22.7 is the OP that would be produced by a one molal solution.

On the other hand, if the solutes are partially dissociated, as is common in soil solutions, not only the concentration of the molecules, but also that of their dissociated ions must be taken into account. This can be done, at least theoretically, by the following equation:

$$OP = \frac{22.7\,mit}{T}$$

where $i = 1 + (n-1)a$ and where n is the number of ions formed by molecular dissociation, and i is the extinction coefficient or the fraction of molecules that dissociate. KCl dissociates to form two ions, so for this solute, $n = 2$. About 76% of its molecules dissociate, so $a = 0.76$. Both a and n will vary with the salt present in the soil solution. The soil solution contains many different solutes dissolved in the water, and the extent of dissociation of their molecules is influenced by the other chemicals present. These complicate any attempt to determine the OP by the above calculations.

A more accurate method of OP determination of a solution is to make use of the cryoscopic properties of the solution. The concentration of particles, such as ions and molecules, will determine the freezing point, boiling point, and vapor pressure of the solution. These can be measured accurately and correlated with the value of the OP of the solution. As an example, the greater the OP of a solution, the lower will be its freezing point, so by determining the freezing point of a solution, we can determine the OP of the solution. This is a good, often-used measurement. The relationship that exists between freezing point and OP can be seen in the following equation:

$$\frac{OP}{22.7} = \frac{1.860}{\Delta}$$

where Δ = freezing point of the solution.

Soil Osmotic Pressures

We can also speak of osmotic pressure in terms of the potential of a solution to produce a pressure, if placed under conditions wherein the pressure can be measured. For instance, it has been determined that a 1 molal sucrose solution can produce a pressure of 22.7 bars under conditions conducive to such measurements, so we say that a 1 molal solution of sucrose has an osmotic pressure of 22.7 bars. (The term atmospheres was formerly used to define pressure, but has recently been replaced by the term bars which is equal to 10^6 dyne/cm or 0.987 atmos.) Since OP depends upon the concentration of solute particles, as water is lost from a solution, such as when plants taking up water from the soil, or, when water evaporates from the soil, the osmotic pressure of the solution increases due to an increase in the ratio of solute to solvent molecules.

Osmotic pressure is important because it is a pressure that plant roots must work against to take up water. The greater the osmotic pressure of the soil solution, the more difficulty the plants will have taking up the water even if it is present. In a good soil, the OP value will be less than 2 bars, even when the water content of the soil is low. However, in many of our Western states, and in other arid regions of the world, evaporation rate is high and rainfall is low. As a result, the salts produced by the decomposition of the soil minerals accumulate in the soil instead of being washed out with the free water. After a period of time, the salts will accumulate to such an extent that the OP of the soil solution will exceed 2 bars. In fact, values in excess of 2000 bars have been reported in such soils. These soils, with a high OP, are known as *saline soils*. With the expansion of irrigation agriculture, *salinity* is becoming more extensive because the salts in the irrigation water are accumulating in the soil. Entire civilizations have been wiped out in the past partly due to this problem. Because of the high OP values, plants cannot take up the needed water even though the water is present in the soil, so crops cannot grow in such soils. Also, the salts are toxic to many plants. These soils support only sparse vegetation limited to a few tolerant plant species, as seen in Fig. 2.5.

Soil Moisture Tension

Water is held in the soil by two forces. One is OP and the other is the force by which water molecules are adsorbed to the surface of the colloids of the soil and the cohesion by which water molecules are held together. Water molecules have a strong tendency for cohesion due to polarity and

Fig. 2.5   Sparsity of vegetation due to salinity. Photo taken near Brigham City, Utah.

hydrogen bond formation. Capillarity is due to this ability. There is a layer of water molecules adsorbed with great force to the colloids and all surrounding molecules are bound to this and subsequent layers of water molecules. The further the water molecule is from this surface layer, the less strongly it is bound, but nevertheless, it is held by some attractive force. The sum of the force holding water molecules to the soil colloids either by adsorption or by cohesion is referred to as the *soil moisture tension* (SMT). The magnitude of SMT can be determined either by tensiometers or a pressure membrane apparatus. Modified tensiometers are often used to automatically turn on sprinkler systems when lawns need irrigation, sometimes much to the chagrin of picnickers.

WATER POTENTIAL

Water is held in the soil by the combined forces of OP plus soil moisture tension. This total force is called the *water potential* (WP) and is also expressed in bars.

Since the water potential of the soil is determined by the OP and the SMT, (WP = OP + SMT), anything that alters these values will alter the ability of the plant to extract water from the soil. The OP of the soil is increased either by adding salt to the soil, or by removing water. As the water content of the soil decreases, the OP increases. As the water content of the soil decreases, the SMT also increases because the water removed first is that held less tightly and therefore that farthermost from the colloidal surface. As more water is removed, the water is held more tightly so the SMT increases. Therefore, as the soil dries out, both the OP and the SMT increases, and if the soil becomes too dry, the water potential will become so great that the plant cannot extract enough water from the soil to supply its needs, and will wilt. This is a common occurrence in many regions during the growing season of the year, and especially in arid regions.

PLANT ROOTS

Water enters the plant primarily through the roots. Therefore, the amount of roots present on the plant and their distribution in the soil, will be important factors in determining the amount of water taken up. Plants have one of two types of root systems namely, the fibrous root system found on grasses, and the tap root system found on many other plants. Examples of these root systems can be seen in Fig. 2.6. Either system is efficient in distributing the roots through the soil, but tap roots usually grow deeper into the soil.

Extent of Root System

The roots of most plants do not go as deep into the soil as the shoots grow in height. In fact, the roots seldom go deeper than ten feet, irrespective of the depth of the soil. Even the roots of trees seldom grow to great depths. They do grow out from the tree for some distance, sometimes 50–60 feet, and then turn down, but seldom grow to depths greater than 10–15 feet. The volume of the root system is usually as extensive as the shoot of the same plant, even in trees. One interesting pattern of root growth is shown by the saguaro tree of the southwest deserts. Its roots grow down into the soil for about two feet and then spread out for some distance, often as far as the tree is high, which may exceed 50 feet.

However, plant species do differ in the depth of their root system even in the same soil. Some have most of their roots in the top two feet of soil, some in the next three feet, and some take up most of their water and minerals from depths greater than five feet. These differences in root

Fig. 2.6 Examples of fibrous and tap root systems. Plants labelled b and g have tap roots, and plants labelled f and p have fibrous root systems. (From *Prairie Plants and Their Environment* by J. E. Weaver, 1968. University of Nebraska Press, Lincoln.)

distribution from one species to another allow species to better compete for water and minerals even when the species exist side by side. Such patterns of root distribution are very important among native plants, but cultivated species are spaced to reduce competition from adjacent plants. They are spaced because their root systems do extract water and minerals from the same depth in the soil, as well as for other reasons to be discussed later. Roots not only compete in obtaining water and minerals but also for oxygen and by giving off chemicals that may be toxic or at least detrimental to competitor species.

Root Growth

Roots grow at their tips due to cell division and subsequent cell enlargement. Therefore, anything that interferes with division or enlargement will interfere with root growth. The physiology of growth at the cellular level will be discussed in a future chapter. Suffice it to say that growth requires energy obtained from respiration within the cell, a favorable temperature, adequate water and other nutrients, and space in which to grow.

Roots grow very slowly during the Winter, primarily due to low temperatures that prevail. Some growth does occur all Winter, but the growth rate is very slow. This effect of low temperature is no doubt due principally to reduced rate of metabolism within the growing cells, although as ice is formed in the soil solution, this may become another limiting factor, due to its mechanical resistance. Water movement from the soil to the root is also slowed down in Winter, being $\frac{1}{3}$ to $\frac{1}{2}$ as rapid at 0°C as at 25°C. Lateral water movement through the soil never occurs over great distances, but water must move to the cell from the soil and through the cell walls and intercellular spaces. Low temperatures also decrease membrane permeability, and are associated with a decrease in the vapor pressure of water, and in the increase in the viscosity of both the water and the protoplasm. These would affect water uptake. However, the greatest decrease in water uptake at low temperatures is due to reduced root growth.

Soil Aeration and Root Growth

Soil aeration is an important factor in root growth and good aeration must be maintained for the growth of most plant roots. This oxygen is needed for root respiration. Root cells carry out respiration just as all other plant cells do, and this respiration is the source of energy needed for root growth. If oxygen is absent, respiration rate is reduced so root growth is decreased. This also reduces the permeability of the membranes to water and mineral nutrients. It is a common occurrence to see plants wilt when one saturates a soil with water. It seems strange that plants should be growing in water and still be suffering from a lack of water, but this happens. Plants differ greatly in the extent of their tolerance for water-saturated soil. Of course *hydrophytes* — those plants normally growing submerged, or partly so, in water — such as water lilies and cattails (*see* Fig. 2.7) are very tolerant and live their life span under such conditions. At least part of this tolerance is due to air channels (aerenchyma) that extend from their leaves to the roots. However, *mesophytes*, those plants such as our crop plants that live in soil with normal water content, are more susceptible, although even these differ from one species to another in regard to their tolerance of such conditions. For example, soybeans are

Fig. 2.7 Hydrophytes, such as those shown in this photo, spend their lifetime with their roots in water-saturated soil.

quite resistant, sorghum moderately so and corn very sensitive to injury in water-saturated soil. Why plants differ so much in this respect is not known. It should also be pointed out that in a saturated soil, not only does a deficiency of oxygen exist, but also carbon dioxide given off during respiration accumulates, and this in itself may be toxic to the plant, causing changes in membrane permeability. In most soils, at varying depths characteristic of the location, if one digs down into the soil, a layer will often be found below which the soil is always saturated with water. This is the *water table*. The depth of the water table below the soil surface is variable from one location to another. However, if the water table rises, plants will often be killed because the roots will not survive in such water-saturated soil. In valleys with poor drainage and in valleys where irrigation has been introduced, this is a frequent phenomenon. Also, the water table can be high especially near the shore of lakes, streams, etc.

High water content is a factor which reduces root aeration, but is not the only factor. A high clay content of the soil leads to poor aeration as does compacting the soil by mechanical pressure. Have you ever noticed the paths across the campus, made by students walking across the lawn, and how the grass does not grow in these paths? If not, *see* Fig. 2.8. Poor aeration due to soil compaction is an important factor although compaction also makes the penetration of the soil by the roots more difficult by increasing mechanical resistance. Anything that reduces aeration of the soil will decrease root growth due either to a deficiency of oxygen for root respiration or to an accumulation of carbon dioxide from respiration or to both.

Fig. 2.8   A photo showing the results of soil compaction on plant growth.

Salinity and Root Growth

A high content of solutes in the soil will also reduce root growth due both to the effects of the solutes on raising the OP of the soil solution and thereby reducing the amount of water that is available for root growth, and because these solutes may be injurious to the cells at high concentrations. Any chemical will be injurious to a plant if present in high enough concentration and soil solutes often do reach toxic concentrations. Some of these effects are due to the direct interference of the chemical with the living activities of the cell and some are indirect, whereby they interfere with the normal function of nutrients needed by the plant. Boron often becomes toxic in *saline* soils of the Western United States, as does sodium in *sodac* soils. In acid soils of the Eastern United States and in other regions of high rainfall, other soil chemicals may be toxic due to their increased solubility associated with the increased hydrogen ion concentration.

Soil Water Content

The water content of soils is also an important factor influencing root growth and water uptake. Just following seed germination is a very critical time in the life of the plant. Since the soil nearest the surface dries out first, following rain or irrigation, the roots of the young seedling must grow down fast enough to maintain a supply of water. In the drier regions of our country, the ability of the seedling to survive depends upon how fast the roots can grow into the soil. Burr oak, for instance, is able to grow well on the Western edge of the Eastern deciduous forest because its seedlings can avoid drought conditions by their rapid root growth the first year following germination. Many desert plants are also so characterized.

WATER ENTERS THE PLANT

Water Moves to the Root

When a root is in the vicinity of water in the soil, and the water potential of the soil solution is less than that of the root, water will move by diffusion through the soil and into the root. Only when the water potential of the root is *greater than* the water potential of the soil will water enter the root. When the water potential of the root is less than the water potential of the soil, the plant can actually lose water to the soil. However, it should be noted that some authors give WP a negative value, in which case water moves toward a more negative WP.

As the water moves into the root, it first moves by diffusion into the intercellular spaces and into the cell walls of the root cells. Most of the water that gets into the plants, goes in through the root hair zone, with lesser but significant amounts entering through the older regions of the root. Nevertheless, somewhere between the soil and the xylem, the water must pass through the living protoplasm of the cell. We do not know in which area of the root this occurs, but some consider the endodermis to be so involved.

Water Enters the Protoplast

Since water enters the living cell, it would be well to review cell structure. A plant is composed of many different kinds of cells. The root cells that are principally involved in water uptake are parenchyma cells. If these cells are studied with the microscope, especially the electron microscope, we find each cell has a definite structure, similar to that shown in Fig. 1.3, but without chloroplasts.

To enter into root cell protoplast, the water in the cell wall must pass through the plasma membrane. It then often passes through the mesoplasm and the vacuolar membrane into the vacuole where it is stored. Such entrance is by osmosis, the movement of water through a differentially-permeable membrane. The plasma membrane, vacuolar membrane, and in fact, most cell organelles, are differentially-permeable membranes. These membranes are composed principally of protein and lipid molecules joined in such a way that they form a double layer. Perhaps the protein forms the basic structure with the lipids on the surface. These membranes are very thin.

Osmosis

Osmosis occurs whenever the water potential of the protoplast is greater than the water potential of the solution in the cell wall. The water potential of the protoplast is equal to the difference between the OP and TP (WP = OP − TP). The water potential of root cells will vary with the salt content of the soil, because the more salt in the soil, the more salt that gets into the plant roots, by the water content of the cell, and by the concentration of the various organic solutes.

Turgor Pressure

The turgor pressure of the protoplast also is a factor in determining the WP value of the cell. When a cell is in pure water, the OP and TP values are the same, but when placed in a solution, the TP value will be less than the OP value, since OP is the maximum potential pressure and TP is the actual pressure. The TP value can equal the OP value, but never exceed it and, in fact, in the plant it is always less than the OP value, since the cell is never surrounded by pure water but always by a solution.

The water potential of the root cell is usually about 15 bars, so the plant can take up water only if the water potential of the soil solution is less than 15 bars. As a result, the water content of the soil that gives a water potential of 15 bars is known as the *permanent wilting point* of the soil and the plant will wilt if the water content is less, or is not increased. Therefore, water in the soil between the range of field capacity and permanent wilting point is considered as *available* water. The actual water content of the soil at permanent wilting will vary with the colloidal content of the soil and with the salt content of the soil.

IMBIBITION

There are incidences when the water potential of the plant may initially be due more to imbibition than to OP. Plant tissues usually have a high water content, with water comprising more than 50% of the fresh weight. However, mature seeds, pollen grains, spores, and some lichens and algae have much lower water contents, with values less than 10% of the fresh weight being reported. In these cases, the initial uptake of water is due to imbibition. *Imbibition* can be defined as the hydration of colloids of the plant cells. Many colloids exist in plants, especially the cell wall materials, and the proteins and enzymes. These adsorb water just as the soil colloids do. When the water content of the cell is low, all of the water will be so bound. The force holding these water molecules to these colloids is great. If one places dried plant material, such as a piece of brown alga stipe in water, the material swells due to the uptake of water associated with imbibition. If this swelling is prevented, the pressure produced can be measured and values as high as hundreds of bars of pressure have been so measured. If the alga is dead, the pressure so produced cannot be due to osmosis since the cell membranes are destroyed when dead, and can only be due to imbibition. When imbibition is occurring, and this happens only when the water content of the cell is low, the water potential can be expressed as being equal to the *imbibitional pressure* (IP) values plus OP values minus the TP values. Such water potentials can and often do exceed greatly the 15 bars associated with osmosis. Thus imbibition is associated with a great tendency of the plant material to take up water from its environment. However, imbibition decreases rapidly as the water content of the cell increases, and is not important in determining the water potential when the water content of the cell is as great as it usually is.

Potometers

The amount of water that enters a plant can be measured by the use of *potometers* (*see* Fig. 2.9), by measuring the water lost from the soil or

Fig. 2.9 A potometer. Water is continuous from the base of the branch and through the calibrated horizontal tube. As water enters the plant, it moves through the horizontal tube where its rate of movement can be observed. Potometers can also be used on roots to measure water uptake by the root.

water in which a plant is growing, by measuring the amount of water lost by the plant, by measuring changes in the water content of the plant, or by the use of water that has been labelled with radioactive hydrogen.

From what we have learned from this chapter it appears that the entrance of water into the plant is a complex phenomenon controlled by many factors. It is important that we learn what these factors are and how to control them because water availability is often the most important factor controlling plant growth, and yields could be greatly increased if more water entered the plant at the time the water is needed. Let us now learn what happens to the water once it gets into the plant.

Chapter Three

Distribution of Water
Within the Plant

Water first enters the plant by diffusing from the soil into the cell walls and intercellular spaces of the plant roots. Such diffusion is a passive process, which means that energy derived from plant metabolism is not needed for diffusion to occur. It involves the passive movement of water from an area of high concentration to an area of low concentration.

Water also diffuses through the soil from an area of high concentration to an area of low concentration, and into the plant roots when the water concentration within the root is less than that in the soil. In this manner, the intercellular spaces and the cell walls have a water concentration about equal to that of the soil. When water is removed from the cell walls or the intercellular spaces the supply is replenished by diffusing in from the soil.

WATER MOVES THROUGH THE ROOT

Where does the water go from the intercellular spaces and the cell wall area? The answer is, into the protoplasm of the root cells. If one looks at the cross-section through the plant root using a microscope, one sees the structures illustrated in Fig. 3.1. The root hairs, epidermis, and cortex are composed of parenchyma cells with abundant cell walls and intercellular spaces. These cells also contain protoplasm, which includes the cytoplasm and a vacuole, both with a high water content. Water from the cell walls and intercellular spaces can and does enter the cytoplasm and vacuoles of these cells to maintain their high water content. However, the total water needs of these cells is very much less than the supply of water that enters the plant during its lifetime.

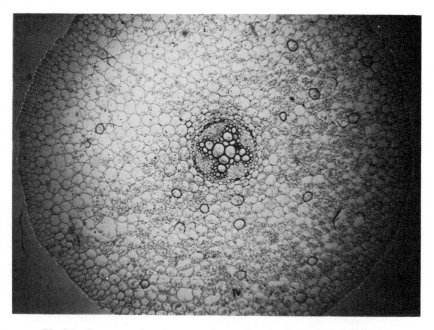

Fig. 3.1   A cross-section of a ranunculus root, as seen through the microscope.

If one looks carefully at the endodermal cells, one can see a strip of material known as the *casparian strip*. This casparian strip is impermeable to water so water cannot move through it. To pass from the cortex into the vascular tissues internal to the endodermis, the water must pass into and through the vacuole of the endodermal cells. To do so, it must pass through cell membranes which are able to control the movement of water through them. Then, the water moves out of these cells, and into the vessel elements of the xylem, through which it is moved (*translocated*) to the tops of the plant. As long as the cell membranes are permeable to water, water will move through the endodermal cells into the vessels when the water potential within the vessels is greater than the water potential in the endodermal cells.

MECHANISMS OF WATER MOVEMENT

Root Pressure

One way of developing a high water potential in the vessel elements is by increasing the solute content of these tubes. This is done by the secretion of inorganic salts from the surrounding root cells. As a result, water moves into the vessel elements, and if it is not rapidly translocated, a pressure will build up, similar to the turgor pressure of an osmometer. This positive pressure is known as *root pressure* and can often be observed in plants when the rate of water loss by the plant is low. Root pressure can be observed either by cutting off the top of the plant near the soil, and observing the exudation of water from the stump, or by the phenomenon known as guttation. However, if the pressure of the water exuded from the stump is measured, the magnitude of this pressure will be only 2–3 bars. If the water potential of the soil is greater than 2 or 3 bars, a situation which is common, root pressure is nil and no exudation occurs.

Root pressure is a means whereby water can be translocated to the tops of the plant. However, since this pressure is low, and practically non-existent in some plants, root pressure would not serve as the mechanism for translocation except under conditions when the plant cells were saturated with water, when water loss was nil, and when the soil was warm with a high moisture content and low salt content. It does occur in some plants, even submerged aquatic plants, but may not be important to the plant.

Guttation

Root pressure is responsible for the phenomenon known as guttation. *Guttation* is the process by which water is lost through the leaves, in the form of liquid water, and is evident when the rate at which water enters the plant is more rapid than the rate of water loss by the plant. Cool, humid mornings following a day of warm temperatures are especially conducive to guttation, if the water content of the soil is high but not at saturation point. Guttation is observed as water is forced out through the pores in the leaves. These pores are often located near the vein endings of the leaves and are called *hydathodes*. The water forced out often appears as droplets near the margins of the leaves, especially in the early morning hours during the Spring, or in mountainous or tropical regions.

The water forced out is not pure water, but is a solution with water as the solvent and with various chemical substances as solutes. These solutes include variable, but important, amounts of gases, inorganic elements, sugars, nitrogen compounds, and various other organic compounds. The mineral elements especially prevalent include calcium, potassium, sodium, magnesium, and often nitrates. Glutamine and other organic nitrogen compounds have also been detected in large amounts if the plants have been recently well fertilized with nitrogenous fertilizers.

These solutes in the water of guttation can have an important effect on the plant. Frequent guttation has been shown to cause a deficiency of mineral nutrients within the plant. Also, as the water evaporates on the leaf surface, the remaining solution becomes progressively more concentrated, increasing the concentration of solutes. This may result in leaf burn either due to high OP or to specific toxicity of one or more of the solutes.

The amount of water lost through guttation varies with the water content of the soil, soil temperature, soil aeration, air temperature, and the vapor pressure of the water in the atmosphere. It also varies with the nutrient status of the plant, being higher in those plants that have adequate amounts of nitrogenous fertilizers available. The rate decreases with increased soil water potential. In the tropics, guttation is much more common than in other regions of the world. Some tropical plants may lose a third of a liter of water each night as a result of guttation. However, in the temperate regions, guttation is not a major way in which water is lost, and much less than 1% of the water lost by plants is actually lost through guttation.

Capillarity

Another mechanism of water translocation in plants is capillarity. In the previous chapter we learned of the occurrence and importance of capillarity in the soil, and how most of the water that a plant gets from the soil is water of capillarity. Well capillarity is also present within the plant. However, its importance in long-distance translocation of water in the plant is questionable.

Transpiration-cohesion Theory

Root pressure does, at times, function to move water upward through the plant, but cannot force water to the heights necessary to reach the tops of trees. To explain how water is translocated to such heights, which may exceed 200 feet, it is necessary to learn the *transpiration-cohesion* theory of water movement. According to this theory, water moves as a result of the high water potential developed as water is lost through evaporation from the plant leaf. This evaporation creates a deficit of water in the vessels and therefore a high water potential. As stated earlier, water then moves from a low water potential to a high water potential and if the potential is great enough, as it often is, water can thus move to the top of the tallest tree. That water does evaporate from the leaves of plants, a phenomenon known as *transpiration*, will be discussed later in this chapter, but such water evaporates from the surface of the mesophyll cells of the leaf and diffuses out through the stomates. This evaporation creates a deficit of water in these cells and therefore increases the water potential. This attracts water from cells closer to the xylem and this attraction continues to the xylem vessel elements.

In the xylem, the water molecules over the entire length of the xylem are joined together like links in a chain. They are joined together because water molecules have the ability to cohese. The molecules are joined by hydrogen bonds as well as by electrostatic attraction. The strength of these bonds is great enough to allow the water to be pulled up the xylem from the roots, so an increase in the water potential causes the water column in the xylem to be pulled up the stem, pulling water from the root cells and thereby creating a deficit within the roots, increasing their water potential. The increased water potential in the root cells causes more water to be taken up from the soil so we have a continuous water column from the soil to the atmosphere near the leaves.

The rate at which water will move up through the xylem to the leaves then depends upon the rate at which water evaporates from, or is transpired by, the leaves, the permeability of the root endodermis to water, and the water potentials of the root cells and soil. This rate may be very low when the air is cool and has a high relative humidity or when the stomates are closed or when the water potential of the soil is very high. Under ideal conditions the rate may be several centimeters per hour.

We can summarize our discussion of the mechanism of translocation of water from the roots to the tops of the plant by saying that water moves from an area of low to an area of high water potential. This movement occurs primarily through the cell walls, intercellular spaces, and empty vessel elements, except where it moves through the living membranes of the endodermal cells. Perhaps root pressure and transpiration-cohesion both explain mechanisms by which this movement occurs and the relative importance of each mechanism depends upon the size of the plant and the prevailing environment at the time.

LATERAL WATER MOVEMENT

Water is translocated laterally as well as vertically. As the water moves up the vessel elements, some water leaks out and moves into other plant cells. This happens, of course, in response to differences in water potential between the water in the xylem and that in the surrounding cells of the stem, root, buds, etc. First, the water moves into the cell wall of the cell needing water, and from there it leaves the cell wall and enters the protoplasm by osmosis. All cell membranes are differentially permeable. This means that they will allow certain chemicals to pass through them but not others. They also have the ability to change their permeability from time to time, being permeable to water sometimes, and less permeable to it at other times. Passage through the endodermal cells is another example of osmosis.

WATER MOVES INTO THE CELL

Although the rate of movement of water through the membrane is controlled by the permeability of the membrane, water moves through it only in response to differences in water potential inside and outside of the membrane. When the water potential inside the protoplasm is greater than that outside the protoplasm, water moves in. An increase in water potential within the protoplasm can be brought about by an increase either in imbibitional pressure or in osmotic pressure of the protoplasmic contents. This may be due to water moving out of the cell, or to an accumulation of solutes such as mineral elements, sugars, etc. Osmosis will cease when the water potential within the protoplasm is equal to the water potential outside the protoplasm. The water potential within the protoplasm can be reduced either by diluting the internal solution or by increasing the turgor pressure. Since turgor pressure is the actual pressure produced within the cell as water moves in, as osmosis occurs and water moves into the protoplasm, the turgor pressure value increases more than the osmotic pressure value decreases until eventually the water potential within the protoplasm is equal to that outside the protoplasm. At this time, osmosis stops, with water moving in as fast as it moves out.

One can view the water within the plant as consisting of water within the cell walls and intercellular spaces, water that is continuous throughout the plant and that moves to that part of the plant having the highest water potential. This water in the cell walls and intercellular spaces is available to move into the protoplasm of each cell as the water potential of the protoplasm, including the vacuole, exceeds that of the water in the cell wall.

LOSS OF WATER FROM THE PLANT

Since vertical water translocation involves water loss, let us study water loss in more detail. The total water content of a plant is less than 1% of the water that enters a plant in its lifetime. Most of the water a plant takes up is lost. In fact, some herbaceous plants lose as much water in one day as can be found in the plant at any given time. Some of this loss occurs in the form of liquid water—guttation—but this represents less than 1% of the total water loss. Most water is lost in the form of a gas, commonly known as water vapor. Water evaporates whenever living cells come in contact with air, just as water will evaporate from an open pan of water if it is exposed to the atmosphere. Loss of water from a plant in the form of water vapor, through evaporation, is called *transpiration.*

There are three kinds of transpiration, named according to the plant structures through which the water vapor leaves the plant, namely, cuticular, lenticular, and stomatal transpiration.

Cuticular Transpiration

Whenever the plant surface comes into contact with the atmosphere, a water-repellent, waxy layer is formed on the surface. This waxy layer is known as the *cuticle.* This cuticle is composed of many substances and is rather complex. The cuticle is found on the surface of stems, fruits, flower petals, and leaves. Since it is water-repellent, it serves to protect the plant from excess water loss, but also may function to minimize mechanical damage to leaf cells, to reduce fungal and insect injury, and to shield the plant from excess ultraviolet radiation from the sun. It has been shown that removing the cuticle from plant leaves will greatly increase the rate of transpiration. Of course, some of this waxy layer on plant structures is naturally rubbed off, but the plant has the ability to regenerate the cuticle so the wax is replaced when removed. This wax, the surface layer of the cuticle, is a complex mixture of long-chain alkanes, alcohols, ketones, aldehydes, acetals, esters, and acids. Each chain commonly has 21–37 carbon atoms in it, so the molecules are rather large. However, it appears that these large molecules are synthesized by the plant from small—two carbon—molecules known as acetate, by joining many of these acetate molecules together to form one wax molecule. Irrespective of the ability of the cuticle to resist water movement through it, some water loss does occur as *cuticular transpiration.* However, the amount of water so lost is less than 1% of the total water loss, a value much less than that found when the cuticle is removed, so the cuticle is efficient in reducing water loss.

Lenticular Transpiration

On the outer surface of plant stems and of some fruits, are found small openings or pores called lenticles (Fig. 3.2). These lenticles are always

Fig. 3.2   The surface of an apple fruit showing the lenticles (light-colored spots).

open so water does evaporate and pass out through them at all times. Such loss is called *lenticular transpiration*. However, lenticular transpiration is responsible for only a small fraction of the water lost from plants.

Stomatal Transpiration

Actually, the amount of water lost either through guttation, lenticular transpiration, or cuticular transpiration is less than 3% of the total water loss by the plant. Over 97% of the water lost is lost through the stomates, a process called *stomatal transpiration*. Therefore, when we refer to water loss or to transpiration, without further qualification, we mean stomatal transpiration.

Stomates are small pores found in the surface of leaves, fruits, and some other plant structures. Each stomate is surrounded by two elongated, specialized cells called *guard cells* (*see* Fig. 3.3). These guard cells form part of the epidermal layer, but differ from other epidermal cells by their elongated shape and by the chloroplasts they contain. Stomates are very numerous on plant leaves, especially on the under surface. If you look at this surface under the microscope you cannot help but see them. Some examples demonstrating the distribution and number of stomates per leaf are given in Table 3.1.

Stomates have two important functions within the plant. One is beneficial, the other is harmful. Their beneficial function is to allow gases to enter and leave the leaf. This is necessary to allow carbon dioxide to move in to be used in photosynthesis, and oxygen produced through photosynthesis to move out into the atmosphere, replenishing the oxygen

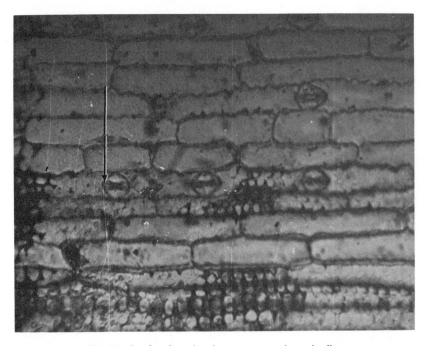

Fig. 3.3   Leaf surface showing stomates and guard cells.

supply of the atmosphere as it is depleted. At the same time, the other function is the loss of water vapor from the leaf via stomatal transpiration. This latter function, though harmful, cannot be prevented. While the gases are diffusing into and out of the leaf, water vapor is moving out with them.

CONTROL OF WATER LOSS

The water content of the plant is very high and must be maintained. In addition, water loss through the plant is the chief means by which soil

Table 3.1   The distribution and numbers of stomates per leaf.

Species	Lower epidermis No./cm²	Lower epidermis No./leaf	Upper epidermis No./cm²	Upper epidermis No./leaf
Hackberry (*Celtis occidentalis*)	22,000	690,000	0	0
Red oak (*Quercus rubra*)	68,000	3,000,000	0	0
Castor bean (*Ricinis communis*)	15,200	2,350,000	7,300	1,125,000
Tomato (*Lycopersicon esculentum*)	10,000	2,125,000	2,300	425,000
Watermelon (*Citrullus vulgaris*)	20,600	387,000	10,600	200,000
Waterlily (*Nymphaea alba*)	0	0	22,000	690,000

moisture is depleted, and in these days of recognized need for the conservation of water within the soil as a source of water for man, any loss of water by the plant must be viewed as undesirable. Indeed, it is probable that transpiration is valuable to the plant, since it is a means by which mineral nutrients are translocated within the plant and may be a means by which plant leaves are cooled, but the rate of water loss by plants is much greater than is necessary for the well-being of the plant. Therefore, let us consider the rate of transpiration to help us to understand how this rate may be reduced.

Temperature and Humidity

The rate of water loss by the plant depends upon many factors, including the rate at which water evaporates from the leaf mesophyll cells and the rate at which it diffuses out into the atmosphere surrounding the plant. The rate of evaporation is greater as the temperature is increased and as the relative humidity in the intercellular spaces decreases. (Why?) The relative humidity within the intercellular spaces depends upon the rate of evaporation and the rate of diffusion of the water molecules out of the leaf. The rate of diffusion of the water molecules out of the leaf depends upon the relative humidity of the atmosphere just outside the stomates, the temperature and the size of the stomate through which the water vapor must diffuse. With an increase in temperature or a decrease in relative humidity outside the leaf, the rate of diffusion and subsequent water loss is increased.

Light and Stomatal Opening

At night the stomates of most plants are closed whereas in the daytime they are open. Some of our desert plants behave just opposite to this but they are exceptions. How does stomatal behavior help conserve water? If the stomates are closed, water loss is stopped since water vapor cannot diffuse out of the leaves, to any great extent, with the stomates closed. However, the rate of water loss through the stomates is not directly related to the extent of opening of each stomate. The stomate does not have to be open wide to allow rapid diffusion of water vapor through it. Even if it is closed half way, diffusion is only reduced about 10%. Stomatal closure can be brought about by either darkness or by water deficits within the plant.

The opening and closing of the stomate is governed by the two guard cells which surround each stomate. When these guard cells have high turgor pressures, they are swollen and the stomates are open. When the guard cells have a low turgor pressure – are plasmolyzed – the stomates are closed. Thus, the opening and closing of stomates are determined by the water relations of the guard cells, and any factor which may alter these water relations will alter stomatal opening. Closure of the stomates during the heat of the day is easily explained, since at this time the plant is losing water more rapidly than it is taking water up, so the whole plant is deficient in water and all the cells may be plasmolyzed. Stomates also close when the soil is dry, since dry soil has a high water potential and therefore will not supply the plant with its water needs, creating a water deficit within the plant. Stomates also close when the plant is growing in water-logged soils, due to a deficiency of water in the plant under such conditions.

Stomates normally open in the morning and close at night, and such diurnal activity is due to the presence or absence of light.

The dependence of stomatal opening on light can be easily demonstrated by floating leaf discs on water. When exposed to the light, the stomates of these leaf discs open. Then when the light is turned off, the stomates close. This opening and closing can be repeated many times, merely by exposing the discs to light or to darkness.

That this light requirement is tied in with photosynthesis is also substantiated, since guard cells devoid of chlorophyll do not react, and since the action spectrum is the same for stomatal opening as for photosynthesis. However, it is not known whether photosynthesis acts by changing the pH of the cytoplasm or by supplying energy carriers, such as ATP, which may act to bring water into the cell. One theory to explain the mechanism of stomatal opening states that photosynthesis is necessary since it supplies energy carriers needed to operate an ion pump mechanism which forces potassium ions into the guard cells, thereby increasing the OP of these cells, and causing water to be taken up, swelling the guard cells and opening the stomates.

Another theory explains the action of light on the stomate as follows: In the light, photosynthesis occurs in the chloroplasts of the guard cells, using carbon dioxide stored in the cell. This use of carbon dioxide reduces the acidity of the cell causing stored starch to be converted to sugar. Since each starch molecule is made up of many sugar molecules when starch is converted to sugar, this increases the osmotic pressure of the guard cells. The high OP causes more water to move into the guard cells by osmosis and thereby increases the turgor pressure of these cells, causing the stomates to open. At night, photosynthesis stops, carbon dioxide from respiration accumulates, the protoplasm becomes more acid converting sugar to starch which reduces the osmotic pressure and the turgor pressure and thereby the guard cells become less turgid, closing the stomates. However, there is no good evidence that carbon dioxide does have any effect on stomatal opening. Therefore, what is known about the mechanism of stomatal opening in the light allows the following generalization: Stomates open when photosynthesis occurs in the guard cells, and this somehow increases the water content of these guard cells, causing them to expand, and this expansion opens the stomates. Stomates close when photosynthesis stops due to an increased loss of water from the guard cells.

Anti-transpirants

Certain chemicals, often called *anti-transpirants*, can be applied to the plant to either cause the stomates to close or to plug up the stomates and thereby stop diffusion of water through them. Phenylmercuric acetate represents the first type of anti-transpirant while oil sprays represent the latter. The use of these and other related anti-transpirants, under field conditions as a means of reducing transpiration and the consequent conservation of soil moisture, is currently being extensively investigated. So far, the results have not been very encouraging.

Wind

The rate at which water vapor diffuses out of the stomates depends too upon the relative humidity just outside the stomate. As water diffuses out through the stomate, water molecules build up their concentration near the mouth of the stomate. This will eventually slow up diffusion through the stomate unless this water vapor is removed. This removal can be greatly speeded up if a wind is blowing. Just a gentle breeze is very efficient in removing the water vapor, but if the wind is too strong the stomates actually close.

Atmospheric Pressure

The atmospheric pressure also has an influence on the rate of diffusion of water vapor through the stomate, but its influence is minor. With an increase in atmospheric pressure, the rate of diffusion decreases.

Water Potential and Transpiration

Since the rate of evaporation of water from the surface of the mesophyll cells is dependent upon the amount of water present in the walls of these cells, the water potential of the plant will also affect water loss. As the water potential increases, the rate of water loss decreases. The increased water potential can be caused by salt or sugar accumulation, by excessive water loss over water uptake or by soil moisture deficits, or other causes of high soil water potential, causes such as low temperature, or high salt content. As previously mentioned, deficits within the plant can also be caused by a high water content of the soil. When the water content of a soil in which a plant is growing is increased through precipitation or irrigation, a series of changes begin in respect to the rate at which transpiration occurs in the plant. If the soil is saturated with water, the rate decreases rapidly until it is nil, and remains low until the water content of the soil in the root zone decreases by gravitational flow. At this time, the rate increases to a maximum, and is controlled largely by the climate. As the soil water content now slowly declines, there is a concomitant increase in soil water potential and in the force necessary to remove the water from the soil so the rate of transpiration decreases. As the soil moisture content approaches permanent wilting percentage, the soil moisture stress increases very rapidly and soon the point is reached whereat the rate of transpiration is nil due to the decrease of available water in the soil.

EXTENT OF WATER LOSS

Just how extensive is the water loss by the plants? We made a statement earlier that the amount of water in the plant at any given time is only a small fraction of the amount of water that enters the plant during its lifetime. We also found that many herbaceous plants replace their entire volume of water each day. If one observes the water level in a stream he will find that during the day it is lower than at night due to transpiration of plants growing along the stream bank. The removal of such plants will usually greatly increase the water level in the stream. Corn plants or apple trees require about 11–15 acre inches of water per year (an acre inch is the water required to cover one acre of ground to a depth of one inch). Forests may require between 5–30 acre inches per year.

Transpiration Ratio

Another way of expressing the water loss by plants is to compare the weight of water that a plant uses while producing 1 lb of dry matter. This value is expressed as the ratio of water used per dry matter produced and is known as the *transpiration ratio*. Such ratios are given in Table 3.2 and show that the values vary considerably depending upon the species of plant studied. Alfalfa requires 800 lbs of water to produce 1 lb of dry matter, whereas a pine tree may require only 50 lbs of water. The lower the transpiration ratio, the more efficiently the plant uses the water it takes up.

Table 3.2   Transpiration ratios of some plant species.

Species	Transpiration ratio
Pine	50
Corn	350
Wheat	450
Apple	500
Alfalfa	800

PLANT WATER CONTENT

One of the easiest determinations to make in the study of the water relations of the plant is its fresh weight. All one needs to do to determine the *fresh weight* is to obtain a plant from the field or greenhouse and weigh it immediately after cutting. After this fresh weight has been determined, the plant material can be placed in an oven at 105°C, left there overnight, and weighed again. This latter weight is the *dry weight*, another important determination often made by plant physiologists. The difference between the fresh weight and the dry weight is the *water content* of the plant, which is usually expressed as percent of the fresh weight

$$\frac{\text{water content} \times 100}{\text{fresh weight}} = \% \text{ water on a fresh weight basis}$$

One need not use the entire plant for these determinations, but can use leaves, fruits, roots, stems, or other plant tissues or organs. However, one must be sure that the plant material is perfectly dry when one obtains the dry weight, and if it is not dry, the sample must be returned to the oven, and later weighed and this continued until the weight does not change with subsequent drying. Bulky materials such as woody stems may require several days or longer until they are dry.

By the above method, one can easily determine the water content of various plant materials. Typical values for the water content, of leaf blades, as expressed as percent of the fresh weight, are reported in Table 3.3. (The percent dry weight can be determined by subtracting the percent water from 100.) From these data, two facts become evident. First, that the weight of leaves is largely water and therefore the leaf blade is composed mostly of nothing more than water. Second, it is evident that the water content of the leaf blade may vary from one species to another. Those species that are trees or woody shrubs have the lower leaf water contents, being about 60–70% water. Those species whose leaves are succulent have the highest water content, about 90% or more water, and the herbaceous plants have intermediate water contents of about 70–90% water.

Table 3.3   The water content of fresh leaves of some plant species.

Species	Water content (% fresh weight)
Iris	90
Dandelion	84
Virginia creeper	75
Osage orange	71
Oak, elm, ash	65

Species Differences

To explain the differences in water content of the leaf blades from one species to another, we must consider the fact that the water content of the leaf, on a fresh weight basis, varies with the amount of water and the amount of dry weight. The plant cell is composed of protoplasm surrounded by a cell wall. This cell wall may be of variable thickness. The water content of protoplasm is in excess of 95% water, and this value is quite constant from one species to another. Therefore, the variation in water content of the leaf blades from one species to another must be due to variations in the thickness and therefore weight of the cell walls. The thicker the wall, the lower the water content of the leaf, on a fresh weight basis. If the cell walls are very thin, the water content would be greater than 90%, whereas if they are thick the water content would be less. Therefore, the leaf blades of succulent plants have less cell wall material than the leaf blades of trees or shrubs.

Although most of the variation in water content of plant organs will be due to differences in the amount of cell wall material, small variations may occur due to differences in the water potential of the cells and to differences in location of the organ on the plant. The latter is evident in Table

3.4 which shows the differences in water content of the upper and lower leaves of an oak tree about 30 feet high. The lower leaves have the highest water content because they are closer to the source of water, namely, the roots. It takes work to pull water up a stem and this results in a lower water content in the higher leaves. However, in herbaceous plants, such differences do not exist due to the smaller size of the plant.

Table 3.4   The water content of upper and lower leaves of an oak tree.

Location of leaves on the tree	Water content (% fresh weight)
Upper	65
Lower	69

Age Differences

As a tree grows taller and older, the lower leaves become shaded so much that they cannot produce sugars through photosynthesis. Therefore, their water potential is less than that of the leaves higher up the plant. As a result of these differences in water potential, the lower leaves lose their water to the upper leaves. Therefore, as the tree gets taller and older, the upper leaves have the highest water content. Senescent leaves – those in old age – lose much of their water potential and, therefore, even on herbaceous plants where we find senescent and young leaves on the same plant we would find that the senescent leaves have a lower water content than the younger leaves. Of course, the young leaves have less cell wall material than the senescent leaves which would also contribute to a difference in water content. The explanations given above should allow you to interpret and predict differences in the water content of leaves of various plant species.

The woody stems and twigs of trees are composed of cells that have very thick cell walls, so these structures should have lower water contents than the leaf blades. Indeed they do. The water content of sapwood is about 50%, and that of bark and heartwood a little less. Again differences occur depending upon the location of the wood on the plant, since the twigs near the top have a higher water content than those lower on the tree. This is probably due to differences either in water potential or in the amount of cell wall material. Perhaps water potential differences are most important here.

The water content of flower petals is often in excess of 90%, whereas that of mature seeds, pollen grains and spores may be less than 10%. In the flower petals the high water content would be due to low cell wall material, but in the mature seeds, pollen grains and spores, the low water content is due largely to the lack of water within the cells, since their cells become desiccated with maturity.

Time Variations

Short-term variations in the water content of plant structures also can be observed. Variations from hour to hour throughout the day have been reported, and usually show a lower water content during the day than at night. One can interpret such variations in terms of the amount of water taken up by the plant compared with the amount of water lost.

Over a 24 hour period, the rates of water uptake and water loss vary, as can be seen in Fig. 3.4. During the daytime, water loss usually exceeds water uptake resulting in a reduction in the water content of the plant. At night, the reverse is true, with water uptake exceeding water loss, restoring the water deficit that occurred during the daytime.

Variations with Weather

The magnitude of the water deficit that occurs over a 24 hour period varies considerably with the weather conditions that prevail. On a rainy day, the rate of water uptake keeps up with the rate of water loss creating

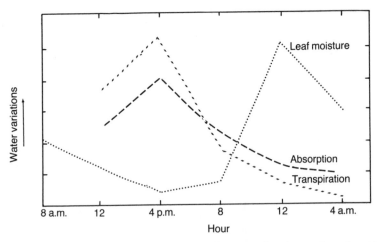

Fig. 3.4 Plant water uptake, loss, and content as these vary from hour to hour throughout the day. (From Kramer, P. J. *Plant and Soil Water Relationships.* Copyright 1949, McGraw-Hill Book Company. Used with permission of McGraw-Hill Book Company.)

no deficit. On a cloudy, cool day with adequate available soil moisture, the rate of loss only slightly exceeds the rate of uptake, creating a slight deficit. On an overcast, warm day, some deficiency may occur, but on a hot, sunny, windy, dry day, the type of weather that frequently occurs during the summer, the rate of loss greatly exceeds the rate of uptake and a severe deficiency results which reduces growth and often causes temporary wilting. Such deficiencies of water may occur on such a day even in the presence of abundant soil moisture, but if the soil moisture content is low, the deficiency is even greater. No doubt you have observed how fresh the plants look after a rain. This is due to the alleviation of water deficiencies within the plant by direct absorption of water through the leaves, and by reduced transpiration.

As seen in Fig. 3.4, on a bright sunny day during the growing season, the maximum water content occurs about 1:00 a.m. in the morning, during the nighttime when the rate of transpiration is nil and the water uptake exceeds water loss. The lowest water content of the leaves occurs about 4 to 5 in the afternoon when the rate of transpiration has been high and has exceeded that for water uptake for some time. Since leaves represent the structures from which most water loss occurs and are located further from the source of water, one might expect and would find that hourly variations in water content are of greater magnitude in the leaves than in other plant parts. As we previously learned, water within the plant acts as a unit, and any deficit in one part of the plant will cause water to move from other parts of the plant to the part where the deficit occurs. Therefore, we might expect that deficits in water in the leaves would result in deficits in water in other parts whereas the maximum water contents of other parts would occur later than in the leaves. Leaves do show the greatest variation in water content and they do recover more quickly from water deficits. In the stems and roots, the maximum water content is found in the morning hours just before sunup. Fruits also show variations in water content similar to those of stems and roots. However, fruits do not seem to be able to hold on to their water as well as other plant parts, so they too show great variations in water content. Fruits rapidly lose water to other plant parts, where deficits may occur, and water loss from these fruits is of such magnitude that marked changes may occur in diameter in response to water deficits in the plant, being of smaller diameter during the day than at night.

It should be pointed out that daily variations in water content also cause daily variation in OP, TP, and therefore in the water potential of plant cells.

If the soil has a high salt content, if frozen, or has a high or low water content, the plant may not be able to renew at night the water lost during the daytime. As a result, the water content of the plant continues to decrease from day to day and may reach such a low level that the plant dies. Some species die when the water content is only slightly lowered, whereas others can tolerate extreme desiccation. The difference between these extremes is a measure of the drought tolerance of plants, and will be considered in the next chapter.

In this chapter, we have studied the mechanisms by which water moves through the plant from the roots, where it is taken up, to the cells that need it or to the leaves and other areas where it is lost largely by evaporation. We have learned that the plant has a high water content, but a water content that varies with the species, plant part, and the environment in which the plant is growing. Let us now consider why this water is important to the plant, how the plant maintains a high water content and what happens to the plant when this needed water is no longer available.

The Functions of Water Within the Plant

PROPERTIES OF WATER

Water is a most important and most interesting biological substance. Chemically the water molecule is composed of one oxygen and two hydrogen atoms with an average *molecular weight* of 18.016. About 99% of the hydrogen and oxygen atoms are ^1H and ^{16}O although traces of heavier *isotopes* of both hydrogen and oxygen are present, and may affect the properties of water. Since three isotopes of oxygen and three of hydrogen may be present, eighteen different kinds of water molecules may be present, each varying slightly in weight.

In the water molecule, the hydrogens are bound to the oxygen to form a *bond angle* of 105°, with the distance between the hydrogen and oxygen atoms at the covalent bond being 0.99 angstroms.

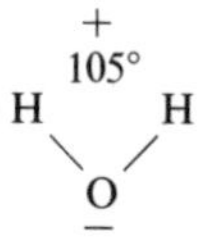

$$+$$
$$105°$$
$$H \qquad H$$
$$\diagdown \diagup$$
$$O$$
$$-$$

Since the hydrogen atoms are not on opposite sides of the oxygen (bond angle is not equal to 180°), the hydrogen side of the molecule has a positive electrical charge and the oxygen side a negative charge. As a result of this unequal distribution of electrical charge around the molecule, the water molecule is *polar*, it has a positive side and a negative side, and therefore will attract and become attached to other atoms or molecules with electrical charges. Therefore, water is a good solvent for other polar molecules such as amino acids or other water molecules, and for ions such as calcium or potassium.

Hydrogen Bonds

Water molecules are found as free molecules largely only in the vapor state. This is due in part to the polarity of the molecules but primarily to the unique ability of water molecules to join together by *hydrogen bonds* (hydrogen bridges). Hydrogen will join together small but strongly electro-negative atoms, such as O and N, as found in other water molecules, nitrates, phosphates, proteins, sugars, organic acids, amino acids, alcohols, etc. Each water molecule has the ability to join with four of its neighbors by hydrogen bonding. Such is the case with ice. But as the temperature rises, these bonds break so that in the liquid state each molecule is joined to but 3, 2, or 1 of its neighbors depending upon the temperature. The hydrogen bond is weaker than ordinary ionic or co-valent bonds but stronger than van der Waals forces. Actually, in a hydro-gen bond, the hydrogen is not located in the center between the electro-negative ions but is attracted more strongly to one than to the other as shown below, where the broken line represents the weaker bond.

$$O—H \cdots\cdots O$$

Density

Due primarily to hydrogen bond formation, water has several very important but unique properties. Solid water (ice) has a lower density—it is lighter—than liquid water due to hydrogen bonding. Most other chemicals become more dense as the temperature decreases and as they become transformed from a liquid to a solid. Not so with water. Its maxi-mum density is near 4°C, and it becomes less dense at temperatures below and above this temperature as can be seen in Fig. 4.1. As water freezes, it expands. This fact is important biologically since if this were not so, ice would form on the surface of lakes, streams, etc. in Winter and sink to the bottom. This would result in the entire lake becoming solid ice during Winter. During the Summer, only the surface of the lake would melt since water is a good insulator. As a result, the hydrophytes—algae and other water plants—would be continually embedded in ice. They could not exist long in this condition. On the other hand, the increase in volume that accompanies freezing can be harmful. For instance, when ice forms in the vacuole of the cell, the expansion can and does mechanically damage the cell.

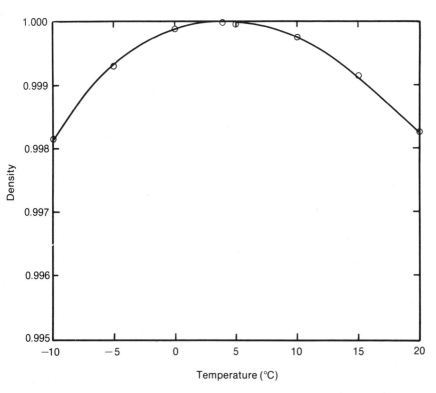

Fig. 4.1 The relationship between temperature and the density of water. (Data of Hadgman, C. D. (Ed.), *Handbook of Chemistry and Physics*, 1955–56. Edition 37. Chemical Rubber Publ. Co., Cleveland, Ohio).

Fusion Temperature

As the temperature of ice increases, the melting point or freezing point is reached at which the ice changes to liquid water (melts). This temperature is 0°, a temperature which is very high compared with other molecules similar to water that do not form hydrogen bonds. In fact, if it were not for hydrogen bond formation, ice would melt or water would freeze near −100°. The conversion of water to ice and vice versa occurs sharply, not gradually, as the temperature changes. However, such a change requires energy or emits energy to the extent of 79.7 cal/g. Thus, as a gram of ice melts, 79.7 calories of energy are absorbed as heat which lowers the temperature of the surrounding air or as one gram of water freezes, 79.7 cal of heat is given off. Thus, the temperature is stabilized.

Specific Heat

Water is well known for its high *specific heat* (heat capacity). About one calorie is required to raise the temperature of water 1° over the temperature range from freezing to boiling. Thus, the temperature of the atmosphere is maintained surprisingly constant by the ability of water to take up or emit large amounts of heat before its temperature changes. It also functions this way in plants to maintain their temperature within the range needed to sustain their living activities.

Heat of Vaporization

As the temperature approaches 100°, water molecules leave the liquid water and go off as a vapor (vaporize or evaporate). When this change from the liquid to vapor state occurs, 538.7 cal/kg of water is required. This is a very high *heat of vaporization* and its magnitude is due primarily to the need to break the hydrogen bonds. In any respect, evaporation requires a great deal of heat and thereby lowers the temperature. Plant leaves are cooled slightly by evaporation. On the other hand, condensation, which is the reverse of evaporation, releases energy to warm the environment. In summary, water is a very good agent for maintaining a rather constant temperature of the plant and its environment due primarily to the ability of its molecules to form hydrogen bonds.

Surface Tension

Within a container of water, each water molecule is attracted by and attracts its neighbors on all sides. However, on the surface this attraction is only in a downward direction since few water molecules exist above the surface molecules. This results in a *surface tension* of a magnitude of 72.8 dynes/cm at 20°, a value which varies considerably with the temperature. As the temperature increases the surface tension decreases until at 100° it is virtually non-existent. Surface tension is very important in the physiology of the plant as seen when we considered capillarity, in Chapter 2.

Radiation Absorption

Water is able to strongly absorb certain types of electromagnetic radiation, especially infrared radiation (heat waves) and red light. That is why lakes and ponds of clear water appear blue or bluish-green. The water molecules absorb the red light from sunlight, leaving the blue light to be reflected and seen.

Dissociation

Water molecules are not the only particles found in water. These molecules have the ability and do ionize or break up to form hydrogen and hydroxyl (OH) ions. In so doing, the hydrogen loses an electron to become a positively-charged proton (H^+) and the hydroxyl (OH^-) gains an electron to become negatively charged. The proton is called a *cation* because it is positively charged and the OH ion an *anion* since it is negatively charged. The proton becomes attached to a water molecule to form hydronium ions (H_3O^-). This can be written as follows:

$$2H_2O \rightleftharpoons H_3O^+ + OH^-$$

Ionization is reversible so an equilibrium is set up. The *ionization (dissociation) constant* of water is 1.008×10^{-14} at 20°, which indicates that only a very small fraction of the water molecules are ions at a given time. However, since the ions are electrically charged they are able to conduct an electrical current, and we can determine the ion concentration by the ability of the water to do so. Pure water has a theoretical *specific conductivity* of 0.038×10^{-6} mhos/cm at 18°. However, water is never pure since it always contains dissolved gases, salts, and organic matter which raises its specific conductivity above that stated above.

IMPORTANCE OF WATER TO THE PLANT

Now that we have been introduced to this remarkable chemical known as water, let us consider why it is so important to the plant. We know that if a plant suffers from a lack of adequate amounts of water, it will die, so water must serve essential functions within the plant. Actually, the functions of water within the plant are many but can, perhaps, be listed in five categories. First, as indicated earlier, water helps to maintain a relatively uniform temperature of the plant tissues. Second, it is an excellent solvent. Third, water is necessary for the translocation of chemical substances through the plant. Fourth, water is a raw material for biochemical reactions and therefore participates in several of the chemical reactions which occur in plant cells. Fifth, water is necessary to maintain turgidity. Now let us consider how water functions in each of these capacities.

HOW WATER FUNCTIONS

Temperature Stability

Water plays an important role in helping to maintain a temperature of the plant structures that will permit the plant to remain alive. As mentioned earlier, water has a high specific heat or heat capacity. As a result, water will absorb 1 cal of heat before its temperature is raised 1°. Water will therefore absorb a great deal of heat before the temperature of the water is raised high enough to be injurious to the plant. Since plant cells are composed mostly of water, there is sufficient water present to prevent the temperature from rising too high. If a healthy leaf and a dry leaf are exposed to the same amount of heat, the temperature of the dry leaf will go much higher than that of the healthy leaf since the dry leaf does not have water to absorb the heat.

Water also lowers the leaf temperature during transpiration. Since water has a high heat of vaporization, for each gram of water that is evaporated from the cell walls of the leaf mesophyll cells during transpiration, 585 cal of heat is used, thereby lowering the temperature. Therefore, one might expect that the more rapid the rate of transpiration, the lower the leaf temperature. This is a valid assumption and does point out one benefit of transpiration. However, this is not the only means of heat loss by plants. When plants are growing in the sun, the heat from the sun strikes the plant. Some of this heat is absorbed by the water due to its high heat capacity, some is used to vaporize water but much of the heat is lost from the plant by the physical processes of conduction, convection, or reradiation. Loss by *convection* occurs by heating the air in contact with the plant, and the heated air, being lighter, then rises carrying the heat away from the plant. Heat is lost through *reradiation* when the heat absorbed by the plant is converted to heat waves which are sent off through the atmosphere away from the plant, just as a heat lamp sends out heat waves that heat anything near it. Heat is lost from the plant by *conduction* or, in other words, by traveling through another material away from the plant. As a result of reradiation, conduction, and convection, most of the heat that strikes a plant is carried away from it, leaving little for transpiration to dispose of. Nevertheless, transpiration does help to keep a plant cool.

The conversion of water to ice also keeps the plant at a more favorable temperature because when 1 gram of ice is formed, 79.7 cal of heat are given off, raising the temperature. This fact is sometimes made use of in the orange groves of California, where farmers irrigate their orange groves when there is danger of frost. As this irrigation water turns to ice, it raises the temperature enough to often prevent frost damage to the oranges.

Solvent

Water is an excellent solvent in plants. Many substances will dissolve in it to form a solution. In fact, it will dissolve so many chemical substances that it has been referred to as the universal solvent. There are three reasons why it is such a good solvent. First, many solutes will combine with it chemically whereby they become dissolved. For example,

$$KNO_3 + H_2O \longrightarrow KOH + HNO_3$$

or

$$CO_2 + H_2O \longrightarrow H_2CO_3$$

Second, since water is composed of polar molecules, each water molecule having a positive and a negative side, it will react with other polar molecules to dissolve them. The reason for this reaction is that if two objects have the same kind of electrical charge they will repel, or move away, from each other. However, if they have opposite charges — one positive and one negative — they will attract each other (come together). There are many polar molecules in plants, molecules such as those of amino acids, organic acids, and inorganic ions of various types. The third reason that water is such a good solvent is because of its ability to form hydrogen bonds. This it does with other water molecules, or with other molecules containing oxygen or nitrogen, molecules such as amino acids, organic acids, sugars, proteins, etc. Being a good solvent is important since these solutes must move through the plant in water, and must perform chemical and physical reactions, many of which occur in water medium.

Translocation

One of the very important functions of water within the plant is to function in translocation. We previously learned how water is translocated through the vessel elements up the xylem from the roots to the tops of the plants, and how water is translocated from cell to cell through the cell walls and intercellular spaces. However, water is not the only chemical translocated in the plant. Sugars produced in the leaves through photosynthesis must move to the root cells and other cells that do not carry on photosynthesis. All plant cells need sugar but only those with chlorophyll synthesize it. Nitrogen compounds, hormones, and other materials as well as the mineral nutrients or fertilizers all need to be translocated from one part of the plant to another.

Translocation can occur from one location within a cell to another location within the same cell, a process known as intracellular translocation. Such movement is either by diffusion or by a process known as protoplasmic streaming. Or translocation can occur from one cell to another, or across cell membranes, or through cell walls. However, when translocation is considered without further qualification, it is assumed that we refer to translocation from one organ to another.

Early in the history of this world, when all plants were unicellular, translocation was no problem. As the plants became more advanced so that they consisted of a few cells, rather than one, but still remained submerged in water, translocation still was not a problem. But, as the plant migrated to a terrestrial environment, specialized structures for translocation became a necessity. Now the plant encountered the problem of moving materials from the cells that absorbed them, through living cells that used them, to cells far removed from the site of uptake. Such movements had to involve specialized cells that did not monopolize the substances being translocated. These are found in the vascular tissues.

One type of vascular tissue is found in the wood, and is called the xylem. We have already seen how water is translocated upward through the xylem. However, mineral nutrients—fertilizers—are also translocated upward through the xylem, from the roots, where they are taken up by the plant from the soil, to the leaves and other aerial parts of the plant. These mineral nutrients move upward in the form of ions or of inorganic salts, dissolved in the water. As the water moves up the xylem, the mineral nutrients dissolved in the water move up with it.

However, the composition of the xylem stream does not remain the same as it moves from the roots to the tops. The higher it moves, the more dilute it becomes. The reason for this is that the mineral nutrients leak out of the vessels into the surrounding cells. Some of these are accumulated by these adjacent cells, some are translocated through the ray cells that traverse the stem, some are precipitated in the vessels and remain there as insoluble particles, some are deposited in the buds along the way and what is left is finally moved to the leaves via the transpiration stream where they accumulate in the leaves. These vessel elements are the functional cells of the xylem and they form a path that goes straight up the stem. If one places mineral nutrients in the soil on one side of a tree, they will move up only to the leaves on the side of the tree at which they were placed. Lateral transfer is much slower than vertical transfer.

It should be noted that the velocity of movement of these mineral nutrients up the stem will be the same as the velocity of movement of the water up the stem. The faster the rate of transpiration, the faster mineral nutrients move upward.

So, we can generalize and say that water and mineral nutrients move up the plant through the vessel elements of the xylem, the mineral nutrients moving dissolved in the water of the transpiration stream.

In the bark, just external to the cambium layer, is another vascular tissue, the *phloem*. This is made up of tubes of living sieve elements, placed end to end. It is through this phloem that nearly all of the organic compounds move either in an upward or downward direction. Through the phloem, carbohydrates, in the form of the sugar, sucrose, are translocated downward or upward from the leaves where they are produced during photosynthesis, to the roots, flowers or fruits where it is needed. There is also a net upward movement of carbohydrates and other organic compounds from the roots to the tops of the plants under certain conditions; in the Spring of the year when growth is resumed but the leaves have not entirely developed; when flowers and fruits are developing, since at this time more sugars are needed than are produced by the leaves; during early stages of seedling development, when the leaves are not producing adequate amounts of sugar so the plant must obtain its needs from the seed; during the early stages of shoot growth from bulbs, rhizomes, tubers, etc.; and finally when the leaves are too young to produce all of the sugars needed by the plant.

Some mineral nutrients are also translocated through the phloem. Usually this is after the mineral nutrient has been built into an organic compound, such as nitrogen in amino acids, but also they can move as inorganic chemicals. Nitrogen, phosphorus, and sulfur move out of the leaves as parts of organic compounds, but potassium, magnesium, and chlorine move as inorganic ions. Such mineral nutrients move out of the leaf at all times during the day and night so that mineral nutrients are moving out of and into the leaves at the same time. However, movement into the leaf is usually faster than out of the leaf. The exception to this is just before the leaf or flower petal falls from the plant, at which time there is a net movement of some of the mineral nutrients out of the leaf or flower petal to the roots, stems, etc. In this way, the plant can conserve these valuable nutrients.

Development of the plant depends upon the translocation of adequate

amounts of organic chemicals when and where they are needed. Therefore, anything that alters the rate of translocation of these materials will
affect plant growth. This is important to us, so let us consider now some
of the factors that will affect translocation through the phloem.

High or low temperatures will affect translocation. At either extremely
high or extremely low temperatures, growth is reduced. A cool temperature (15–20°C) seems to be about optimum for translocation through the
phloem. One can actually stop translocation through the phloem by
heating the bark with steam or by applying freezing temperatures. In
fact, such treatments can tell whether translocation is occurring through
the xylem or through the phloem because translocation through the
xylem is not stopped by such treatments.

A lack of oxygen stops or reduces transport through the phloem. We
might expect this since, in contrast with the xylem, the phloem cells are
living cells at maturity, and living cells need oxygen to remain alive. This
oxygen is needed for cell respiration. Treatments, such as chemical inhibitors, which will inhibit respiration of the phloem cells will reduce phloem
translocation, but not xylem translocation.

Boron deficiency also reduces the rate of transport through the phloem.
The exact mechanism for this effect is not known, although there is some
evidence that boron alters cell membrane permeability to sugars and
other organic compounds.

It appears that translocation through the phloem involves living cells
and their continued metabolic activity. Large amounts of materials move
through the phloem, for example about $\frac{1}{2}$ gram of sugar may enter a
pumpkin fruit per hour for over a month. The velocity of movement is
rapid too, being about 10–150 cm per hour. In addition, movement can be
bidirectional. How can we formulate a theory to explain the mechanism
of movement through the phloem that will satisfy all of these facts?

One answer to this was proposed by a German worker, Munch, over
35 years ago. He formulated what is known as the *Munch* or pressure-
flow *hypothesis* which is still the best theory to explain translocation
through the phloem.

According to the Munch hypothesis, movement through the phloem is
due to differences in turgor pressure between the source of the organic
compound and the destination. If we look at Fig. 4.2, we can see an
illustration that shows how the Munch hypothesis works. If two osmo-
meters, one containing 1 bar of sucrose solution and the other $\frac{1}{2}$ bar of the
same solution, the solution level will rise to a greater height in the osmometer with the greatest sucrose concentration, due to the greater pressure

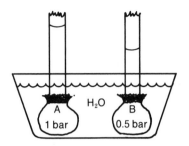 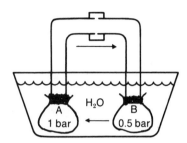

Fig. 4.2 Illustration of the mechanism of the Munch hypothesis which explains how organic molecules, such as sucrose, move through the phloem.

(TP) produced. Now if we join these two osmometer tubes, solution will flow from (a) to (b), because of the greater TP in (a) than in (b). This will continue until the concentrations in (a) and (b) become equal. However, if they do not become equal, water will be forced out of (b), travel through the tray to (a) and a cycle will be formed. This cycle will continue as long as the sugar concentration in (a) is greater than in (b). If we continuously remove sugar from (b) and add sugar to (a), the cycle will continue indefinitely. Now let us look at this model as it relates to the plant. If (a) is the leaf cells, producing sugar through photosynthesis, and (b) is the root cells, removing sugar by converting it to starch, and if the phloem is the osmometer tubes and the xylem the tray, we can see how solution can flow from one part of the plant to another in response to changes in TP. The sugar and other organic compounds move then along with the water, dissolved in it. Now then, why would the metabolic activities of the phloem cells alter the rate of flow? Well, the answer is that such activities alter the permeability of these cells to the solution.

Raw Material

Water is also important because it serves as a raw material for many biochemical activities within the plant and many of these are very important. We shall study them later, but let us list the role of water in photosynthesis and in many hydrolysis reactions as important functions of this type.

Maintain Turgidity

Perhaps one of the most obvious functions of water within the plant is to maintain turgidity. This is obvious because if water fails to maintain turgidity, the plant wilts. We have all seen plants with their leaves wilted. Perhaps this was due to the fact that we failed to water the plants as often as we should have or it failed to rain for an extended period of time so the plants in our yards or farms or out in the field, dried out and wilted. How much water a plant can lose without wilting depends somewhat upon the type of plant. If it is a plant that normally grows in the sun—a *sun plant*—it may be able to delay wilting until 20–30% of its normal water content is lost, whereas if it is a plant that normally grows in the shade—a *shade plant*—it may wilt after losing only 3–5% of its normal water content. The cause of wilting is a loss of turgor pressure. If turgor pressure is present, the pressure pushes the protoplasm against the cell wall, making the cell firm, much like an inner-tube makes a tire firm when the tube contains air. When turgor pressure is lacking, the cell is soft and pliable (becomes *plasmolyzed*), much like a tire goes flat when it loses its air. Leaves remain firm and expanded on plants only so long as their cells are *turgid*, or, in other words, have turgor pressure. When the cells of the leaf lose their turgidity, the cells become pliable and soft and the leaf collapses.

As seen earlier, the loss of turgidity also causes the stomates to close. This decreases the rate of transpiration, raising leaf temperature, and reduces the rate of photosynthesis by shutting off the carbon dioxide supply. Without photosynthesis the plant cannot grow and increase in yield and dry weight. Therefore, a loss of cell turgidity reduces the growth of the plant or stunts its growth. Turgor pressure is not only necessary to increase the dry weight of the plant, but also to increase its size. The growth of the plant is due to an increase in cell division and/or an increase in cell enlargement. Both are reduced when turgidity is lost. More will be learned about these two important growth processes in a later chapter.

So you see, water has some very important functions to play within the plant. Is it any wonder that the plant soon dies if deprived of water. Mention should now be made of the physical state of water within the plant.

STATES OF WATER WITHIN THE PLANT

Water within the plant is either free water or bound water, as is water in the soil (Chapter 2). Most of our discussion so far has been concerned with *free water*; that water present in the liquid form, and not bound to the physical or chemical structure of the plant. Free water is therefore free to move within the plant from one location to another. The *bound water* is that water which is bound to the plant constituents and is not free to move from one part of the plant to another. This water may be bound in any one of four ways. It may be bound to ions and/or polar molecules where it hydrates these chemicals, or it may be bound with hydrogen bonds to other molecules including many non-ionized or non-polar molecules, or it may be bound to colloids, especially to the protein of the protoplasm or to the cell wall materials, whereby it hydrates these colloids, a phenomenon important in imbibition. Also it may be bound through capillarity.

The ratio of bound to free water varies with the water content of the plant. In protoplasm about 5% of the water is bound when the protoplasm has its normal water content. As the cell loses water, the water that is lost is mostly free water, so the percentage of bound water increases. Therefore, as the plant dries out, its bound water percentage increases and its free water percentage decreases. A plant cannot normally survive if it contains only bound water.

DROUGHT TOLERANCE

Plants vary considerably in their ability to survive as their water contents are reduced. However, the lack of adequate amounts of water within the plant is one of the most frequent causes of reduced plant growth in the field. We seldom realize how often plant growth is reduced in the field, due to the lack of an optimum water content. This frequently occurs even in plants growing in areas of abundant rainfall, and is caused by a more rapid loss of water than water uptake, which results in a below-optimum water content of the plant. The ability of the plant to resist the undesirable effects of water deficits within the plant is known as its *drought tolerance*.

Physiologists and others have tried for many years, without much success, to find a mechanism that is common to all plant species to make them drought tolerant. No doubt there is no one mechanism associated with drought tolerance. Life is not that simple. Each species has one or more mechanisms for protection against water loss and its subsequent effects, and the number of such mechanisms is so great that all species need not have the same mechanisms. Let us now consider what some of these mechanisms of drought tolerance are.

To protect itself from drought, a plant can take four approaches. It may develop a more efficient system for water uptake from the soil, it may have structural or physiological modifications to prevent water loss, and subsequent injury, it may store water, or it may normally grow only during the season of maximum precipitation. Since a loss of water is common in plants, each species has various modifications in its structure and physiology to protect it from death by water loss.

Root System

A study of desert and other drought resistant plants reveals that some of them, such as the saguaro cactus or the creosote bush, have very extensive root systems, root systems that do not go deep into the soil, but that branch profusely and cover large areas just under the soil surface. This allows more root surface for water absorption in the region of the soil where the water is most likely to be found. This is also the explanation for the drought tolerance of sorghum.

Some species, such as alfalfa, send their roots down very deeply into the soil, and therefore are able to extract water from the soil at depths below that available to other species.

Cuticle

One common structural modification increasing drought resistance is a very thick cuticle on the leaf surface. Since the cuticle is impermeable to water, this reduces water loss by cuticular transpiration. All species have a cuticle on their leaves, but the thickness of this cuticle varies from species to species. The thicker the cuticle, the less water loss. Pinon pine and creosote bush are examples of species with extremely thick cuticles.

Leaf Loss

Ocotillio (*Fouquieria* sp.) and *Euphorbia splendens* are often without leaves. Most of the growing season, their leaves are absent. Only during the short rainy season do they develop leaves. A month or so later, when the water becomes deficient in the soil, their leaves abscise, not to be formed again until next year. This is a good mechanism for reducing water loss, since over 90% of the water lost by plants is through their leaves, but this also greatly reduces growth rate since the photosynthetic area is also reduced.

Small Leaves

Many of our desert plant species have very small leaves, for example the palo verde, or few leaves, as in desert hackberry. These being small or few in number, give less surface for the loss of water to occur from via stomatal or cuticular transpiration.

Stomatal Variations

Some species have leaves with few stomates through which water can be lost, and these leaves sometimes abscise rapidly under conditions of moisture deficiency. In fact, moisture deficiency causes the leaves of many plants to abscise, thus reducing the surface from which transpiration can occur.

Some species have stomates that are modified structurally to reduce water loss through them. Such modifications include placing the stomates below the epidermis, a condition known as sunken stomates; covering the stomates with hairs to reduce water loss; closing the stomates during the daytime when water is deficient; and the presence of small stomates.

Water Storage

Some plant species found in the desert are able to survive long periods of drought because they store water in large cells within their roots, stems, or leaves during the rainy season. Such species also have modifications to reduce water loss, but can survive for many months or even years with no rain, using the water stored in their large parenchyma cells. Barrell cactus and saguaro cactus are two good examples.

Protoplasmic Resistance

In spite of all of these possible modifications which make the plant more able to obtain or keep water, the really important factor in drought tolerance is the ability of the protoplasm of the cell to survive desiccation. Structural modifications merely postpone the time when the water content of the plant tissues will decrease. Then the important factor will be how much water loss can the protoplasm endure.

Selaginella sp. can be dried out and remain so for an extended period of time, only to be revived and to continue to grow when water again becomes available. Creosote bush, a common inhabitant of our desert regions of the United States, retains its leaves all year and still can survive well under drought conditions. Seeds, spores, and pollen grains can also be dried out and survive extended periods of desiccation. What changes occur in the cell to make the plant resistant to such severe treatments? We cannot answer this question accurately, but let us look at some of the reported changes that take place as a cell dries out. Often there is an increase in osmotic pressure of the cell under drought conditions. One might expect this since with a reduction of water content, the solution remaining would be more concentrated. However, this is not due entirely to water loss. With drought, there is often an increase in the soluble solute content. Normally, when adequate water is available, the OP of the cell is due to solutes of both inorganic and organic chemicals. Under such conditions, about $\frac{1}{4}$ of the OP is due to organic solutes, and $\frac{3}{4}$ to inorganic solutes. With drought, less inorganic molecules are taken up, but there is an increase in water-soluble organic molecules, especially sugars and amino acids. These arise from a conversion of starch to sugars, and from protein breakdown which releases amino acids, amides, and soluble proteins. Such an increase in OP would allow the plant to retain its water longer, due to the higher water potential of the plant.

As drought conditions persist, soluble nitrogen compounds are further broken down, releasing their nitrogen in the form of ammonia. As this ammonia accumulates, it soon becomes toxic, killing the plant. Some of this ammonia is removed by the plant in the formation of proline and amides, but as the water deficit continues even these cannot keep the ammonia content below the toxic level. Indeed, the ability of the plant to keep its ammonia accumulation low may be related to its ability to tolerate drought.

In this chapter, we have had the opportunity to learn the properties of water, how water functions within the plant, and finally some of the consequences of the lack of adequate amounts of water within the plant. Much remains to be learned about water as a plant nutrient, but we certainly can now appreciate its value to the plant and the frequency with which water limits the growth of our plants. Let us now move on to a consideration of other nutrients, particularly the inorganic elements of the soil that serve as nutrients to the plant.

Chapter Five

Mineral Nutrition of Plants

The soil is the primary source of water for plants, and it also supplies the inorganic elements that are found in plants, many of which are essential as nutrients. This chapter is primarily concerned with these elements in multicellular plants.

PLANT ASH

Previously, we learned that if plant material was dried in the oven, the water that formed the main constituent of the plant was driven off, and the dry matter that contained all other plant constituents except the water was left. If this dry matter is now placed in an oven at a higher temperature, about 500°C, and left there for several hours, the organic compounds decompose, converting them to carbon dioxide gas and to water, both of which are volatile and will pass off into the atmosphere. What are left are the inorganic components of the plant which comprise the *ash*. The ash then contains the inorganic chemical elements that were taken up and accumulated within the plant. The amount of ash found in plant tissues varies with the species of plant sampled, the organ of the plant sampled, the composition of the soil in which the plants were growing, and the amount of water previously transpired by the plant.

Species Variations

The ash content will vary from one species to another, as shown in Table 5.1. Here it can be seen that the ash content of hydrophytes—those plants living in or near water—is lowest, being about 2-3% of the dry

Table 5.1   The ash content of some plants.

Type of plant	Ash content (% dry wt.)
Bacteria	6–14
Algae	11–21
Fungi	1–10
Mosses	2–3
Hydrophytes	1–3
Cultivated plants	7–18
Halophytes	4–24

weight of the plant. Cultivated plants have a higher ash content, about 10% of the dry weight, and halophytes – those plants living in a saline soil – with the highest content of ash, being about 10-22% of the dry weight. Therefore, the ash content is often about 1% of the fresh weight of the plant, which is not very high, but, as we shall see later, many of the ash constituents are necessary for the plant, and the plant will not survive if these nutrients are not present in adequate amounts, even though the amount may be small.

Organ Variations

The ash content also varies from one organ to another, of the same species, as can be seen in Table 5.2. The leaves have the highest ash content with the ash contents of the other plant organs being less but quite similar. As revealed in the last chapter, inorganic elements that enter the plant through its roots become dissolved in the water of the transpiration stream and move into the leaves with this water, as water is lost through transpiration. The inorganic elements do not evaporate as water does, so as the water is lost, these elements remain behind and accumulate in the leaves. This is why the leaves have the highest ash content of all plant organs. There is a direct relationship between the available inorganic elements content of the soil and the ash content of the

Table 5.2   Ash content of various plant structures.

Plant structure	Ash content (% dry wt.)
Wood	Less than 1
Bark	1–5
Roots	3–5
Fruit and Seed	2–8
Leaves	2–20

plant. As more inorganic elements become available to the plant in the soil, they are taken up and accumulated by the plant. This is why halophytes have a higher ash content than hydrophytes.

Variation with Transpiration

As discussed in the previous paragraph, the ash content of the plant increases with the rate of transpiration. If two individual plants of the same species are growing in the same soil, but one is transpiring more rapidly than the other, the one with the highest rate of transpiration will have the highest ash content.

Ash Composition

As mentioned earlier, the ash is composed of inorganic elements that enter the plant. If a chemical analysis of this ash is performed to determine its composition, all of the chemical elements present in the ash that are present in the soil in which the plant is growing will be found. Therefore, the composition of the ash depends upon the composition of the soil in which the plant is growing. However, the amounts of each of these chemical elements in the ash will not be the same as in the soil, nor will the ratio of one chemical element to another be the same in the plant as in the soil. In the first place, the plant only takes up those chemical elements that are dissolved in the soil solution so some may be present in the soil in the insoluble form which the plant cannot use. In the second place, plants are selective as to what chemical elements they take up, as will be discussed later. Nevertheless, the ash is composed of two groups of inorganic chemical elements, the *mineral nutrients* and the *inert elements.*

THE MINERAL NUTRIENTS

The mineral nutrients are those inorganic chemical elements that are essential for the plant and with which the plant cannot do without. They comprise part of what is known as the essential elements of plants. These essential elements are composed of the mineral nutrients plus those inorganic elements that make up the organic compounds, namely carbon, hydrogen, and oxygen. In higher plants there are currently considered to be sixteen essential elements, thirteen of which are mineral nutrients.

Criteria for Essentiality

To be classed as a mineral nutrient, an inorganic element must satisfy three criteria. First, the plant must not be able to complete any phase of its life cycle in the absence of or lack of adequate amounts of the element. Second, no other element can substitute for this element to satisfy the first criterion. Third, the elements must be directly involved in the plant, and not act through another element or condition to satisfy the first criterion. If a chemical element satisfies all of these criteria we consider it to be a mineral nutrient. If it fails to satisfy any one of these, we call it an inert element. As shall be seen later, it is possible that some of the inert elements are, in reality, mineral nutrients, but it has not been possible to demonstrate their essentiality due to inadequate experimental techniques.

The thirteen inorganic elements presently considered to be essential for higher plants and therefore called mineral nutrients include nitrogen (N), phosphorus (P), potassium (K), magnesium (Mg), sulfur (S), calcium (Ca), iron (Fe), copper (Cu), zinc (Zn), boron (B), manganese (Mn), molybdenum (Mo), and chlorine (Cl). All other inorganic elements, such as sodium, aluminum, silicon, etc. are inert elements.

According to the criteria that must be fulfilled to be considered as mineral nutrients, each of the thirteen mineral nutrients is essential for the survival of the plant. In the absence of adequate amounts of any one, the plant will die without completing its normal life cycle. Therefore, it is perhaps not valid to say that any one is more important to the plant than any other.

Macronutrients

For convenience, the mineral nutrients are often subdivided on the basis of the relative amounts of each found within the plant, into macronutrients and micronutrients. The *macronutrients* are those found in greatest concentrations within the plant, and include nitrogen, phosphorus, potassium, magnesium, sulfur, and calcium. There are, therefore, six macronutrients. As stated above, the macronutrients are no more important to the plant than are the micronutrients.

Micronutrients

The *micronutrients* are those mineral nutrients that are found in lesser amounts within the plant, and include iron, copper, zinc, boron, manganese, molybdenum, and chlorine. Some of these micronutrients are needed only in very small amounts, in fact, in such small amounts that it is often difficult to even demonstrate their presence.

Since the mineral nutrients represent the function of the ash that is of most importance, we shall be concerned primarily with these throughout the remainder of the discussion of mineral nutrition. However, it should be noted that in a few species, especially among the lower plants such as the fungi and algae, but also rarely in the higher plants, an inert element will be found to be essential and therefore will have to be classed with the mineral nutrients, or a macronutrient for higher plants may be a micronutrient in lower plants.

ENTRANCE OF MINERALS INTO PLANTS

Now that we know something about the presence and composition of ash within the plants, let us consider how the mineral nutrients get into the plants. As previously stated, they enter the plant through the root. This is generally true, but not always. They can enter through the leaves also when present on the leaf surface. However, they are not usually so located. More and more, though, use is being made of supplying the plants with mineral nutrients that the plant cannot normally obtain through the soil—*fertilizers*—by spraying these fertilizers directly on the leaf. There are some advantages to doing this. Perhaps the soil is sandy so that the fertilizers, if applied to the soil, would wash away (*be leached out*) and thus be lost to the plant and wasted. Perhaps the fertilizer is tied up or precipitated within the soil so that the plant cannot use it even when it is present. In these cases, foliar application of the fertilizer would be advantageous, and no doubt such methods of fertilizing will increase in the future.

Rate of Foliar Uptake

If the same amount of several mineral nutrients is sprayed onto the leaf, this does not mean that each one will enter the plant with equal rapidity. They will not. Actually it is not known how these mineral nutrients enter the leaf, whether it be entirely through the cuticle or through the stomates, but perhaps both pathways are used. Nevertheless, different mineral nutrients enter at different rates. Usually N and K enter very rapidly, whereas P, S, Mg, and the micronutrients enter more slowly. The rate of foliar uptake also varies with environmental conditions, with the rate of entrance being greatest when the leaf is moist and light intensity is high.

Nutrients in the Soil

By far the majority of the mineral nutrients in the plant are supplied by the soil and enter through the roots. In the soil, the mineral nutrients are found in three conditions. First, there are the *fixed elements* — those that are not available to the plant. Some of these are constituents of the rock. They must await weathering before they can be released and used. Some form insoluble precipitates in the soil. Second, there are those mineral nutrients that are exchangeable, and third, those that are dissolved in the soil solution and are readily available to the plant. The proportion of each mineral nutrient present in each of these three conditions varies from one soil to the next, and all three are in equilibrium, as seen below.

$$\text{soluble} \rightleftharpoons \text{exchangeable} \rightleftharpoons \text{fixed}$$

A few words should be said here about the exchangeable mineral nutrients. In Chapter 2, it was revealed that the soil colloids — clay and humus particles — have the ability to bind water. These particles also have a negative electrical charge on their surfaces and because of this negative charge, the positively-charged mineral nutrients (*cations*) are attracted to and bound to the soil colloids. The ability of these colloids to bind cations is referred to as the *base-exchange* or *exchange capacity* of the soil. Soils high in clay or humus content will have a higher exchange capacity than soils low in these components.

The cations that are held by base-exchange, especially K, Ca, Mg, Na, and H, can easily be removed by another cation, and it does not have to be another cation of the same element. Cations may have one, two, or three charges, being called mono-, di-, or trivalent cations respectively. What is actually exchanged by the soil colloid is positive charges. Thus a monovalent cation can replace another monovalent cation, or a divalent cation can replace either another divalent cation or two monovalent cations, and a trivalent cation can replace either another trivalent cation, three monovalent cations, or a monovalent and a divalent cation. The cations include the monovalent H, K, Fe, Cu, Zn or the divalent Ca, Mg, Cu, Zn, Fe, or Mn and the trivalent Al which is not a mineral nutrient but is common in soils. The released, or exchanged, cations return to the soil solution from which they can be removed by the plant. There is an equilibrium set up between the cations bound to the colloids of the soil and those present in the soil solution, as seen above.

Entrance into the Free Space

In the soil solution, the mineral nutrients diffuse to the plant root as the root grows into the area where these mineral nutrients are located. The mineral nutrients do not move far laterally in the soil, so the roots must grow to them in order to obtain them. Along with the soil water molecules, the mineral nutrients diffuse into the cell walls and intercellular spaces of the root. This area of the root that the mineral nutrients of the soil are able to diffuse into is known as the *free space* of the root, and may comprise as much as 20% of the root volume. Eventually, the mineral nutrient either diffuses back out of the root or it enters the protoplasm of the cell, either to be accumulated in the vacuole of the cell or to be secreted from the endodermis into the xylem vessels to be translocated to other parts of the plant.

Any part of the root can take up the mineral nutrients and does. However, it appears that most of those translocated enter in the area of the root hairs.

Passive vs. Active Uptake

The entrance of mineral nutrients into the free space of the root by diffusion is known as *passive uptake*, and no metabolic energy is involved. However, mineral nutrients cannot enter into the protoplasm by diffusion. There is a barrier at the plasma membrane, and entrance through it can occur only at the expense of metabolic energy furnished by the cell into which the mineral nutrient is entering. There are several bits of evidence to support this contention. First, the cell cannot accumulate mineral nutrients if respiration is stopped or the cell is dead. Second, not all mineral nutrients are accumulated as rapidly or to the same extent as another. Third, the extent of accumulation of a given mineral nutrient within the cell is great, with the concentration being as much as a thousand times as great inside the cell as outside. If diffusion occurred, the concentration inside would be similar to that outside. This uptake of mineral nutrients by the cell at the expense of metabolic energy is called *active uptake*. Much research has been expended, over many years, to find a mechanism of active uptake. So far, no infallible theory has been advanced to explain how active uptake works.

THEORIES OF ION UPTAKE

Perhaps two theories are most actively being considered at the present time, the carrier theory and the anion respiration theory. Both have been studied so extensively that they should be considered here.

Carrier Theory

The carrier theory of ion uptake is illustrated in Fig. 5.1. A cation (K^+) diffuses to the surface of the membrane, where it combines with a carrier (C). The K^+ cannot move through this membrane, but this combination (CK) can move through and does so. When it reaches the inner surface (CK) breaks down, releasing the cation (K^+) into the interior, which may

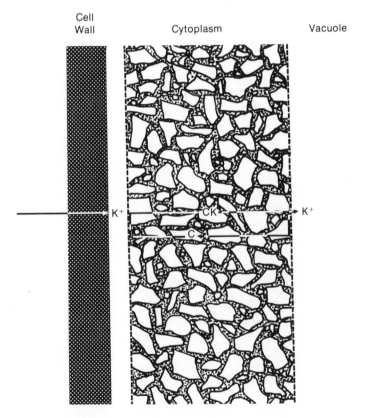

Fig. 5.1 Illustration of the carrier theory of ion uptake. K represents a cation. C represents a carrier, and CK represents the carrier-cation complex.

be the vacuole, cytoplasm, etc., and freeing the carrier which is either destroyed or can then return to the outer surface of the membrane to again participate in ion transport. This theory assumes that the membrane is impermeable to the ion, but not to the ion-carrier complex. It explains the role of metabolism by stating that metabolic energy is required either for the synthesis of the carrier, or for the movement of the carrier across the membrane. It explains the role of oxygen, in ion uptake, as being essential for the release of metabolic energy.

It has been known for many years, that not all mineral nutrients are taken up by plants to the same extent. Monovalent ions enter more rapidly than divalent, and divalent more rapidly than trivalent, etc. Also, even among ions with the same valence, differences in the rate of uptake exist. For example, most plant species will accumulate potassium to a great extent, but not sodium. Both are monovalent cations. This *selectivity* is explained, according to the carrier theory, as due to the presence of *binding sites* on the carrier – sites to which the ions become attached. Each site is specific, allowing only certain ions to attach there. If binding sites for sodium do not exist, the sodium will not pass through the membrane.

The carrier theory will explain many phenomena associated with the uptake of mineral nutrients. However, a chemical substance that functions as the carrier has not been found. We have attributed such properties to such molecules as proteins, enzymes, RNA, amino acids, sugar phosphates, and others, but none of these have definitely been accepted as the carrier. If such a carrier could be found, this would certainly strengthen the carrier theory. The carrier theory does not explain how the carrier-ion complex moves across the membrane, either.

Anion-respiration Theory

Another theory that is popular today is based upon the explanation put forth by Lundegardh. This theory states that within the membrane a series of chemicals that act as chains for electron transfer exist, with electrons being passed along the chain from the outer surface of the membrane to the inner surface. As an electron is passed inward, this leaves a positive charge which can be satisfied by an anion taking the place of the electron. As the electron moves further inward, the anion follows it. This continues until the anion reaches the interior of the membrane at which time it is released into the vacuole, etc. Cations follow the anions inward to maintain electrical neutrality. The membrane is assumed to be impermeable to anions, but not to cations.

Electron transport chains have been found in plant cells, usually associated with respiration, as will be discussed in Chapter 8. However, they have not been found in all membranes. Also, it is difficult to explain selectivity with this theory.

Ion Concentrations

Irrespective of the mechanism employed, mineral elements do get inside the plant cell, and do become accumulated there in concentrations often much higher than their concentration in the external solution, as seen in Table 5.3. Here it can be seen that some mineral nutrients are present in concentrations over 1000 times their external concentration. These are free ions within the cell too, and not bound chemically or physically to other cell constituents.

Table 5.3   Ion accumulation by a green alga. (Adapted from Spector, W. S. *Handbook of Biological Data.* © Federation of American Societies for Experimental Biology. With permission.)

Ion	Ion concentration		
	In cell	In culture medium	In cell/culture medium
Ca	10.2	0.8	12.8
Mg	17.7	1.7	10.4
Na	10.0	0.2	50
K	54	0.05	1080
Cl	91	0.9	100
SO_4	8.3	0.3	28

Once inside of the cell, the mineral nutrients may remain in the solution of the vacuole, or they may enter the cytoplasm, including various cell organelles where they may participate in various metabolic activities for which they are needed. However, they may also leak out of the cell. Previously, we stated that the cell membranes were impermeable to ions. This is true, but cells do have the ability to secrete ions into the external medium. They have *ion pumps* that force ions out of the cell. This activity requires energy from metabolism and therefore is an active process.

The mechanisms of uptake and secretion or leakage set up an equilibrium within the plant whereby mineral nutrients are present both within the protoplasm of the plant cells, and in the free space. Studies have indicated that perhaps as much as 40% of the mineral nutrients within the plant may be in the free space, with the other 60% being found within the protoplasm.

LOSS OF MINERALS FROM PLANTS

Once the mineral nutrients get into the plant, they do not necessarily remain there for the life of the plant. As previously stated, some of these are transported from one part of the plant to another, but some are also lost from the plant. Some are lost by leaching, some by the abscission of plant parts, some by diffusion out of the roots or by decay of the roots, some are lost by secretion from the plant, and some even by being converted to a gas within the plant and diffusing out into the atmosphere. Since mineral nutrients are important to plants, and since it would be desirable to reduce the amount of these that must be added to plants as fertilizers, their loss by the plant is of great economic importance.

Leaching

About 80–90% of the potassium and 50–60% of the calcium of a plant may be lost by leaching. This represents a considerable loss. Fortunately, not all mineral nutrients are lost with equal facility through leaching. Mn is easily lost, Ca, Mg, S, and K are lost less easily and Fe, P, Zn, or B lost only slowly.

Leaching is a process whereby the mineral nutrients are washed out of plant tissues. It is particularly important with leaves and leaves may lose a great deal of the mineral nutrients they have accumulated, through this mechanism. Apparently, loss by leaching is brought about by the mineral nutrients in the free space diffusing to the surface of the leaf or other tissues and there dissolving in the rainwater which falls off the leaf carrying the mineral nutrients with it to the soil. In young leaves, little loss occurs, but as the leaf matures, and especially when it reaches senescence (old age) the amount of loss increases greatly and may be very significant, as indicated above. In fact, it may be so extensive that a new generation of plants can be cultured in the *leachate* without need for added fertilizers. This leachate may have important ecological implications, especially in soils that are deficient in certain mineral nutrients, such as tropical soils. Field crops often become deficient in certain mineral nutrients after a heavy or extensive rain, due to the leaching of these minerals from the plant.

Loss from Roots

Loss of mineral nutrients often occurs from the roots. This may be due to diffusion of mineral nutrients from the free spaces of the roots back into the soil, or it may be due to the death and subsequent decay of the roots or root hairs. The importance of such loss is difficult to evaluate.

Loss by Abscission

Plants lose a great deal of their mineral nutrient content through *abscission*. Abscission is the process whereby plant structures fall off the plant. Leaves abscise from deciduous plants in the Fall, and fruits and flower petals abscise from plants during the growth process. It is true that some of the mineral nutrients are translocated out of these structures prior to abscission, going to the roots or stems where they are conserved or to the young leaves, etc., but still a considerable amount is lost from the plant due to abscission. This may be very important. For instance, in the tropics, due to the high rainfall, there are few mineral nutrients in the soil. All are tied up in the growing plant life. If they remained here, there would be no nutrients for new plant growth. However, when leaves, and other parts abscise, they fall to the ground and decay, slowly releasing their mineral nutrients to the soil solution to be immediately taken up by new plants before these mineral nutrients can be washed out of the soil. Therefore, there is a cycle of mineral nutrients whereby they move from the soil into the plant and from the plant into the soil. This is important in continuing the propagation of plants in any area. If all of the vegetation from a given area in the tropical rainforest is removed, it would not be possible to continue to grow crops in that area without the use of considerable amounts of fertilizers to compensate for those mineral nutrients removed with the native vegetation.

Loss by Secretion

Plants also lose some mineral nutrients by secretion by glands. These glands, such as the nectaries, secrete solutions that do contain mineral nutrients. Once outside the plant, these mineral nutrients can then be lost by being washed off the plant surface by rain.

A small amount of nitrogen is lost by plants by being converted to ammonia or some other gaseous form. Once in this form, the nitrogen is volatile and will diffuse out of the plant into the atmosphere.

Loss by Translocation

As previously learned, plant organs, such as leaves and flower petals, can lose mineral nutrients by translocation to other plant parts. It is also possible for plants to lose mineral nutrients by translocation to other plants. It is not often realized that many adjacent plants are actually joined together by *root grafts*. Root grafts occur when two roots from neighboring plants come together and join, making their vascular tissues continuous from one plant to another. Diseases are often transported to neighboring plants in this way. The extent of transfer of mineral nutrients from one plant to another through these root grafts has not been determined and is in need of further study.

All in all, it is evident that several mechanisms exist through which plants do lose mineral nutrients that they have previously accumulated. Therefore, the mineral content of the plant or plant part is determined by the amount of minerals taken up and the amount lost. This balance must maintain an adequate amount of each of the mineral nutrients in each cell of the plant or the plant suffers from a deficiency of one or more mineral nutrients, a deficiency that will reduce plant growth and yield, or even kill the plant.

PLANT CULTURE TECHNIQUES

To determine what the optimum content of each mineral nutrient should be within the plant and to determine the essentiality of a given mineral element, it is necessary to culture plants in such a manner that we can control or determine the amount of each mineral element which enters the plant or is lost by it. These culturing techniques for mineral nutrition studies can be grouped into three categories, namely, soil culture, slop culture, or water culture. Each of these cultures has certain advantages and disadvantages for mineral nutrition studies, which shall now be considered.

Soil Culture

Soil culture is growing plants in soil. This soil may be in the field or in pots in the greenhouse or living room. It has the advantage that the roots are growing in the natural environment and the plants are therefore as one would find them in the field so the application of the results of such studies is quite reliable. Also, less attention to the cultures is required during the growth of the plants, than with other techniques. On the other hand, the soil is always contaminated with many chemical substances so you cannot be sure what chemical elements are available to the plant or in what concentration. Some mineral nutrients added to the soil may be tied up or in some other way made unavailable to the plant by the soil. Soil composition and texture may influence your study. However, such studies are useful and should be used wherever possible.

Slop Culture

Slop culture is sometimes called sand culture. It consists of growing plants in containers of sand, and applying solutions as often as needed. This culture technique does give the root a solid support and is somewhat similar to a soil environment and artificial methods for aerating the roots are not needed. The disadvantages include the fact that sand too is contaminated with chemical elements, and frequent irrigation is necessary due to the inability of the sand to hold the water. Also, the sand is not buffered, so a close control of the pH of the solution is necessary to prevent it from becoming too acid or alkaline. The sand can be made chemically purer than the soil, it can be more efficiently sterilized to remove soil organisms, and the chemical composition of the solution available to the plants can be more accurately controlled. More attention must be paid to these cultures, and there is some evidence that plants behave differently in the extent of mineral nutrient uptake in slop cultures, as compared with soil cultures.

Water Culture

For a more accurate control of the mineral nutrient composition of the solution at the root surface, *water culture* or *hydroponics* is used. With this method, the plants are grown in a container, often a glass or plastic jar, with the roots in the solution within the jar and with their tops extending above the jar into the air, as shown in Fig. 5.2. The advantages of this method are that the chemical composition of the solution at the root surface can be accurately controlled or known and easily altered as the need arises. Many plants grow just as well in solution cultures as in a good soil. However, there are certain disadvantages associated with the use of this method. The jar must be darkened to prevent the growth of algae in the solution, the plant must be supported, the pH of the solution must be determined regularly and controlled, a favorable osmotic pressure of the solution must be maintained, aeration must be supplied by bubbling air through the solution continuously, and iron must be added every day or two or it soon becomes unavailable to the plant. Also, this is a very unnatural environment, and roots do not behave the same in water culture as in soil culture. Therefore, the interpretation of results and subsequent application of results to field conditions is limited.

Fig. 5.2 A setup for the water culture of plants. The plant is held in place with cotton, and the glass tube is connected to an air pressure line for aeration. Normally the jar would be covered to prevent algal growth.

The culture of plants either in slop culture or in water culture, requires the use of an aqueous solution that contains all of the mineral nutrients required by plants except the one studied, and their presence in adequate amounts. Fortunately, it has been found that most plant species grow equally well in the same solution. However, the concentration given in Table 5.4 is for crop plants. Native plants require a much lower concentration, so often diluting the solution recommended by a factor of three or four will be best for these.

UPTAKE OF SPECIFIC IONS

Since most of the mineral nutrients enter the plant through the root from the soil solution, it would be well at this point to consider those factors within the soil that will influence the availability of the mineral nutrients. Generally we can say that soil water content, soil aeration,

Table 5.4   Composition of a nutrient solution for the culture of cultivated plants. (Modified from Hoagland and Arnon in *The Water-culture Method from Growing Plants without Soil*, 1950. Calif. Agr. Expt. Sta, University of California, Division of Agricultural Sciences, Circ. 347.)

To prepare 1 liter of nutrient solution, add to 985 ml of distilled water, the following:	
1 ml	1 M KH_2PO_4
5 ml	1 M KNO_3
5 ml	1 M $Ca(NO_3)_2$
2 ml	1 M $MgSO_4$
1 ml	Fe-EDTA
(Prepared according to Steiner and Winden, *Plant Physiol.*, **46**: 862–863, 1970.)	
1 ml	Micronutrient solution
Micronutrient solution is prepared by adding to 1 liter of distilled water, the following:	
2.86 grams	H_3BO_3
1.81 grams	$MnCl_2\cdot4H_2O$
0.22 grams	$ZnSO_4\cdot7H_2O$
0.08 grams	$CuSO_4\cdot5H_2O$
0.02 grams	MoO_3

temperature, pH, soil compaction, types and size of soil particles, and the kinds and concentrations of other inorganic elements in the soil may all influence the rates at which a mineral nutrient enters the plant, but now let us be more specific, and consider those factors that may influence the uptake of the individual ions.

Nitrogen

Nitrogen is very abundant on this earth. In fact, it comprises 78% of the gases in the atmosphere. However, it occurs in the atmosphere as nitrogen gas, a form which most plants cannot use. It is easy to demonstrate that a plant will die from the lack of nitrogen even when bathed in it. To be available to the plant, the nitrogen must be converted into either nitrates (NO_3), ammonia (NH_3), or one of the nitrogen-containing organic compounds of low molecular weight, such as an amino acid. In the soil, most of the nitrogen that is available to the plant is in the form of either nitrate or ammonia; ammonia predominating in poorly aerated soils, such as swamps and bogs, and nitrates predominating in most other soils. Even when nitrogen is added to a soil as ammonia, it is rapidly oxidized to nitrate, with the nitrate entering the plant. So, in most of the soils with which we are familiar, nitrate predominates as the form of nitrogen taken up and available to the plant. Its uptake is reduced when boron is deficient.

Nitrate is a monovalent anion, so it is not adsorbed to the soil colloids. Therefore, it is easily leached out of the soil, and rarely accumulates in high concentrations.

It appears that the uptake of nitrogen by the plant occurs readily whenever available nitrogen is in the soil, so keeping the plant supplied with nitrogen is principally a matter of keeping usable nitrogen in the soil. There is a good correlation between the amount of organic matter in the soil and the amount of nitrogen. Perhaps, stored nitrogen in the soil is that tied up in organic compounds.

Phosphorus

Phosphorus is also an anion. It can exist as either the mono-, di-, or trivalent anion, although it appears to be taken up mostly as the mono-valent H_2PO_4 anion. Nevertheless, an equilibrium is always set up among the three forms, as seen below:

$$H_3PO_4 \rightleftarrows H_2PO_4^- \rightleftarrows HPO_4^= \rightleftarrows PO_4^=$$

Which form predominates depends upon the pH of the solution. There-fore, the amount of phosphorus that enters the plant will depend upon the pH of the soil solution, with the uptake decreasing at values above pH 6.

In soils, much of the phosphorus is a component of the soil minerals with 25–60% being present as organic phosphorus. As this organic phosphorus is broken down to release phosphates, these enter the plant. Although phosphorus is an anion, it is exchangeable. However, it is not present in high concentrations in the soil solution.

Phosphorus uptake is reduced in the presence of high concentrations of nitrates, calcium, or iron and is increased by higher concentrations of magnesium. Calcium and iron form insoluble precipitates with phos-phorus. Magnesium is associated with phosphorus metabolism within the plant.

Potassium

Potassium is a monovalent cation. It is present in adequate amounts in most soils, but often not in a form that can be taken up by the plant. Its entrance into the plant, once present in the soil solution, is decreased by high nitrate or phosphate concentrations, and often by high sodium, calcium or magnesium concentrations. However, the problem of getting adequate amounts of potassium into the plant is usually a problem of

potassium availability, since it is readily and strongly adsorbed and fixed in the soil, even when added as a fertilizer. It does not form an insoluble precipitate in soils, as phosphorus often does, and all its salts are readily soluble.

Calcium

Calcium is a divalent cation, and is taken up by the plant as such. It is the major ion in the soil solution and on the exchange complex of most soils, except sodac or tropical soils, but is largely fixed as minerals or crystals. Generally, soils of humid tropical areas are low in calcium, while soils of arid or semiarid regions are rich in calcium. Usually it is present in adequate amounts in the soil to supply plant needs, although it may be a problem keeping the calcium present in adequate amounts in the soil solution. Its solubility increases with the carbon dioxide concentration, and decreases with increased temperature and increased pH.

Magnesium

Magnesium is also a divalent cation and is taken up by the plant in this form. It is usually present in adequate amounts in the soil, as is calcium. Like calcium and potassium, magnesium interacts with calcium and potassium so an excess of either or both calcium or potassium will reduce magnesium uptake, and their deficiency will result in an increase in magnesium uptake.

Sulfur

Sulfur is an anion, much like nitrogen. It enters the plant largely through the soil as the divalent sulfate anion (SO_4^{--}), although near industrial areas it may enter as SO_2 gas to a limited extent. Either form can be used by the plant, as can some organic sulfur compounds. It is normally present in the soil solution, but is not exchangeable. It enters the soil solution either by the breakdown of organic sulfur compounds from plant and animal remains or as fertilizers. Gypsum and triplephosphate both contain sulfur and are often used for soil treatments, a use which also adds sulfur to the soil.

Iron

Iron is found in most soils, and enters the plant as the divalent cation. However, its concentration in the soil solution is usually very low. It is precipitated by phosphates and at high pH values. The divalent form is found in waterlogged or otherwise poorly-aerated soils, whereas the trivalent form predominates in normal well-aerated soils, so the *oxidation potential*—the ease with which oxidation can occur—of the soil is an important factor in iron availability. Also, the amount of iron entering the plant is influenced by the presence of other cations in the soil solution. Manganese, zinc, or copper, when present in excessive amounts, will cause iron uptake to be reduced, perhaps by antagonism at the site of uptake.

Copper and Zinc

Copper and zinc are present in most soils as the divalent cation, and enter the plant as such. However, such entrance is reduced by carbonates or phosphates, which precipitate the copper, by high pH, or by high concentrations of soluble zinc, iron, or manganese which are antagonistic to it. It might be worth noting here that deficiencies of some of the micronutrient cations, copper, zinc, iron, and manganese, are caused by farmers applying too much phosphorus fertilizers to their fields. Copper and zinc may also be deficient in sandy or peaty soils.

Manganese

Manganese is found principally in the soil solution as the divalent cation form, being in low concentrations in neutral or alkaline soils and high in acid or waterlogged soils. Its uptake by the plant decreases at pH values above 7, and it is influenced in its availability by the concentrations of iron, zinc, and copper.

Molybdenum

Molybdenum is taken up in the mono- or divalent anionic form. Its availability increases with increased pH of the soil solution. In recent years, increasingly more areas in which molybdenum deficiency occur are being found.

Boron

Boron is taken up as the borate anion. It is frequently deficient in acid or coastal plain soils, but not in prairie or desert soils. However, its uptake by the plant is reduced during drought or at high pH values of the soil solution, or by high calcium contents of the soil.

Chloride

Chlorine is taken up as the monovalent chloride anion. It is readily soluble in the soil solution and probably is never deficient in the soil.

Uptake Control

It can be seen, from the previous discussions, that there are many factors in the soil that influence the rate at which the mineral nutrients are taken up by the plant. Some of these factors influence the amount of the nutrient dissolved in the soil solution, some influence the valency and therefore the ability of the plant to take up the mineral nutrient, and some interact with the site of active uptake at the plant protoplasmic surface. The interaction of all of these factors controls the rate of uptake and therefore the mineral nutrient content of the plant.

Since mineral nutrients are essential to the plant, they must play important functions within the plant. Indeed they do. However, since these functions are usually closely associated with plant metabolism, the consideration of the function of each of the mineral nutrients within the plant will be postponed until plant metabolism is discussed.

Now that we have learned which mineral elements are needed by the plant, how the plant gets them and loses them, and techniques used for studies of plant mineral nutrition, we shall consider how the plant obtains and uses its carbon.

Chapter Six | Photosynthesis and Carbon Dioxide Fixation

Water and mineral nutrients are important to the plant, but even if these are present in adequate amounts the plant will not long survive without the final inorganic nutrient, carbon dioxide. Chapter 4 showed that if 100 grams of plant material were dried, 10 grams of dried plant material resulted. If this was then placed in an oven at high temperature, perhaps as much as 1 gram of plant ash would remain. The difference between the dried plant material and the ash represents the amount of organic compounds present in the plant tissues, namely, about 10% of the fresh weight, or 90% of the dry weight of the plant. Cells are constructed from these organic compounds, obtain their energy from these organic compounds, and store and even secrete these organic compounds. These compounds are all composed largely of carbon, and the ultimate source of this carbon is the carbon dioxide present in the atmosphere.

CARBON DIOXIDE IN THE ENVIRONMENT

Carbon dioxide is present in the atmosphere as carbon dioxide gas. Although its concentration does vary somewhat, it is fairly constant at 0.03% by volume at the earth's surface. This constancy is maintained by the rate of removal of carbon dioxide from the atmosphere being balanced by a similar rate of its entrance into the atmosphere. Carbon dioxide is removed from the atmosphere by its fixation into organic compounds by living organisms, primarily by green plants, and is released from plants and other living entities through the process of respiration and by other decarboxylation reactions as will be seen in Chapter 8. A *decarboxylation reaction* is one in which a *carboxyl* (–COOH) group is removed from an organic compound and is converted into carbon dioxide.

The concentration of carbon dioxide in the atmosphere is also maintained constant through activity at the water-atmosphere boundary which exists where the atmosphere is in contact with the water of ponds, lakes, streams, rivers, and oceans, as seen in Fig. 6.1. Actually, at 20°C, there is about as much inorganic carbon in the waters of the earth as in the atmosphere, so in total these amount to a great deal of carbon available for carbon dioxide fixation by plants.

Carbonic Acid System

Carbon dioxide is a species of the *carbonic acid system* illustrated in Fig. 6.1. In pure water, about 99% of the species is present as carbon dioxide, and only about 1% as the other components. However, this relationship varies considerably under natural conditions since the total amount and the ratio of these species present will vary with temperature, atmospheric pressure, the carbon dioxide partial pressure of the atmosphere, the pH of the water, and the amount of excess base (those cations

Fig. 6.1 Species of the carbonic acid system including carbon dioxide gas of the air, and dissolved carbon dioxide, bicarbonate, and carbonate of the water.

associated with carbonates or bicarbonates). Therefore, if temperature and gaseous carbon dioxide partial pressure are constant, more inorganic carbon will be found in sea-water and in hard fresh waters than in water that is acid or in soft fresh water. Why?

pH Effects

Since hydrogen ions are involved in equilibria within the system, water with a high pH would have a greater proportion of carbonate and bicarbonate ions than one with a low pH value. One can estimate the relative amounts of carbon dioxide and bicarbonate in water by use of the Henderson–Hasselbalch equation, which is derived from the equilibrium constant. Since the amount of carbonic acid varies with the carbon dioxide concentration, carbonic acid can be omitted from the calculations to get the following equation:

$$pH = -\log K_1 + \log \frac{(HCO_3^-)}{(CO_2)}$$

or

$$pH = pK_1 + \log \frac{(HCO_3^-)}{(CO_2)}$$

where $pK_1 = -\log K_1$

K_1 = first dissociation constant (concentrations are in mols/liter)

The latter equation is the Henderson–Hasselbalch equation. Since pK_1 equals about 5.7 at 25°C, then at this temperature, equal amounts of bicarbonate and carbon dioxide would be present when the pH is 5.7. At higher pH values, more bicarbonate than carbon dioxide would be present. The relative concentrations of carbon dioxide, bicarbonate, and carbonate at various pH values are illustrated in Fig. 6.2.

As indicated in Fig. 6.1, the species of the carbonic acid system are in equilibrium so if the carbon dioxide concentration of the atmosphere increases, much of this increase goes into the formation of more dissolved carbon dioxide, carbonic acid, bicarbonates, and carbonates in the waters. Inversely, as the carbon dioxide concentration of the atmosphere decreases, carbon dioxide enters the atmosphere from the waters by the conversion of carbonates, bicarbonates, carbonic acid, and dissolved carbon dioxide into carbon dioxide gas. This is a very efficient system for maintaining a constant carbon dioxide concentration of the atmosphere, and reminds one of the pH buffer systems.

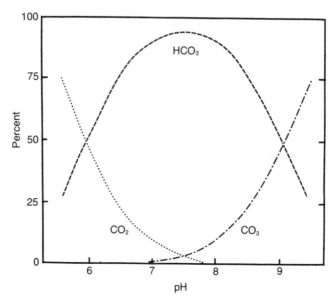

Fig. 6.2   Relative concentrations of species of the carbonic acid system in sea water at 18°. (Redrawn from Paasche (1964) *Physiol. Planta.* Suppl. III with permission.)

CARBON DIOXIDE ENTERS THE PLANT

Carbon dioxide enters the multicellular plant by diffusion through the stomates and lenticels of the leaves, stems, flowers, fruits, and roots, into the intercellular spaces, as seen in Fig. 6.3.

Passage through the Stomates

The process of diffusion was reviewed in Chapter 3. Since the lenticels are always open, there is a continuous diffusion into the plant through them as long as the concentration of carbon dioxide inside the plant is less than that of the plant's environment. In the case of the stomates, the same can be said to be true except when the stomates are closed. When the stomates are open, they do not restrict carbon dioxide diffusion when the light intensity is low. In fact, at low light intensity, the stomates can be nearly closed, with no effect on carbon fixation. But at high light intensity, carbon fixation is limited by the rate of diffusion through the open stomates. Indeed, the rate of carbon fixation is limited more by light intensity than by carbon dioxide availability, except at high intensities.

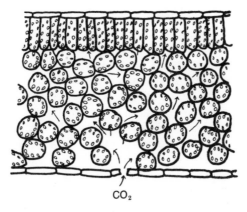

Fig. 6.3 Drawing to illustrate the path of diffusion of carbon dioxide into the leaf and chloroplasts.

Complete closure of the stomate stops the entry of virtually all carbon dioxide into the leaf. These stomates are usually closed at night and open during the daytime, as indicated in Chapter 3, although water deficits and high temperatures often cause them to close during the daytime. Of course, at night, photosynthetic carbon dioxide fixation does not occur so the carbon dioxide requirement is nil.

Movement within the Leaf

Movement of carbon dioxide gas through the intercellular spaces of the leaf is perhaps due largely to diffusion, but such movement will be greatly enhanced by mass movements of the air due to differences in temperature and pressure as well as by mechanical deformation of the leaf. Such mechanical deformation occurs regularly due to bending of the leaf by wind.

Within the intercellular spaces, the carbon dioxide concentration is surprisingly constant. Whether the stomates are open or closed, it seldom falls below 0.01% by volume.

Movement into the Cell

When the intercellular carbon dioxide comes in contact with the cell walls, it dissolves in the water which impregnates the walls. This dissolution may be a physical phenomenon with the amount being dissolved depending upon the carbon dioxide concentration of the intercellular space near the cell wall, the chemical composition of the water, the prevailing temperature, the pH of the water, and the rate of carboxylation and decarboxylation reactions occurring in the cell. Since carbon dioxide is a gas, its solubility decreases as the temperature increases, so less is dissolved at higher temperatures than at lower temperatures. That which does dissolve often becomes another component of the carbonic acid system and moves into the cytoplasm of the cell.

The diffusion of carbon dioxide or bicarbonate through water, such as that of the cytoplasm, is a much slower process than its diffusion through air. In fact, the rate of the former is only about 1/10,000 or less than that of the latter. This might indicate that the rate of diffusion of carbon to the site of fixation may be a rate limiting step in fixation reactions. However, such transport is usually greatly speeded up by protoplasmic streaming and by mass movement caused by temperature and pressure differences within the cytoplasm.

Look back to Fig. 6.2, and compare the relative amounts of carbon dioxide and bicarbonate at the pH value normally attributed to cytoplasm. It is evident that at pH 6.8, carbonates are essentially absent and about 90% of the carbonic acid species consist of bicarbonate, and approximately 10% as carbon dioxide. Studies have indicated that, although some specie differences can be observed, plants can use both carbon dioxide and bicarbonate for fixation reactions, but probably not carbonate. Therefore we shall not be too concerned about the species of the carbonic acid system that actually moves to the site of carbon dioxide fixation. Also, the conversion of carbon dioxide is normally a slow process but this would not eliminate one species or another in plant cells, since such cells contain an enzyme, carbonic anhydrase, which catalyzes this conversion.

Sources of Carbon Dioxide

It should also be mentioned that the atmosphere is not the only source of carbon for carbon dioxide fixation in plants. In addition, small amounts of carbon dioxide may be taken up by the plant from the soil solution where it is present as bicarbonate salts, or it may be supplied from the respiratory activities of the root and other cells. However, if these respiratory or other decarboxylation reactions represented the only source, there would be no increase in dry weight of the plant, (Why?) so the increase in such dry weight largely represents the results of carbon dioxide fixation using atmospheric carbon dioxide as the carbon source.

DARK FIXATION OF CARBON DIOXIDE

There are actually two broad categories of reactions by which carbon dioxide is fixed in plants. One is by *dark fixation*, which is common to all living cells, both plant and animal, and the other is by *photosynthetic fixation* which occurs only in the chloroplasts of green plant cells and is therefore found only in plants, and only in the green cells of plants at that. In plants, the rate of dark fixation totals less than 1% of that of photosynthetic fixation and represents a loss of energy rather than energy conservation.

Crassulacean Acid Metabolism

The occurrence of dark fixation of carbon dioxide is most obvious in the case of what has been called crassulacean acid metabolism. At certain seasons of the year, the leaves of certain succulent plants such as species of the genera *Bryophyllum*, *Kalanchoe*, *Sedum*, *Kleinia*, and *Crassula*, and even the green stems of *Opuntia*, will accumulate high concentrations of organic acids during the night. Malic acid is the organic acid which is present in highest concentrations, but citric and isocitric acids may also increase. This high acidity decreases during the daytime as the acids are destroyed. The explanation for this phenomenon is that during the night carbon dioxide accumulates in the cells, since it is not being used in photosynthesis but is being released through respiration. As it accumulates, it is reacting with some of the products of respiration to form the acids. The predominant reaction associated with this phenomenon is believed to be either the carboxylation of pyruvic or of phosphoenolpyruvic acids to form either malic acid directly or through oxalacetic acid which is then reduced to malic acid or oxidized to form citric or isocitric acids, as shown in Fig. 6.4.

A.   pyruvic acid $+ CO_2 +$ NADPH$_2$ $\rightleftharpoons$ malic acid $+$ NADP $+ H_2O$

B. α-ketoglutaric acid $+ CO_2 +$ NADPH$_2$ $\rightleftharpoons$ isocitric acid $+$ NADP $+ H_2O$

C.   phosphoenolpyruvic acid $+ CO_2 +$ ADP $\rightleftharpoons$ oxalacetic acid $+$ ATP

      phosphoenolpyruvic acid $+ CO_2 + H_2O$ $\rightleftharpoons$ oxalacetic acid $+$ P

D.   acetyl-CoA $+ CO_2 +$ ATP $\rightleftharpoons$ malonyl CoA $+$ ADP

Fig. 6.4   Mechanisms by which dark fixation of carbon dioxide occurs.

Mechanisms of Dark Fixation

Some of the reactions that could participate in this and similar dark fixation of carbon dioxide are listed in Fig. 6.4. In all cases, the raw material becomes carboxylated during the reaction. In other words, the carbon dioxide becomes attached to the end of the molecule to form a carboxyl group (–COOH). These reactions are catalyzed by a group of enzymes called *carboxylases*, enzymes whose prosthetic groups contain the vitamin biotin. They are also reversible reactions which presents the possibility of their function in decarboxylation reactions under certain conditions.

Dark fixation of carbon dioxide occurs in all living cells at all times, both day and night. The reason the organic acid content of succulent leaves is reduced during the day is that the carbon dioxide is used primarily for photosynthetic carbon fixation leaving little for dark fixation. Also, at this time there is a net breakdown of the organic acids, to form carbon dioxide, which results in a release of the carbon dioxide and its subsequent availability for photosynthesis.

The reaction forming malonyl coenzyme A is a very important reaction in plant cells because it is the means whereby lipids and related materials are synthesized. It is not known how essential the other methods of dark fixation are for higher plants, but certainly this reaction should be essential since cells cannot survive without lipids. Dark fixation is known to be important for fungi for the synthesis of lactic, citric, fumaric, and succinic acids. In fact, many fungi cannot survive in the absence of exogenous carbon dioxide.

Dark fixation of carbon dioxide appears to occur in the *mitochondria* of the cell and does require energy derived from metabolism to carry out these reactions. This energy is derived largely from respiration, as will be shown in Chapter 8.

CARBON DIOXIDE FIXATION THROUGH PHOTOSYNTHESIS

By far the most valuable carbon dioxide fixation process to the biological world is that of photosynthetic carbon dioxide fixation. This is so, not only because it represents a mechanism whereby carbon is made available for the synthesis of organic compounds, but also a mechanism through which energy from the sun, which is captured during the light reaction of photosynthesis, is stored in these organic compounds. Thus, valuable metabolic energy is not expended as in the dark fixation processes.

Photosynthetic carbon dioxide fixation is but one of two processes of photosynthesis. The second, namely the light reaction, will be studied in Chapter 7. Photosynthesis is the process whereby the green plant uses carbon dioxide and water in the presence of light to form organic compounds plus oxygen, and can be summarized by the reaction shown in Fig. 6.5. In so doing, energy is transduced from the radiant energy of light to the chemical energy of organic compounds, oxygen is released into the atmosphere, and carbon dioxide is used to fix carbon into organic compounds.

Fig. 6.5   A summary of the entire photosynthetic process.

IMPORTANCE OF PHOTOSYNTHESIS

Photosynthesis is of vital importance to all life here on earth for four reasons. First, it is the essential source of energy for all living organisms, both plant and animal. Second, plant growth and yield are determined by the rate of photosynthesis. Third, it is necessary for the synthesis of many organic compounds used by man and other animals. Fourth, it replenishes the atmosphere of the earth with oxygen that is so vital to life, and without which we could not survive.

WHERE PHOTOSYNTHESIS OCCURS

Photosynthesis occurs in the chloroplasts. Evidence to support this contention comes from many sources, including the observations that the action spectrum of photosynthesis is the same as the absorption spectrum of chlorophyll, that chlorophyll is found only in the chloroplasts, in higher plants, and that isolated chloroplasts can carry out the complete photosynthetic process.

Not all plant cells contain chloroplasts, (Which do not?) but they are found in the mesophyll and guard cells of the leaf, in the epidermal cells of many herbaceous plant stems, in the subepidermal cells of the calyx of flowers, and in the subepidermal cells of unripe fruit. These cells that contain chlorophyll are sometimes called *chlorenchyma* cells. Also, blue-green algae and a few bacteria contain chlorophyll but no chloroplasts.

Chloroplast Structure

If you look at chlorenchyma cells through the microscope, you can see the chloroplasts in them. These chloroplasts appear as dark, spherical bodies suspended within the cytoplasm of the cell. There may be only one or there may be more than 100 per cell. They are ellipsoidal, with diameters of 4–10 microns and are 1–2 microns thick. Each chloroplast is limited by a double membrane, called the *chloroplast membrane*. Circumvented by this membrane, in most chloroplasts, there are a number of membranous, dark bodies, about 0.4 microns in diameter, called the *grana*. These grana are suspended in a solution called the *stroma*, but are joined together by lamellae. The chloroplast then consists of the chloroplast membrane, stroma, and numerous grana. The light reaction of photosynthesis occurs on the grana, and photosynthetic carbon dioxide fixation reaction in the stroma. An electron micrograph of a chloroplast is shown in Fig. 7.6.

Chloroplast Chemicals

The stroma of the chloroplast is a colloidal system, containing salts, and proteins and starch as well as other organic molecules.

Chloroplasts may make up 20–30% of the dry weight of plant leaves. Each chloroplast contains 75% water, about 80% of which is free water in the stroma of the chloroplast. Each also contains about 37% of the total protein of the leaves of higher plants, which make up about 50% of the dry weight of the chloroplast or about 12% of the fresh weight of the chloroplast.

About 30% of the chloroplast is made up of fat-soluble materials which includes the lipids and the chloroplast pigments. Chlorophyll is one of the most abundant of these pigments, making up 6% of the dry weight of the chloroplast. Other important chemicals found in the chloroplast include several cytochromes including cytochromes f and a b-type cytochrome, both of which are found only in chloroplasts, vitamins C, E, and K_1, and carbohydrates of various types. Details of grana structure will be given in Chapter 7.

Chloroplast Formation

Chloroplasts are formed either from colorless proplastids or by the division of mature chloroplasts. In higher plants, formation from proplastids seems to be most common. Proplastids are much smaller than chloroplasts and consist of the limiting membrane enclosing the stroma, but have no grana. The grana form as the chloroplast develops from the proplastid.

The chloroplast is a very complex cell organelle, with the machinery to carry out many syntheses found in the living cell. In fact, it is so complete, that one theory states that originally the chloroplast was a living, intact, cell that invaded another cell and took up a symbiotic relationship with it that has continued even today. In fact, some scientists are working now to culture chloroplasts free from the cell. It may be possible to do this and even to get them to reproduce under such conditions. They do have some nucleic acids needed for self-duplication. However, they seem to be limited in their capacity for aerobic respiration although recent evidence indicates that respiration does occur in chloroplasts, to some extent, at the same time photosynthesis is occurring. This is called *photorespiration.* This needs further investigation, since other evidence indicates the chloroplasts do not take up oxygen, do not exhibit Krebs' cycle activity, have no cytochrome oxidase system and possess no hexosemonophosphate shunt system.

MECHANISM OF PHOTOSYNTHETIC CARBON FIXATION – CALVIN CYCLE

The carbon fixation reaction of photosynthesis occurs in the stroma – the aqueous solution of the chloroplast. Dissolved in the stroma are globular proteins called fraction I protein. These proteins are constituents of enzymes needed to catalyze the carbon fixation reaction. In this reaction, as summarized in Fig. 6.6, a five-carbon sugar phosphate named ribulose-1,5-diphosphate combines with carbon dioxide to form a six-carbon compound.

$$CO_2 + \text{ribulose-1,5-diphosphate} + H_2O \longrightarrow (6C) \longrightarrow 2(\text{3-phosphoglyceric acid})$$

Fig. 6.6   The photosynthetic carbon fixation reaction.

Carbon Dioxide Fixation

Whenever ribulose diphosphate is present, it will rapidly combine with any carbon dioxide (or bicarbonate?) present irrespective of the presence of light. Metabolic energy is not needed for this. The six-carbon compound formed is very unstable. It is so unstable that we have not yet been able to isolate it, and therefore have not been able to identify it. This compound rapidly breaks down to form two three-carbon molecules of 3-phosphoglyceric acid (3-PGA). Therefore, whenever ribulose diphosphate and carbon dioxide come together, two molecules of 3-PGA are formed, and whenever one molecule of carbon dioxide is fixed, two molecules of 3-PGA are formed. This 3-PGA is the first stable product of photosynthetic fixation.

3-PGA Metabolism

One of two things can happen to this 3-PGA. It can react with ATP to form 1,3-diphosphoglyceric acid and then be reduced to form sugar phosphates or it can be oxidized to form 2-PGA and such chemicals as amino acids and fats. One of the contemporary fields of study is to determine the control mechanism which determines which pathway will be followed. Certainly both species and environmental control mechanisms exist. At one time, all carbon fixed during photosynthesis was believed to be reduced to form sugars or starch before being utilized in the formation of other chemicals, and indeed in some species such as sunflower, under certain conditions, this is true. Not too many years ago, it was believed that sugars were the sole products of photosynthesis. This is not true, although sugars and starch are the principal products of photosynthesis. In many species of plants, much of the carbon will be used directly in the formation of amino acids and fats without forming carbohydrates first. Mature leaves of certain species will produce, almost exclusively, sugars, as a result of photosynthesis, whereas their young growing leaves will produce largely proteins and fats. More must be learned about what determines the end product of photosynthetic fixation.

When sugars are formed first, the two 3-PGA molecules are *phosphorylated*, using ATP formed during the light reaction of photosynthesis as the energy source, to form two molecules of 1,3-PGA. These are then reduced, using the $NADPH_2$ from the light reaction of photosynthesis, to 3-phosphoglyceraldehyde. Such reduction results in the transfer of energy from the energy carriers, ATP and $NADPH_2$, to the aldehyde molecules. For the reduction of two molecules of 3-PGA to two of 3-phosphoglyceraldehyde requires about 8–10 quanta of energy. One of the aldehyde molecules, is transformed into one dihydroxyacetone phosphate molecule following which this molecule joins with the remaining aldehyde molecule to form the six-carbon fructose-1,6-diphosphate. From this, sugars or starch can be formed.

For continuation of the photosynthetic fixation reaction, it is necessary that ribulose diphosphate be regenerated. By referring to Fig. 6.7, one can continue to trace the pathways by which eventually another ribulose diphosphate molecule is formed or by which carbon is diverted to form other organic compounds. To maintain this supply of ribulose diphosphate, five of every six 3-PGA molecules must be so used, but the sixth forms sugars, organic acids, proteins, etc.

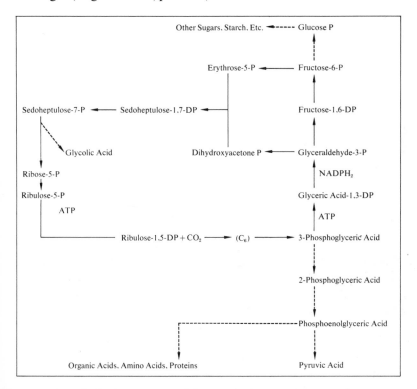

Fig. 6.7 The carbon reduction cycle of photosynthesis. Dotted arrows indicate reactions not in the cycle, but which result in carbon loss from the cycle.

A technique used for studying carbon fixation is to expose the cells to radioactive carbon dioxide ($^{14}CO_2$), kill the cells after a few seconds, extract the radioactive chemicals with alcohol, and separate and identify the radioactive organic compounds by paper chromatography. When exposure is made in the dark, radioactivity is found in certain organic acids and amino acids. If done in the light, organic acids and amino acids retain little of the radioactivity and most of it is found in the 3-PGA and in sugar phosphates. The organic acids and amino acids become labelled to about the same extent in both the light and the dark, but in the light, carbon fixation is much more rapid and results largely in other compounds, as indicated in Fig. 6.8.

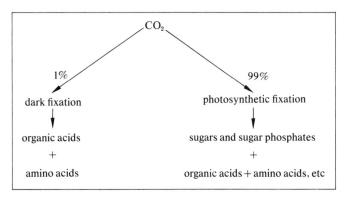

Fig. 6.8   Extent of carbon dioxide fixation in plants.

THE C-4 DICARBOXYLIC ACID PATHWAY

Not so long ago, it was thought that the Calvin cycle of photosynthetic carbon fixation was the one and only mechanism by which this fixation occurred. Now evidence indicates that this is not true. Another mechanism has been found, namely the C-4 dicarboxylic acid pathway. This mechanism, as outlined by Hatch and Slack, two Australian workers, is shown in Fig. 6.9. This pathway consists of two interconnected cycles. The first occurs in the mesophyll chloroplasts and the second in parenchyma sheath chloroplasts in the same leaf. The first begins with pyruvic acid and yields oxalacetic, malic, and aspartic acids. A few of the oxalacetic acid molecules have their carboxyl groups transferred to ribulose diphosphate to form 3-PGA and pyruvic acid. The 3-PGA then proceeds, as in the Calvin cycle, to produce sugar phosphates, sugars, and starch. Actually, the second cycle is much less active than the first.

Plants that use the C-4 dicarboxylic acid pathway have some interesting characteristics which differ from species using the Calvin cycle. First, most of the carbon dioxide fixed appears in malic and aspartic acids rather than in 3-PGA. Second, they also differ in maximum rates of photosynthesis, light-saturation values, photorespiration (which occurs only with

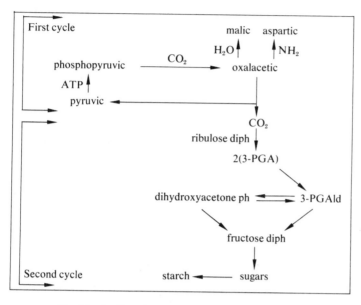

Fig. 6.9 Outline of the C-4 dicarboxylic acid pathway.

the Calvin cycle), compensation point, and the effects of oxygen and inhibitors on the rate of photosynthesis. Species of plants with this type of photosynthesis include the grasses native to the tropics, such as sugar cane, corn, and Johnson grass, as well as some species of dicots, such as species of *Amaranthus*, *Atriplex*, and *Cyperus*. However, it does appear that the type of photosynthesis that prevails can vary from one organ to another in the same plant, since sugar cane leaves use the C-4 dicarboxylic acid pathway, whereas cell cultures derived from their stems use the Calvin cycle. Wheat, oats, and many other grass species, those native to temperate climates, use the Calvin cycle.

Since most of the dry weight of plants is composed of organic compounds, and these arise from carbon fixation reactions, and since this increase in dry weight is important for plant growth and also because it is the basis of controlling plant yield, it would be well to consider factors which influence the rate of carbon dioxide fixation. First, a few words might be said about dark fixation since the chemical reactions responsible for this are endergonic reactions requiring energy, it might be expected that the rate of fixation decreases as energy becomes less available. Since this energy is derived from respiration, then anything which interferes with respiration would reduce the rate of fixation. Indeed, this rate of fixation is decreased by anaerobic conditions such as a lack of oxygen or decrease in oxygen availability, or by the presence of respiration inhibitors such as HCN. The rate of dark fixation is stimulated by certain mineral nutrients especially nitrogen compounds. However, since dark fixation adds only about 1% or less of the total carbon to plants, perhaps the rate at which it occurs is not significant. Nevertheless, studies should be made to learn additional factors which control this rate and also to learn factors which determine the product produced via dark fixation, or the ratio to such carbon ending up in organic acids as compared with its presence in amino acids.

Factors which influence photosynthetic carbon dioxide fixation are much better known since they have been studied much more extensively, often to shed more light on how to increase crop yield.

HOW PHOTOSYNTHESIS IS MEASURED

Although photosynthesis involves the light reaction, to be considered in the next chapter, in addition to photosynthetic carbon dioxide fixation, the rate of photosynthesis is normally measured by measuring the rate of carbon fixation. Therefore, we can talk about the rate of photosynthetic carbon fixation as the rate of photosynthesis, and shall do so in this section. The rate of photosynthesis is determined in several ways. One can measure the net increase in dry weight of the plant, neglecting that due to inorganic elements, or measure the amount of carbon dioxide which enters the plant, or measure the changes in total energy content of a plant and correlate this with photosynthesis. None of these methods are infallible, due primarily to the inability to accurately measure the evolution of carbon dioxide by respiration which is occurring at the same time as photosynthesis or the release of energy during respiration or the movement of solutes from the tissues studied to other parts of the plants. Nevertheless, what measurements can be made are accurate enough to reveal the magnitude of photosynthesis.

EXTENT OF PHOTOSYNTHESIS

The extent of photosynthesis occurring on this earth is tremendous It has been estimated that the net amount of carbon fixed each year through photosynthesis *on land* is 3×10^{10} metric tons, or about 2 tons per hectare per year. A surprising amount of photosynthesis also occurs in the waters of the earth. The extent *in the waters* is about 0.8 metric tons of carbon per hectare per year, which is less than $\frac{1}{2}$ that on land. However, since the waters cover a much greater area of the earth's surface than does the land, the total fixed in waters is about equal to that on land. The total yield for the earth is therefore about 5×10^{10} metric tons of carbon or 22.5×10^{10} metric tons of organic matter per year, and remains rather constant from year to year. This amounts to about 200 times the food consumed each year by humans.

FACTORS WHICH ALTER PHOTOSYNTHESIS

In spite of the constancy of photosynthesis on the earth from year to year, great variations do occur in the rate of photosynthesis and such variations are associated with many factors. Let us consider what these factors are.

Light Intensity

The characteristics of incident radiation or light, are very important factors. As previously mentioned, the rate of photosynthetic carbon fixation is about 15 times that of dark fixation, so light must be important since it is necessary for the light reaction of photosynthesis, the reaction that makes energy carriers available for photosynthetic fixation. Without this energy, photosynthetic carbon fixation could not occur. In the dark, photosynthetic carbon fixation does not occur, obviously. As the light intensity increases, the rate of photosynthesis increases. When the light intensity is such that the rate of photosynthesis produces organic compounds only as fast as they are used in respiration, it is said that the light intensity is at the *compensation point.* When the light intensity is lower than this, the plant cannot long survive without an additional source of organic carbon, since respiration will continue to deplete its reserve organic compounds until they are insufficient for survival. The light intensity at the compensation point will vary from one species to another, as indicated in Table 6.1.

As the light intensity increases above the compensation point, the rate of photosynthesis continues to increase, eventually leveling off at some intensity below that of full sunlight. The light intensity above which no increase in photosynthesis occurs also varies with the plant species, as indicated in Table 6.1, and with the percentage of the leaves exposed to full light intensity. For example, apple tree leaves carry out their maximum rate of photosynthesis at $\frac{1}{3}$ to $\frac{1}{4}$ full sunlight, but intact apple trees

Table 6.1   The compensation point of some tree seedlings. Values are in percentages of full Winter sunlight. (Data of Burns, G. P., 1923. Studies in tolerance of New England forest trees. IV. Minimum light requirement referred to a definite standard. *Vt. Agr. Expt. Sta. Bul. 235.*)

Species	Light intensity
Pinus ponderosa	30.6
Pinus sylvestris	28.7
Thuja occidentalis	18.6
Larix laricina	17.6
Pseudotsuga taxifolia	13.6
Pinus contorta var. *latifolia*	13.6
Quercus borealis	13.3
Celtis occidentalis	11.5
Picea engelmannii	10.6
Pinus strobus	10.4
Picea excelsa	8.7
Tsuga canadensis	8.4
Fagus grandifolia	7.5
Acer saccharophorus	3.4

require full sunlight for maximum photosynthesis. The difference is due to shading of leaves on the intact apple tree by leaves above or near them, which cuts down the light intensity to which each leaf is exposed. Generally, algae and plants that normally grow in the shade reach their maximum rate of photosynthesis at a lower light intensity than do those plants that normally grow in the sun. Grasses and sunflower plants are examples of the latter.

The intensity at light saturation increases with the elevation of the normal habitat of the plant, and with the temperatures. As an example of the latter relationship, *Mimulus* sp. exhibit a ten-fold increase in light intensity saturation values as the temperature increases from 0 to 40°C.

Light intensity is probably the factor that most often limits the rate of photosynthesis, so it would be desirable to find ways to increase it. Light intensity in the field can be increased by proper spacing of plants, removing competing plants, or by thinning to remove those leaves and plant parts that are no longer efficient in carrying out photosynthesis. Light intensity is also reduced by water vapor, and therefore by a high relative humidity, by fog, smog, and dust. Also, the effective light intensity will vary with the angle at which the light hits the leaf, being maximum at a 90° impact, and less at smaller angles of incidence. Therefore, one might expect some hour to hour variation in light intensity due to the fact that as the sun approaches the horizon, sunlight must pass through more atmosphere with more water vapor, dust, smog, etc., and as the angle of the sun changes, the angle of the incident radiation striking the leaf surface changes. Light intensity also increases with elevation, so alpine species are exposed to greater intensities than are seashore species. This increase in intensity with increased elevation is due to less atmosphere through which the sun's rays must pass. As sunlight traverses the atmosphere, some of the light is absorbed by the water vapor and ozone, as well as by dust particles, and some is reflected back into space. Therefore, the longer the path the light must take through the atmosphere before it gets to the plant, the lower its intensity will be when it gets there. Conversely, the higher the elevation at which the plant is growing, the less atmosphere the light must traverse so the greater its intensity. Also, not all wavelengths of light are absorbed at equal rates by the atmosphere. Ultraviolet and infrared are removed more extensively than others. In fact, high in the mountains, plants often show characteristic symptoms of exposure to high intensity of ultraviolet light and this may be one reason why such plants are usually dwarfed in size.

Light intensity is, of course, related to light duration. At night, the intensity is near zero. Generally, the longer the days, the more extensive the photosynthesis. However, it must be realized that at a given light intensity the rate of photosynthesis is not constant. The stomates may close reducing the rate, due to carbon dioxide deficiency, and even if this does not occur, there appears to be some internal mechanism associated with photosynthesis which approaches saturation, so the rate decreases with duration of continuous exposure to sunlight.

Light Quality

Light quality is also a factor that can alter the rate of photosynthesis. In the field, light quality will vary between that to which the leaves in the sunlight are exposed and to that to which leaves in the shade are exposed. Since leaves remove blue and red light predominantly, and transmit farred light, then leaves and other plant parts growing in the shade would be exposed to small amounts of red or blue light and excessive amounts of farred light. This would alter the rate of photosynthesis and possibly other physiological activities, as shall be considered later.

When the light intensity is great enough to cause the maximum rate of photosynthesis, it is usually much below the light intensity present when the plant is exposed directly to the sun. The reason increased light intensity does not result in a greater rate of photosynthesis is that some other factor becomes limiting.

Carbon Dioxide

The second most important limiting factor is the carbon dioxide concentration of the atmosphere. In both fresh and marine waters, the concentration of carbonic acid species is too low to give a maximum rate of carbon dioxide fixation at high light intensity by submerged plants, and therefore becomes a limiting factor under these conditions. Plant growth, as a result of photosynthetic carbon fixation, can be greatly enhanced by increasing the carbon dioxide concentration of the atmosphere above 0.03%. This is not practical in the field, but can be, and sometimes is, done in the greenhouse. Such enhancement is not without limits, however, and at a value that is often less than 0.5%, carbon dioxide concentration may become toxic to the plant.

Although the carbon dioxide concentration of the atmosphere is quite constant at 0.03% by volume, there may be some decrease in this concentration in the vicinity of the plant during the daytime when photosynthesis is rapid. However, such decrease is not great since air movement keeps the atmosphere mixed and therefore the carbon dioxide concentration constant. The concentration could be low if the air movement was low, but it rarely is, as anyone can attest who has tried to photograph flowers outdoors and has had to wait for the wind to subside so the photograph could be taken.

Temperature

Perhaps the next most often limiting factor would be temperature. Since carbon fixation involves metabolism and since such metabolism has a Q_{10} near 2–4, one would expect the rate of photosynthesis to increase by doubling, tripling, or even to exhibit a four-fold increase as the temperature increased 10°C or a similar decrease when the temperature decreased 10°C. Such magnitude of temperature change is common in most areas during the *growing season*. The growing season is considered to be the number of days in which the temperature exceeds 5°C. Therefore, temperature could be a factor causing hourly or even day to day changes in the rate of photosynthesis.

Air temperatures also vary from season to season and with elevation. At higher elevations, the air temperatures generally decrease 0.5°C for every 100 meters increase in elevation, so alpine plants are normally exposed to lower temperatures than seashore plants, which could lower their rate of photosynthesis. However, the maximum rate of photosynthesis for plants in the arctic is about that in more favorable regions.

The lowest temperatures at which there is a net rate of photosynthesis lies between -2 and $-5°C$. At increased temperature, the rate of photosynthesis increases. However, in the case of evergreen plants in cold climates, frost in the Fall results in a change in the photosynthetic mechanism whereby the Summer rate of photosynthesis cannot be reached rapidly, but only returns with gradually warmer temperatures in the Spring. The cause of this is unknown.

The highest temperatures at which photosynthesis occurs have been reported to be near $85°C$ by some algae that are normal inhabitants of hot springs. However, in temperate region plants, photosynthesis does often occur at $35°C$ or even higher.

The optimum temperature for photosynthesis is extremely variable and controlled by many factors. In order to determine the optimum temperature, the duration of the exposure must be considered. In one experiment, conducted at 25, 30, 35, and $40°C$, photosynthesis increased with temperature for 30 min, but only at $25°C$ was a good rate maintained for long periods of time. At higher temperatures, the values soon declined to values lower than the original. Also, remember that leaf temperature may be $2-10°C$ higher than air temperature, unless the leaf is shaded, due to direct thermal effects of sunlight.

Water

Perhaps the next most important factor which controls the rate of photosynthesis is water. A reduced water content of leaves reduces the rate of photosynthesis. Such reduction is due largely to a reduction in the amount of carbon dioxide fixed per unit of leaf area. As previously indicated, reducing the water content causes the stomates to close. Such closure serves to conserve water but also cuts off the exogenous supply of carbon dioxide, and thereby greatly reduces the concentration of carbon dioxide available for photosynthesis. At a permanent wilting point, the plant loses so much water that the rate of photosynthesis may be reduced to about 90% of its value under conditions of good soil moisture.

Looking back to Fig. 6.4, at the overall reaction of photosynthesis, it may appear that water would limit this reaction because it is necessary as a substrate. Nevertheless, less than 1% of the water in the cell is used in biochemical reactions, including photosynthesis, and the leaf water content never gets this low without the cells dying, so this direct involvement of water in photosynthesis is probably never limiting. However, when photosynthesis has been reduced greatly by water deficits, it often takes several days to fully recover, which cannot be explained on the basis of the deficiency causing the stomates to close. The reason for this phenomenon is not known but may be due to alterations in chloroplast structure. Indeed, it is common to observe that plants growing in the field will have closed stomates during the middle of the day in Summer due to water deficits, but photosynthesis recovers when the deficit is alleviated. Perhaps the inconsistency of these observations is due to the length of time the deficiency prevails.

The fact that water does often become limiting during the growing season in the temperate and desert regions partially explains the observation that the highest rates of photosynthesis are found in plants of the tropical rainforest, marsh, and swamps where water deficiencies rarely occur, or in irrigated crops where soil moisture is kept favorable.

Mineral Nutrition

A deficiency of any one or more of many mineral nutrients can greatly reduce the rate of photosynthesis. Such deficiencies do occur in the field and are probably more frequent than suspected. Nitrogen and magnesium are constituents of the chlorophyll molecule and therefore are necessary for photosynthesis, although chlorophyll concentration is usually in excess of that needed so chlorophyll deficiency perhaps seldom becomes limiting unless the leaves are senescent or very chlorotic. Many of the deficiency symptoms associated with mineral nutrients are caused by effects of these nutrients on chlorophyll synthesis, and breakdown. We have seen how phosphorous is needed for phosphorylation reactions in photosynthesis and therefore is needed for these both during the light reaction and during the photosynthetic carbon dioxide fixation reaction. Potassium, calcium, and iron are needed for protein synthesis and these minerals, along with other micronutrients, are often needed for enzyme synthesis and operation. This indicates then how mineral nutrient deficiencies can reduce the rate of photosynthesis.

Sprays

Although not always suspected, sprays applied to plants may often limit the rate of photosynthesis. When I was living on an orange grove in

California, we used to spray the trees with an oil-based spray to control scale insects. Such oil-based sprays often reduce the rate of photosynthesis because the oil plugs up the stomates reducing the availability of exogenous carbon dioxide. Dust on plants growing near unpaved country roads, or mud on leaves of plants that have been submerged during floods will do likewise, as will a number of anti-transpirants which were discussed in Chapter 3. Other sprays, such as some anti-transpirants and herbicides have poisoning effects on photosynthesis since they inhibit the mechanism, at toxic concentrations.

Wind

Perhaps it would not be proper to end a discussion of environmental factors which influence the rate of photosynthesis without mentioning the role wind may play. It maintains a constant carbon dioxide distribution in the atmosphere, it causes leaves to bend and thus speeds up the movement of carbon dioxide within the leaf and it moves the leaves which alters continuously the pattern of shading. It may transport dust and sprays which could be injurious, as well as smog and smoke. It also plays a role in temperature maintenance of the leaf. Wind is an important but variable factor in the plant's environment.

As a result of interactions and combinations of the factors which influence the rate of photosynthesis, as just discussed, the rate of photosynthesis is not constant from hour to hour, from day to day, or from month to month. However, since the causes of such variations sometimes involve chloroplast and grana structures let us defer further discussion of such variations until the structural characteristics of the chloroplast, as they affect the light reaction of photosynthesis are considered, which will be done in the next chapter.

This chapter, in essence, terminates considerations of plant and cell nutrition. It has revealed what chemicals must be supplied to the plant for its continued development, how the needed chemicals get into the plant and, to some extent, what the plant does with them. Now, the subject of the energy relations of the plant will be considered since all living entities, including plants, need energy just as much as they need chemicals, and they need a continuous supply of this energy.

| *Chapter* *Seven* | Photosynthesis and the Entrance of Energy into the Cell |

All living plants and cells require a continuous supply of energy just as much as they require a continuous supply of nutrients. This chapter describes how this energy gets into the plant and its cells. A discussion of how the plant uses, stores, and loses this energy is given in Chapter 8.

The physicists define energy as the capability to do work.

FORMS OF ENERGY

Kinetic

Energy can exist in many forms. It can occur as *kinetic* or *heat energy*, which is the energy of molecular motion. Objects move because they possess kinetic energy. Objects also have temperature because of their kinetic energy. At $-273°C$, kinetic energy does not exist, but as the temperature increases, so does the kinetic energy. A thermometer measures the kinetic energy of the molecules around it, and our bodies detect this energy as heat.

Potential

Energy can occur as *potential energy*, which is that energy preserved in an object because of its position relative to something else. A leaf at the top of a tree has more potential energy than one near the base. This text shall be little concerned with potential energy.

Chemical

Much of the energy in plants, at least that necessary for life, is present as *chemical energy*, or that energy associated with, and tied up in, the bonds of chemical molecules, especially in the organic compounds, but also associated with atoms, or with groups of atoms, and with ions. This is the energy that joins atoms together to form the molecule. More will be said about this later, both in this chapter and in the chapters that follow.

RADIANT ENERGY

Most of the energy studied in this chapter, occurs as *radiant energy* or *radiation*. This is the energy associated with the electromagnetic spectrum, and includes gamma rays, cosmic rays, ultraviolet rays, X-rays, light, infrared radiation, radio and television waves, sound waves, and electricity.

Energy may enter the plant in many forms, including entrance as chemical energy of organic nutrients, radiation of various wavelengths, temperature changes, etc. This chapter shall not be concerned with entrance as the chemical energy of nutrients, but shall be restricted to other means, the most important perhaps being radiation.

Thermodynamics

The study of energy is known as *energetics*. The study of the interrelationships of various forms of energy is known as *thermodynamics*. According to the first law of thermodynamics, energy can be neither created nor destroyed, but energy can be changed from one form to another. Radiant energy can change into chemical energy and vice versa, and many other forms of energy are interconvertible. According to the second law of thermodynamics, all energy will eventually degrade into heat, so heat represents a form of wasted energy, and especially is this so in plants and in their cells. The conversion of energy into heat occurs continuously in all living organisms and such transformations are associated with, and necessary for, metabolism. However, heat can only partially be converted into other forms of energy. Therefore, some means must be available to replenish the supply of usable energy.

Most of the usable energy in the living world is available because of the transduction of radiant energy of sunlight into the chemical energy of plants through the process of photosynthesis. This occurs by means of the light reaction of photosynthesis, and therefore makes this reaction so very essential to all life on the earth.

Photochemistry

The light reaction of photosynthesis occurs on the surface of membranes of the grana of the chloroplast. It begins with the absorption of light by the chloroplast pigments, and ends with the formation of two energy carriers, namely, $NADPH_2$ and ATP.

The light reaction is a *photochemical* reaction — a chemical reaction that occurs only in the light. Light itself, being a form of energy, is a very narrow portion of the electromagnetic spectrum (*see* Fig. 7.1.) — that portion to which our eyes are sensitive and therefore that we can see.

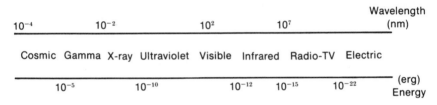

Fig. 7.1   The electromagnetic spectrum, illustrating its wavelength-energy relations.

To understand the nature of radiant energy, consider it as being composed of waves of particles. These particles are called *photons*, and each photon can be visualized as a bundle of energy traveling in waves. The amount of energy in each bundle, or photon, is called a *quantum*. The energy content in each quantum varies from one photon to another as the wavelengths vary, as seen in the equation below:

$$E = hv = h\frac{c}{\lambda}$$

where E = the amount of energy in a photon or quantum, and is expressed
   in ergs (1 erg = 0.239×10^{-7} cal)
 h = Planck's constant = 6.62×10^{-27}
 c = the velocity of light = 3×10^{10} cm/sec
 λ = the wavelength of the radiation
 v = the light frequency = $\dfrac{c}{\lambda}$

From this relationship, we can see that the shorter the wavelength, the greater the energy value of the photon. Light has wavelengths that vary from about 400–760 nm. Therefore, substituting these values into our equation, we find that violet light (400 nm) is composed of photons that have a higher energy content than those of far red light (760 nm). If we determine the energy values for light of various colors, or wavelengths, we find the relationships as expressed in Fig. 7.2. Certainly, the shorter the wavelength of the radiation, the greater the energy content of its photons.

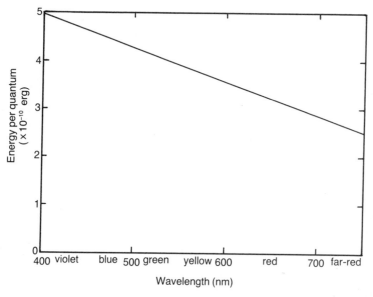

Fig. 7.2   Energy content of each quantum as related to color and wavelength of the light.

A photochemical reaction is a chemical reaction that is not spontaneous, but one that requires energy before it can occur, and this energy comes from light. When light strikes a plant's surface, such as the surface of a leaf, part of the light is reflected, part is *transmitted* through the leaf, and part is absorbed by the leaf. Needless to say, only that which is absorbed can be used in the photochemical reaction. That which is reflected or transmitted is lost and represents wasted energy as far as that leaf or other surface is concerned.

The absorbed radiation is absorbed by certain chemicals. These chemicals are known as *pigments*. Pigments, or dyes as they are sometimes called, are colored molecules that absorb light of their complementary color. A ball is red, not because it absorbs red light, but because it absorbs green light. A leaf is green, not because it absorbs green light, but because it contains the pigments known as chlorophyll which absorb red and blue light, and the green light is left to be reflected into our eyes, allowing us to see the leaf as green. We see only the reflected or transmitted light, not that which is absorbed. Autumn leaves are yellow because their carotenoids absorb blue light, the complementary color of yellow.

This light is absorbed because the pigment molecule is able to absorb its photons. The pigment does so because it is able to use this energy to raise one or more of its valence electrons from their *ground state* to another orbit, known as the *excited state*. In a normal molecule, the valence electrons assume the orbit which conserves the greatest amount of energy, just as we do when we avoid work, or as a satellite does when it is in orbit around the earth. To move the electron to another orbit of a higher energy level requires energy. This energy comes from the photons. However, chemical molecules are specific as to which photons they can capture. Only those of a certain energy content can be captured by each type of molecule.

The electron cannot long remain at this higher energy level. Something has to happen. Either the absorbed energy destroys the pigment molecule or the molecule transmits the energy, lowering its electron back to the stable, ground state or orbit. Colored cloth fades when hung in the sun because the dye molecules absorb energy from the sunlight and cannot transmit all of this energy, so the absorbed energy destroys the pigment, causing the cloth to fade. Perhaps before learning more about this intriguing photochemical reaction of photosynthesis, it would be well to illustrate the importance of light to plants, and to study the pigment associated with photosynthesis, namely chlorophyll.

ETIOLATION

It is not difficult to demonstrate the importance of light to the plant, particularly to the multicellular plant. All that is necessary is to compare two individual plants of the same species, after one has been grown in the light and the other in darkness, as seen in Fig. 7.3. The plant grown in the dark is *etiolated.* Its color is whitish or yellowish. It has very long internodes with spindly stems, due to more and larger cells being present. It has very small leaves that are colorless, in the case of angiosperms, and are weak and narrow. The root system is poor, with small roots being

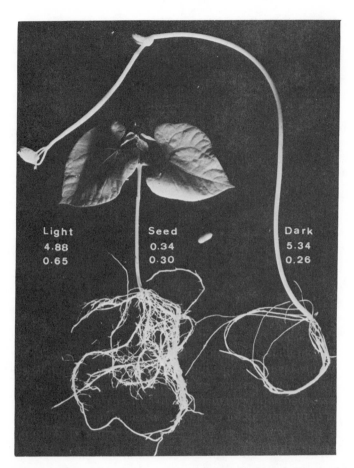

Fig. 7.3 Bean seedlings of the same age, grown in light or in the dark. Top numbers equal fresh weight and bottom numbers equal dry weight, in grams per plant.

present. Also, the plant grown entirely in the dark from seed germination contains less dry weight than did the seed from which it originally developed. This latter observation is difficult to believe since the volume of the plant is so much greater than that of the seed, but is not difficult to demonstrate, since all one has to do is weigh the seed before it germinates and weigh the resulting plant after it has been dried in an oven. A comparison of the fresh and dry weights of such plant structures is also given in Fig. 7.3.

CHLOROPHYLL SYNTHESIS

Two chlorophylls are always present in the green leaves of higher plants, namely, chlorophyll a and chlorophyll b. Their chemical molecular formulas are $C_{55}H_{72}O_5N_4Mg$ for chlorophyll a, and $C_{55}H_{70}O_6N_4Mg$ for chlorophyll b, and their chemical structures are illustrated in Fig. 7.4. You can see that the chlorophylls are composed of carbon, hydrogen, oxygen, and of the mineral nutrients nitrogen and magnesium. Structurally the two molecules are very similar. (How do they differ?)

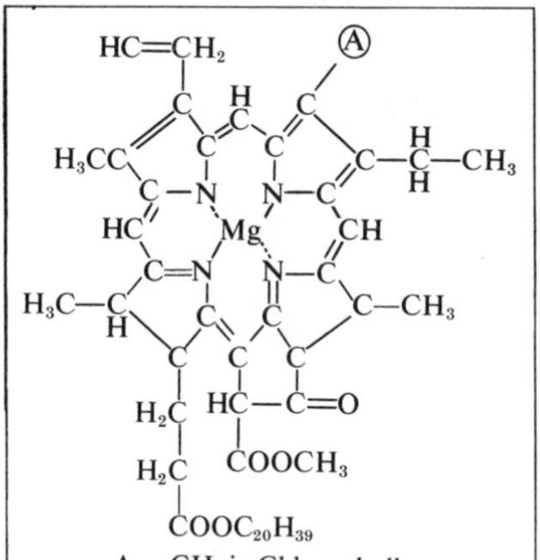

A = CH_3 in Chlorophyll a
A = CHO in Chlorophyll b

Fig. 7.4   Structure of the chlorophyll molecules.

Both chlorophylls are synthesized within the chloroplast using the five essential elements carbon, oxygen, hydrogen, nitrogen, and magnesium. The mechanism of synthesis is identical for both molecules until the final stages. To begin this synthesis, sugars are converted into the organic acid, succinic acid (*see* Chapter 8) and the amino acid, glycine (*see* Chapter 9). These form the four rings, giving the ring structure to the molecule, after which a phytol molecule, synthesized from sugars (*see* Chapter 8), becomes attached. The details of this synthesis are given in Fig. 7.5 which shows that many *intermediates* are involved. If the formation of any one of these intermediates is prevented, chlorophyll will not be synthesized.

The final stages in chlorophyll synthesis involve the addition of hydrogen to the protochlorophyll molecule, to form chlorophyll a, a

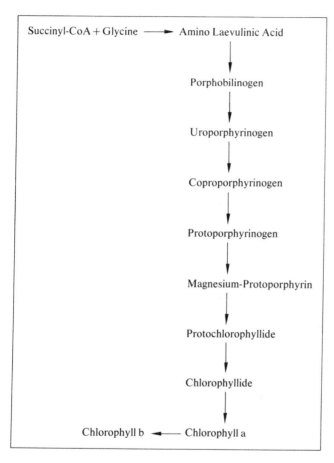

Fig. 7.5 Pathway of chlorophyll synthesis.

reaction that requires light in angiosperms but not in the gymnosperms and lower plants. For this reason, pea plants that are grown in the dark are colorless or light yellow in color but will take on their normal green coloration a few hours after being placed in the light. You may have noticed how weed seedlings, growing under a board or small swimming pool in the backyard, are nearly white in color. However, the next day after this cover is removed, these same seedlings will be green, because now they can convert their protochlorophyll to chlorophyll. Chlorophyll *b* is synthesized from chlorophyll *a*.

THE CHLOROPLAST

Chlorophyll synthesis also requires the presence of lamellae within the chloroplast. The proplastid is very small and does not contain lamellae. It is also colorless, since it does not contain chlorophyll. Chlorophyll is not formed until the lamellae are formed. Therefore, anything that prevents lamellae formation will prevent chlorophyll synthesis. Since lamellae are composed of membranes, their formation must involve protein and lipid synthesis.

CONTROL OF CHLOROPHYLL SYNTHESIS

Now that the mechanism of chlorophyll synthesis is understood, let us consider the conditions that prevent or reduce this synthesis. A knowledge of such conditions is most important since chlorophyll is necessary for photosynthesis and photosynthesis is necessary for plant growth as well as for life itself.

Enzymes

Each step in the pathway of chlorophyll synthesis requires an enzyme, so the absence of any one of these enzymes would prevent chlorophyll synthesis. The formation of these enzymes is controlled by genes through the action of nucleic acids, so a mutation of any one of these genes would stop chlorophyll synthesis. Albino plants occasionally appear among seedling growing in the light. These are white in color due to a mutation that has occurred; they cannot synthesize chlorophyll. Of course, albinos do not live long but soon die due to a lack of sugars. Such plants can, however, sometimes be kept alive to produce fruit and seeds if grown in solution culture, as long as sugars are present in the nutrient medium.

One of the interesting problems in differentiation is why some cells of an individual plant contain chlorophyll whereas others do not. For instance, leaf mesophyll and guard cells do, whereas other leaf epidermal cells and cells of the leaf vascular system do not. Cells of most flower petals do not, as is also true of most stem cells. No doubt these cells do not have the enzymes needed for the synthesis of chlorophyll, but we do not know why they do not have these when other cells of the same individual do.

Iron

Plant leaves often become *chlorotic* — are yellow instead of green in color — due to a lack of iron within the plant. Iron is not a constituent of the chlorophyll molecule, but is essential for chlorophyll synthesis. Iron is necessary for the synthesis of chlorophyll because it is necessary for protein synthesis. In the absence of adequate amounts of iron, proteins that are needed to form the lamellae and to form the enzymes that participate in chlorophyll synthesis are not present. Iron may be lacking in the soil, or it may be fixed in the soil or precipitated in either the soil or the plant so that the plant cannot use this iron.

Nitrogen

When nitrogen is deficient, the leaves turn yellow, due to the lack of chlorophyll synthesis. One might expect this, not only because nitrogen is part of the protein molecule and therefore is needed for protein synthesis, but also because it is a constituent of the chlorophyll molecule.

Magnesium

Since magnesium is also a constituent of the chlorophyll molecule, leaves will turn yellow when magnesium is not present in adequate amounts. However, magnesium is not often deficient in most soils, so this is not an important factor to consider in maintaining a high rate of chlorophyll synthesis within the plant.

Actually, a deficiency of any one of many mineral nutrients may eventually result in a decrease in the chlorophyll content of the leaf due to their essentiality for protein and enzyme synthesis or for energy availability within the cell.

Water

Water is also of some importance for chlorophyll synthesis. Following a heavy rain, the chlorophyll content of plant leaves will often increase, but during a drought, the chlorophyll content will decrease. On the other hand, if the soil is saturated with water, this results in a decrease in the chlorophyll content of the leaves. The water content of the leaf must be high for maximum chlorophyll content.

Temperature

Temperature may also be an important factor in chlorophyll synthesis. The optimum temperature seems to be near 20–30°C, with the rate of synthesis and accumulation dropping off at higher or lower temperatures. On the other hand, some species are able to form chlorophyll even at very low or very high temperatures. Chlorophyll is present in conifer leaves during the Winter and in algae that grow and survive in hot springs, where the temperature may be over 80°C.

The chlorophyll content of plants is due to a balance between chlorophyll synthesis and chlorophyll destruction. If the rates of these two are the same, the chlorophyll content does not change with time. If the rate of synthesis is faster than the rate of destruction, the chlorophyll content increases with time. If the rate of synthesis is less than the rate of destruction, the content decreases with time. The same can be said for any organic chemical substance within the plant, except for the few that are stable, such as DNA, cellulose, lignin, and sporopollenin.

CHLOROPHYLL DEGRADATION

Chlorophyll destruction normally occurs through three mechanisms. It may be destroyed by high light intensity, by an enzyme called chlorophyllase, and by destruction of the grana protein. Plants growing in direct sunlight have a lower chlorophyll content than those growing in the shade, due to *photodestruction*, by the high light intensity. The ultraviolet radiation seems to be the chief cause of this destruction. One might expect therefore, that plants growing near the mountain tops would have less chlorophyll per plant than the same species growing near sea level. Chlorophyll content should decrease with increased elevation. In very strong light, many chloroplasts orient themselves within the cell on a plane parallel to the incident light, an act which reduces their photodestruction. (Why?)

All chlorenchyma cells contain *chlorophylase*, an enzyme that destroys the chlorophyll molecule. This enzyme acts by removing the phytol chain from the rest of the molecule. Its activity is very low in some species, such as grasses and high in *Datura* and *Heracleum*.

There is no doubt that chlorophyll content is influenced by the stage of development of the plant. Mature leaves often have a high chlorophyll content, but as they become senescent, chlorophyll disappears. This is particularly evident when fruits lose their green coloration during ripening, or in the Fall of the year when the leaves of deciduous trees lose their green coloration and become yellow or red. These examples are due to the faster rate of chlorophyll destruction than chlorophyll synthesis.

In the leaves, in the Fall, this condition is brought on principally by short days and cool weather.

The chlorophyll content of plant leaves also varies from hour to hour over each 24 hour period. This variation is not due to variations in any of the known environmental conditions but is associated with an endogenous rhythm in plants, which will be discussed in more detail in a later chapter.

OTHER CHLOROPLAST PIGMENTS

In addition to chlorophylls a and b, the grana contain other pigments that function as secondary pigments by being able to capture light energy and transfer this energy to the primary pigment, chlorophyll a. The most abundant of these include chlorophylls b, c, and d, the carotenoids, β-carotene and lutein, and the quinone compounds. α-tocopherol, and vitamin K_1. The relative concentrations of these major pigments of the chloroplast grana would be about as follows:

chlorophyll a/chlorophyll b = 2.3
chlorophyll/carotenoids = 4.8
chlorophyll/quinone compounds = 5
chlorophyll/phospholipids = 2
chlorophyll/P_{700} = 250

The pigment P_{700} seems to be a special form of chlorophyll a, which absorbs light of a wavelength of 700 nm. It is believed to be one pigment which actually participates in the light reaction of photosynthesis and is said to be located in the *reaction center* of the photosynthetic unit. Thus for each P_{700} molecule there would be 250 chlorophyll molecules, 175 chlorophyll a, 75 chlorophyll b, 50 carotenoids, 50 quinones, and 120 phospholipids, as well as numerous protein, glycolipid, sulfolipid, and sterol molecules.

The relative concentrations of the above molecules in the grana as given above, are average values under normal conditions, but these values can vary considerably. For example, the relative concentrations of chlorophyll a to chlorophyll b would be much higher in species with yellow-green leaves than in those with dark green leaves, and in shade plants as compared with sun plants. Also, as the leaves turn from green to yellow with senescence or in the Fall of the year, the ratio of carotenoids to chlorophyll will increase tremendously.

CHLOROPLAST STRUCTURE

Since the synthesis of chlorophyll is dependent upon lamellae formation, and since the surface of the grana represent both the sites of pigment location and of the light reaction of photosynthesis, let us look at the structure of the grana. The grana are so covered with chlorophyll, that they appear only as dark, saucer-shaped structures under the light microscope. To study their structure in more detail, the use of an electron microscopy is required. A photograph taken under the electron microscope is seen in Fig. 7.6. Here it can be seen that the internal structure of the

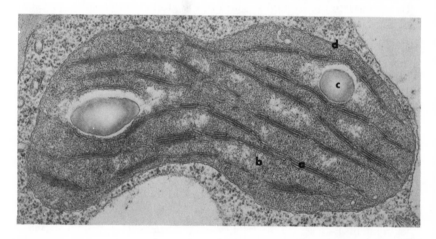

Fig. 7.6 Electron micrograph of a leaf chloroplast, showing granum (a), stroma (b), starch grain (c) and chloroplast membrane (d). (Adapted from *Plant Physiology,* 1968, **43**: 495–503. Courtesy of Dr. E. Gantt and the Smithsonian Institution.)

chloroplast, that inside the membrane, consists of a lamellar system bathed in the stroma. In numerous places, these lamellae come together to form the *grana*. Therefore, the grana can be considered as piles of membranous lamellae (thylakoids). The granum (singular for grana) has a layered structure formed by stacks of these membranes, with some of the membranes being continuous among the grana of a given chloroplast. It is on the surfaces of these grana membranes where chlorophyll and the other chloroplast pigments are located, and where the light reaction of photosynthesis occurs. In the stroma surrounding the grana many physiological phenomena occur, such as photosynthetic carbon fixation, phosphate transfer, protein synthesis, carbohydrate metabolism, starch, fat, and protein accumulation, chlorophyll synthesis and RNA synthesis, and possibly even DNA synthesis.

The grana membranes, like other cell membranes, are composed basically of lipids and proteins. On the surface of the granum, perhaps attached to the protein, is found the chlorophyll of the cell, and this is the only location within the cell where such chlorophyll is found. However, these chlorophyll molecules are not randomly distributed over the surface of the granum, but are found as groups of chlorophyll molecules, grouped together with about 250 chlorophyll molecules in each group. These groups are known as *photosynthetic units*, and each group functions separately to carry out the light reaction of photosynthesis. Not only do the various chlorophylls function in this reaction, but all of the chloroplast pigments act together as a unit.

THE LIGHT REACTIONS OF PHOTOSYNTHESIS

In the photosynthetic unit, many pigment molecules are found. When the light strikes these pigments, they absorb the photons. This increases the energy content of the molecules, and rather than be destroyed, they pass the added energy off to neighboring molecules. This continues until the energy is absorbed by a special pigment near the center of each photosynthetic unit. One of these pigments is P_{700}. What the other pigment is, is not known.

Photosystem II

The light reaction has been referred to as though there was only one mechanism involved, but this is not the case. Since two pigments act as energy acceptors and are active in photosynthesis, it appears that two photosystems actually function in the light reaction. Indeed, there are two systems. One is called *photosystem I*, and the other, *photosystem II*. The energy acceptor at the reaction site of photosystem II is unknown but is referred to as pigment "X". Photosystem II begins when pigment "X" acquires sufficient energy for the photochemical reaction to occur, at which time it loses an electron which moves to and reduces another unknown molecule, called "Y". This leaves "X" short one electron, which it now gets from a water molecule, thus splitting the water molecule and giving off oxygen. This represents the source of oxygen produced during photosynthesis. Somewhere in this *photolysis* of water, the micro-nutrients manganese and chlorine are involved, but their exact role is unknown. With the reduction of "Y", photosystem II is complete. However, now the electron is removed from "Y" and passed on to P_{700}. This electron transfer is similar, in many respects to the electron transfer of the respiratory chain, discussed in an earlier chapter. It appears that a plastoquinone removes the electron from "Y" and passes it on to cytochrome f, or to cytochrome B_6 which then passes it on to cytochrome f. It is cytochrome f which transfers the electron to P_{700}. During this electron transfer between the two photosystems, ATP is synthesized, a synthesis known as *photosynthetic phosphorylation*.

Photosystem I

With the energy from photosystem II plus that transferred to it from accessory pigments, P_{700} becomes excited and transfers an electron to ferredoxin, an iron-containing protein. This reduces ferredoxin, which can now use this additional energy to form the energy carriers ATP and $NADPH_2$, as illustrated in Fig. 7.7, or it can use it directly to supply

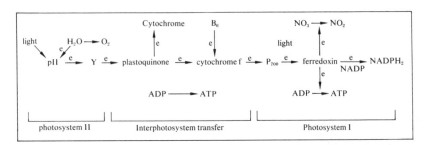

Fig. 7.7   Summary of the light reactions of photosynthesis.

energy to certain energy-requiring metabolic activities, such as nitrate reduction. The energy carriers formed permit storage of the energy released during the light reaction until this energy can be used in the dark reaction of photosynthesis.

Redox

In past studies of chemistry, you have learned about oxidation and reduction. A molecule is oxidized when it loses an electron. When the electron leaves, it carries energy with it, so a molecule that has lost an electron, has been oxidized, has less energy than it previously had. When a molecule is reduced, it receives an electron, so it has more energy than it previously had. The light reaction of photosynthesis involves oxidation and reduction reactions. For instance, P_{700}, after receiving energy from the accessory pigment, is oxidized by passing an electron off to ferredoxin. The ferredoxin is reduced by receiving the electron. Since the electron has energy associated with it, when the ferredoxin receives the electron, and is reduced, it has more energy than it originally had. The ferredoxin then passes this electron off, becomes oxidized, to combine with a hydrogen ion and both become attached to NADP, oxidizing the NADP to $NADPH_2$. Since $NADPH_2$ is the reduced form of NADP, it contains more energy than NADP. Due to this extra energy, which it can later transfer by reducing one of the intermediates of the dark reaction, it is called an energy carrier. Remember, when one molecule is reduced, another must be oxidized.

The rate at which the light reaction occurs in plants is determined by the light intensity and quality, and the amount of pigments and grana. It can be inhibited experimentally by the use of certain chemical inhibitors, such as 3-(p-chlorophenyl)-1, 1-dimethylurea, m-chlorocarbonylcyanide phenylhydrazone 2, heptyl-4-hydroxyquinoline-N-oxide, and desaspidin, which inhibit the production of ATP and $NADPH_2$. Temperature has little effect.

Actually, the light reaction occurs very rapidly, with its rate being measured in milliseconds. The complete light reaction occurs in less than one thousandth of a second, which is very fast.

Although the light reaction has been studied by many scientists, much remains to be learned about it. It is possible to remove chloroplasts, or even fragments of chloroplasts, from the plant, place them in a test tube or flask, and, when exposed to light they will carry out the light reaction, emitting oxygen gas. This activity is known as the Hill reaction, and has been observed for many years. This indicates that only chloroplasts or their fragments, and not the intact cell, are needed for oxygen evolution. Such materials have also been shown to function in photosynthetic phosphorylation, additional evidence that the light reaction of photosynthesis occurs entirely on the membranes of the chloroplast grana, and not in the stroma. However, such fragments cannot complete the photosynthetic process. This can be done by isolated, but intact, chloroplasts, demonstrating that chloroplasts are the only cell structures needed for photosynthesis, but the chloroplasts must be intact.

The light reaction of photosynthesis is the photochemical reaction of plants that requires the greatest light intensity to saturate it. It is also, perhaps, the only such reaction that stores energy for general use by the cells of the plant and for use by most other living entities. Nevertheless, it is not the only photochemical reaction in the plant cell.

THE HIGH ENERGY REACTION

Two photochemical reactions that are of common occurrence in plants and that function, separately or jointly, to control morphogenesis, are the high energy blue reaction and the phytochrome reaction.

When one compares the appearance of individuals of the same species grown in either full sunlight or in the shade, one finds that the sun plants are smaller than the shade plants. The sun plants do have a greater dry weight, as one might suspect since photosynthesis would have been more extensive, and more and larger roots. However, the shoot size is decreased with the internodes being shorter and the leaves smaller but thicker. Also, there is an increase in the number of flowers formed on sun plants as compared with shade plants, as many gardeners have observed. If you want the plant to flower profusely, keep it in the sunlight, not in the shade. The phenomena related to shoot and leaf size and to flowering are associated with the high energy reaction and phytochrome reaction.

The *high energy reaction* has been demonstrated many times in many plant species but not much is known about it. A relatively high light intensity is required, and for a relatively long period of time, but not as high as that required for the light reaction of photosynthesis. The pigment has never been isolated or even identified, which indicates either that it is present in extremely low concentration within the cell or that it is very unstable, breaking down whenever attempts are made to study it. The former reason seems to be very logical, namely, that it is indeed present in very low concentrations, but it is also likely that it may be very unstable.

The absorption and action spectra for the high energy light reaction is in the blue and far red regions of the spectrum, indicating that the pigment is green in color (Why would it be green?). It is not a reversible pigment so the high energy reaction is not light reversible. However, the pigment does seem to be very important for some aspects of morphogenesis.

Currently, knowledge concerning the way by which the pigment controls morphogenesis is lacking. It is not understood how the pigment can function in this capacity. However, the results of this high energy reaction are known.

One function of the high energy reaction is to reduce stem growth. If one measures periodically the height of a species growing in the dark, in the shade, and exposed to full sunlight, one invariably finds that the shortest plants will be those grown in full sunlight. This may seem strange, since under such conditions maximum rate of photosynthetic carbon fixation would be occurring, which might be expected to give greatest growth, but the plant is dwarfed. The light intensity under full sunlight is great enough to allow the high energy reaction to occur, and this reaction reduces stem growth. For this reason, plants growing in the shade are taller than those growing in full sunlight.

The high light intensity reaction also has other manifestations relative to morphogenesis, but most of these are also controlled by the phytochrome reaction so let us study the phytochrome reaction and then return to the study of the photoresponses brought about by the two reactions working together.

THE PHYTOCHROME REACTION

The *phytochrome reaction* is a low energy reaction, as contrasted to the high energy reaction. Low light intensities are all that are required for this photochemical reaction.

The Pigment

The pigment, called *phytochrome*, that is the photoreceptor for the phytochrome reaction, has been isolated and identified. It is a blue colored protein and is found in cells in green plants. In some cases, it is soluble in the ground substance of the cytoplasm, and in others, it is a constituent of cell membranes. It is always present in very minute concentrations, so the cell never appears blue, even when phytochrome is present.

Phytochrome is a reversible pigment with action spectra in the red (660 nm) and far red (720 nm). In plants kept in the dark, it is present in the Pr form, or the form that can absorb red light. Upon exposure to red light, it is converted to the Pfr form, or the form that can absorb far red light. When exposed to far red light, it is converted to the Pr form. Such a conversion also takes place slowly in the dark, so a Pfr form of phytochrome will become a Pr form in the dark even without exposure to far red light. That is why plants kept in the dark always have their phytochrome in the Pr form irrespective of the form of its existence prior to transferring the plants to the dark. This reversible photochemical reaction is shown below:

$$\text{Pr (660 nm)} \underset{\substack{\text{far red light} \\ \text{or darkness}}}{\rightleftharpoons} \text{Pfr (730 nm)}$$

The phytochrome reaction has been studied extensively for about the past twenty years, but still little is known about how the pigment brings about the morphogenetic responses associated with it. However, there is some indication that the phytochrome can directly alter the permeability of the cell membranes.

Seed Germination

One classical example of a role of the phytochrome reaction in morphogenesis, is its effect on seed germination. Seeds of many plant species will not germinate in the dark. Freshly-harvested, Grand Island lettuce seeds are an example. However, if they are exposed to even a brief flash of red light, they can be returned to the dark and will germinate very well. However, if, after exposure to red light, they are exposed to a flash of far red light and then returned to the dark, germination will not occur. One can alternate red and far red light treatments many times, but whether or not the seeds germinate, after returning them to the dark, will depend upon whether the last treatment was red light or far red light.

High Energy vs. Phytochrome

As previously indicated, many morphogenetic activities are controlled either by the high energy reaction or by phytochrome or, more often, by both working together. Such responses include anthocyanin synthesis, inhibition of hypocotyl lengthening, hair formation along the hypocotyl, enlargement of the cotyledons, increase of the negative geotrophic reactivity, opening of the plumular hook, formation of tracheary elements in the hypocotyl vascular bundles, stem elongation, and flowering. Usually these responses are manifestations in differences in the extent of enlargement of the cells on opposite sides of the plant structure, but how these reactions alter this enlargement is not fully understood and awaits further research to explain.

In the case of the effects of the photochemical reactions on the inhibition of isolated wheat roots, both cell enlargement and cell division are inhibited, and in the case of fern prothali stimulation by light, cell division is stimulated.

Other morphogenetic activities, probably altered by one or both of these photochemical reactions, include the viscosity of protoplasm, chlorophyll synthesis, movement of cellular organelles, such as mitochondria and chloroplasts, effects on plasmolysis, and the appearance of the ectodesmata.

In view of the wide variety of responses to these photoreactions, and the fact that they can function separately or together indicates that there is no immediate relationship between these two reactions but that they function independently within the cell. However, both independent activities do at times supplement each other.

FUNGAL RESPONSE TO LIGHT

Fungi are heterotrophic plants and therefore do not contain chlorophyll. Also they do not possess the pigments necessary for either the high energy or the phytochrome reactions. Nevertheless, many of them do depend upon the presence of light for their normal development. In nearly all cases, only the short wavelengths of light, the blue light, and the longer ultraviolet radiation satisfy their requirements. This indicates that their pigment is a flavoprotein, and is probably similar to the phototrophic mechanism in other multicellular plants. Such light is necessary in fungi for carotenoid formation, spore germination, mycelial growth, conidia formation, sclerotia and spore formation, formation of sporodochia and pionnotes, and formation, growth, and differentiation of fruiting bodies in ascomycetes and basidiomycetes, as well as the "light growth response" and phototropism.

PHOTOTROPISM

Another interesting and important photochemical reaction which occurs in higher plants and appears to be similar, as far as mechanism is concerned, to that just discussed in fungi, is *phototropism.* Many examples can be cited of phototropism among plants. Leaves of ivy, and other plants change their orientation during the day as the sun moves across the sky so that they remain exposed to maximum sunlight. The flowers of some species, including sunflower, remain perpendicular to the rays of sunlight during the day due to phototropism. Seedlings, in the dark will bend toward unilateral light, and roots often grow away from light due to their phototrophic response.

In Grass Coleoptiles

Perhaps the greatest amount of research on this subject has been done using grass coleoptiles. Coleoptiles are sheaths which grow out from the germinating seed and enclose the leaves of the young plant. After these coleoptiles have attained a length of about 2–3 cm, they stop growing and the young leaf pushes through the coleoptile. The coleoptile then dies and disintegrates, having served to protect the young leaf during early growth. The coleoptile has been shown to grow in length by enlargement of its cells near the middle of the coleoptile. This enlargement is caused by the synthesis of a hormone, auxin, which is produced only at the tip of the coleoptile to the cells near the middle, where these cells are stimulated to enlarge. Without this auxin, no enlargement occurs.

If grass seeds are germinated in the dark, and after the coleoptiles have partially grown, these young plants are exposed to light coming from only one direction (unilateral light) the coleoptiles can be seen to bend toward the light. This bending takes place only at the middle of the coleoptile. However, if the tip of the coleoptile is covered with a piece of aluminum foil, or some other light-proof material, no bending occurs. This indicates that the tip is the light-sensitive site where the photochemical reaction occurs that results in the bending. However, since the bending occurs in the middle, there must be some influence which is produced in the tip that moves down to the middle to control this bending.

The Pigment

The pigment associated with the phototropic phenomenon is believed to be a flavoprotein. It is most receptive to blue light, of about 400–500 nm and is present in very minute amounts. However, the flavoprotein does not move through the coleoptile so it must produce some other hormone that does move. This hormone has been shown to be auxin, a nitrogen-containing chemical produced in all growing parts of the plant. Actually, auxin is produced in all meristems of the plant, so the flavoprotein does not alter its production. However, it has been shown that when the coleoptile is in the dark or is receiving light from all directions, auxin moves from the tip down the coleoptile to the growing cells, in the middle, uniformly on all sides of the coleoptile. However, when light shines on the coleoptile from only one side, more auxin ends up on the side opposite that exposed to the light, as shown in Fig. 7.8. This causes more enlargement of the cells on the dark side; the side with the most auxin, and therefore causes the coleoptile to bend toward the light. Therefore, unilateral light is absorbed by flavoprotein which causes a lateral movement of auxin, by some unknown mechanism, to the opposite side. Thus, more auxin moves down the dark side and causes the cells on this side to elongate more than the cells on the lighted side, thus causing the coleoptile to

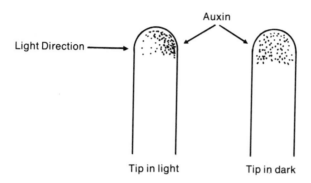

Fig. 7.8   The effect of unilateral light on the distribution of auxin in the coleoptile.

bend toward the light. The extent of bending is proportional to the amount of light received (to the light intensity plus the duration of exposure).

The tips of coleoptiles are not the only locations of the pigment responsible for phototropism. This pigment also functions in leaves or petioles, root tips, and in other plant structures. The photochemical reactions discussed above are summarized in Table 7.9.

Table 7.9   Summary of important photochemical reactions in plants.

Reaction	Pigment	Absorbed light color	Intensity required	Function
Photosynthesis	chlorophyll *a*	red + blue	high	energy storage
High energy	?	blue + far red	high	morphogenetic
Phytochrome	phytochrome	red + far red	low	morphogenetic
Phototrophic	flavoprotein	blue	low	organ orientation

GEOTROPISM

Phototropism is, in many respects, similar to another tropism, namely, geotropism. Geotropism refers to the bending of a plant structure under the influence of gravity. It can either be positive, (grow toward the earth) or negative (grow away from the earth). As in the case of phototropism, the bending is due to an unequal distribution of auxin on the two sides of the stem or root.

Geotropism can be easily demonstrated by laying a potted plant on its side, in the dark. After a few hours, the stem will bend upward, and the roots will bend downward. In both cases, this bending is associated with an unequal distribution of auxin on the upper and lower sides. In the stems, this increase in auxin causes the cells on the lower side to elongate more than those on the upper side so the stem bends up until it is growing vertically, at which time the auxin moves from the tip equally down all sides of the stem, allowing vertical growth to continue. In the case of the root, the root cells are more sensitive to high auxin concentration, so the concentration of auxin that causes the cells of the stem to elongate more, actually inhibits the growth of the cells on the bottom of the root. Therefore, the cells on the upper side elongate more causing the root to bend downward, until it is growing vertically, at which time the auxin moves back from the tip uniformly on all sides, so vertical growth continues.

Importance of Geotropism

How important is geotropism? If it were not functional, seeds would have to be placed in the soil with the site of the root down and the site of the shoot up and in no other position. The same may be said for bulbs, tubers, etc. Otherwise, roots would grow up out of the soil and shoots would grow down into the soil.

ULTRAVIOLET RADIATION

Perhaps it would be well to consider some effects of how energy enters the plant in the form of ultraviolet radiation. This radiation has wavelengths between about 10 and 400 nm. It is a constituent of sunlight, making up about 1% of that natural radiation. Fortunately, atmospheric gases, such as nitrogen and oxygen, as well as ozone in the upper atmosphere, absorb much of the short wavelengths of ultraviolet radiation and therefore prevent such radiation of less than about 300 nm from reaching the earth's surface. However, as the elevation above sea level increases, so does the ultraviolet intensity. This is one cause for the stunted growth of plants of the alpine vegetation of our higher mountains. Ultraviolet radiation will not penetrate glass, as one knows since ultraviolet radiation is the cause of sunburn and we cannot get sunburned sitting in front of a closed window or in a greenhouse. Ultraviolet radiation is often used to kill microorganisms, such as bacteria and viruses. It alters their enzyme activity perhaps by protein denaturation.

HEAT

Heat is another form of energy that has significant effects on plants. As is true of energy generally, heat enters the plant or its cells either through radiation, convection, or conduction. *Conduction* occurs when one substance comes in contact with another whereby their molecules bump together, such as when one end of an iron rod is placed on a hot plate, and one feels the movement of heat up the rod. *Convection* involves movement of the heated material from the site of heating into an area of cooler temperature. For instance, air or water will move in such a direction due to differences in density of the heated and cooled molecules. Hot air rises, and cool air settles. Orchards are sometimes planted on slopes to reduce frost injury. (Why would this reduce injury?) Have you ever noticed how the temperature of water in a pool is hotter near the surface than at greater depths? *Radiation* is heat gain in the form of heat waves, or infrared waves. In this chapter, how the heat enters the plant and its cells will be considered, and heat loss will be studied in Chapter 9. Also, some heat appears as a result of energy wasted during chemical reactions within the plant, but this is probably an insignificant amount.

Heat may enter the plant through conduction. If the air has a higher temperature than the plant, heat will enter the plant as the air molecules transfer some of their kinetic energy to the molecules of the plant as they bump into the plant molecules. Heat may also be transferred by conduction from the soil when the soil temperature is greater than that of the plant. Plant stems will sometimes be burned or scalded near the soil surface due to such heat transfer.

Some heat enters by convection as air is heated near the soil surface and moves up the plant due to the lighter density of such heated air. However, perhaps radiation is a more important means of entry.

Radiation is the form in which much heat enters the plant, raising its temperature. Sunlight contains a great deal of infrared radiation, especially infrared radiation of longer wavelengths. This radiation is very efficiently absorbed, particularly by the water molecules of the cell. As a result, the plant absorbs over 95% of the incident infrared radiation of the longer wavelengths. Of course, these may arise not only from sunlight directly, but also from reradiation from other structures of the environment. Infrared radiation is very efficiently converted to heat so can rightly be considered to be heat waves.

Whether heat enters the plant through conduction, convection, or radiation, it does enter, and as a result this raises the temperature of the plant and its cells. What such temperature changes signify as far as the physiology of the plant is concerned, will be discussed in Chapter 8.

In summary, this chapter has revealed how energy enters the plant either in the form of conduction, convection, or radiation. Once it is in the plant, it may or may not remain. Either it is lost by transmission or reflection or it is absorbed. If it is absorbed, this means that a chemical molecule has energy transferred to it which either increases the motion of the molecule, as in the cases of heat, or activates one or more valence electrons of the molecule in the case of light. Such activated molecules are either destroyed or the energy is passed off to another molecule and used or conserved within the plant for subsequent use in metabolism. How the absorbed energy is used will be the topic for discussion in the next chapter.

Chapter Eight

Energy Storage, Utilization, and Loss

The previous chapter showed how energy entered the plant – how radiant energy activated the pigment molecule, and how this captured energy sometimes destroyed the molecule, and sometimes was used for photochemical reactions. Such photochemical reactions resulted in physiological changes in the plant, such as during the phytochrome reaction, the high energy reaction and during phototropism – changes that were evident primarily as visible manifestations of growth. However, one of the photochemical reactions, namely, the light reaction of photosynthesis, resulted in energy storage within the plant in the form of the chemical, energy-rich bonds of the energy carriers, ATP and $NADPH_2$. Later in the dark reaction of photosynthesis these energy carriers transferred their energy to certain chemical compounds in the form of chemical energy – that energy associated with their chemical bonds. This chapter will consider what happens to this energy after it has been introduced into these molecules.

MECHANISMS OF METABOLISM

What is Metabolism?

A plant is composed of thousands of different types of organic compounds, such as sugars and proteins, and an undetermined number of these are essential for life. During the life of the plant, some of these compounds are continuously being constructed and others are being broken down. Such dynamic activities are characteristic of the living plant. This synthesis and degradation of organic molecules in living entities is known as *metabolism*. Metabolism involves enzymes, and all such reactions are catalyzed by these enzymes. Metabolism also involves various chemicals to act as *substrates* (raw materials), and metabolism involves energy. Synthesis, often referred to as *anabolism*, requires energy to form the bonds between the atoms that join together to form the molecule. Degradation, often referred to as *catabolism*, involves breaking these bonds to release energy. Therefore, in any metabolic activity, energy is either required or released. If it is required, the reaction is said to be *endergonic*, and if energy is released, the reaction is *exergonic*.

Heat Content

To hold atoms together to form a molecule requires energy. The sum total of this energy is known as the *heat content* of the molecule. In the reaction, $A \rightarrow B$, the chemical A is the substrate and B is the product. Since each chemical reaction involves a gain or loss of energy, the energy of B will be greater or less than A depending upon whether the reaction is endergonic or exergonic. This difference in the energy of the product and that of the substrate is known as the *change in heat content*, and is designated as ΔH. This relationship is shown below:

$$\Delta H = \text{energy of products} - \text{energy of substrate}$$

A study of the above relationship reveals that the magnitude of ΔH tells the amount of energy lost or gained during the reaction, and that ΔH can be either positive or negative in sign. If it is positive, energy must be added for the reaction to occur, as in any photochemical reaction. This energy is needed to form chemical bonds. If ΔH is negative, energy will be released during the reaction as bonds are broken so the reaction will occur spontaneously. The energy lost in a spontaneous reaction, is lost either as heat or as usable energy which may be conserved to be used in later synthesis.

Free Energy

The energy lost as heat is designated as $T\Delta S$ (absolute temperature times entrophy) and that conserved as usable energy is designed as *free energy* (F). Therefore, the following relationships exist:

$$\Delta H = \Delta F + T\Delta S$$

and

$$\Delta F = \Delta H - T\Delta S$$

The relationship of ΔF to the free energy of the substrate and product molecules can be seen in Fig. 8.1. One can determine the ΔF value if either the equilibrium constant of the reaction or the redox potential of the reaction are known, as follows:

$$\Delta F = -RT \ln K = -4.606 \, T \log K$$

or

$$\Delta F = n23,000 \, \Delta E$$

In these equations, R = the gas constant, T = the absolute temperature, K = the equilibrium constant of the reaction, n = the number of electrons exchanged, and ΔE = the change in electrical potential during the reaction. ΔS is determined by observing the energy change when the reaction is carried out at a temperature near absolute zero ($T = 0°K$).

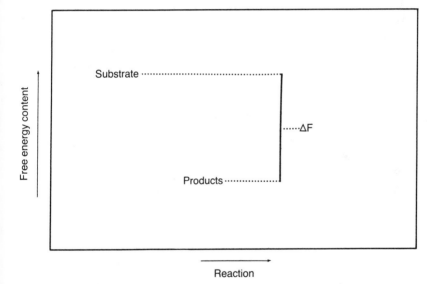

Fig. 8.1 An illustration of the relationship between ΔF and the free energy content of substrate and products.

Use of ΔF

Like ΔH, the ΔF value can be either positive or negative. If it is negative, the reaction evolves energy and therefore will occur spontaneously; if positive, the reaction will occur only if energy is applied, as in any photochemical reaction. (*See* Fig. 8.2.) Therefore, the sign of the ΔF value will tell the likelihood of a reaction occurring.

To determine the likelihood of a reaction occurring, not only the sign of ΔF, but also its magnitude must be known. At equilibrium, the magnitude of ΔF is zero, and this magnitude increases in value with the distance of the reaction from equilibrium. Therefore, ΔF can be considered as the driving force of the reaction or at least as a measure of this driving force.

$+\Delta F$ = ENERGY OF PRODUCTS – energy of substrate (Energy added)

$-\Delta F$ = energy of products – ENERGY OF SUBSTRATE (Energy evolved)

Fig. 8.2   Free energy change, illustrating sign determination and related energy status. Capital letters indicate high free energy content and lower case letters represent low free energy content.

Since ΔF varies with concentration, in many cases, its value can be changed, and a chemical reaction can be driven in one direction or another by changing the relative concentrations of the substrate or the products.

The ΔF value will also reveal if the reaction is significantly reversible, since this too is related to the equilibrium constant. Theoretically, every reaction is reversible but the reverse may be so slow that it is not considered to be reversible for any practical purposes. If the ΔF value is greater than 4 or 5 kcal and negative, the reaction is considered to be irreversible. The ΔF of photosynthesis is about $+690$ kcal, so the reaction is endergonic and reversible, while that of respiration is -690 kcal, so it is exergonic and irreversible, unless energy is added.

Reaction Rates

Consideration of the free energy change reveals the likelihood of a reaction occurring but reveals nothing about the rate at which this reaction will occur. Nor will the concentration of the molecules involved help in this respect. The rate at which a chemical reaction occurs does not depend upon the total concentration of the molecules of the substrate but rather upon the concentration of those molecules which have sufficient kinetic energy to participate in the reaction. Not all of the substrate molecules have this much kinetic energy at any given time. Although atoms and molecules are continuously moving about due to the kinetic energy they possess, not all are moving at the same speed, since each has a different kinetic energy content than its neighbors. However, each substrate molecule must possess a minimum amount of energy to take part in the reaction. This minimum amount of energy is designated as the *activation energy*. The activation energy requirement differs from one reaction to another.

Activation Energy

Since kinetic energy is related to temperature, one can increase the energy content of the substrate molecules by increasing their temperature. Indeed, in this way one can speed up the rate at which a chemical reaction occurs. However, due to the necessity of a tremendous increase in the rate of metabolic reactions in plants, often in excess of 500,000 × increase, and since reaction rate is related to absolute temperature, and since living cells are damaged by high temperatures, it is not possible to increase the temperature enough to give the necessary increase in reaction rates without killing the cell long before this needed increase in rate is realized. However, there is another method by which reaction rate can be increased, and this is by lowering the activation energy requirement of the reaction.

The rate at which a reaction is related to the activation energy (Ea) requirement, and the absolute temperature as follows:

$$\text{reaction rate} = Ae^{-Ea/RT} = \log A - \frac{Ea}{2.303\,RT}$$

where $A = $ a constant which is essentially the same for all reactions, and $R = $ the gas constant $= 1.99$ cal/mole. Since the reaction rate is an exponential function of Ea and T, a small change in either Ea or T will result in a large change in the reaction rate. For most metabolic reactions, a 10°C increase in temperature will cause a two- to four-fold increase in the reaction rate. Since this is not nearly enough of an increase to allow plants to survive, the Ea requirement must be considered as a possibility of speeding up the reaction rate.

If the Ea requirement can be lowered, the rate will be speeded up. Just by decreasing the Ea value to one half, will result in increasing the reaction rate of perhaps 500,000 times. This is the mechanism the plant cell uses for reaction rate increase. It lowers the activation energy requirement of each reaction. It does this by the use of catalysts – organic catalysts called *enzymes*. How the enzymes are able to lower the Ea requirement is not fully understood, but will be considered again in Chapter 11.

Now that we have survived this brief, but, I hope, understandable revelation about the intricacies of the role of energy in metabolic reactions, let us get on with the subject of the storage, use, and loss of energy by the plant and its cells.

ENERGY CARRIERS

The reactions by which the plant synthesizes the many organic compounds found in it are energy-requiring reactions. The energy needed for these reactions is supplied by the energy carriers which couple the energy-releasing reactions to the energy-requiring reactions, so that the free energy released during one reaction is used to carry out another reaction that requires free energy. Such energy carriers include the ATP and $NADPH_2$ which are formed during the light reaction of photosynthesis. This reaction was coupled to the photosynthetic carbon fixation reactions through these energy carriers. Sometimes these energy carriers move from one part of the cell to another, and sometimes they remain fixed in one position but from this fixed position they engage in energy transfer activity. These energy carriers are synthesized primarily in the chloroplast, during the light reaction of photosynthesis, and, as shall be seen later, in the mitochondrion or cytoplasm during respiration. A list of some common energy carriers in plant cells is given in Table 8.1, and includes the reduced coenzymes, nucleotides, thioesters of coenzyme A, and formyl-THFA.

Formation

The anhydrides are formed from ATP, and ATP is formed during photosynthetic phosphorylation, during oxidative phosphorylation, by substrate-linked phosphorylation, from ADP whereby two molecules of ADP give one molecule each of AMP and ATP, as well as by inter-nucleotide phosphate transfer. Chapter 7 revealed how they are formed during photosynthesis, so let us learn something about their formation during respiration.

RESPIRATION

Respiration is a form of oxidation and is often referred to as biological oxidation. Oxidation is defined as a loss of electrons by a molecule, which

Table 8.1   Common energy carriers in plant cells.

Reduced coenzymes
  NADPH$_2$
  NADH$_2$

Nucleotides
  ATP
  CTP
  UTP
  GTP

Thio esters of coenzyme A

Formyl-THFA
Flavin-adenine dinucleotide
Riboflavin phosphate (flavin mononucleotide)

---

may or may not be associated with a loss of hydrogen atoms or a gain of oxygen atoms, and a gain in positive valence. However, usually one hydrogen atom is removed from a molecule when it is oxidized and two electrons are lost at the same time, as given in the example below:

$$NADH + H^+ \xrightarrow{\text{oxidation}} NAD^+ + 2H$$

The two electrons are lost one at a time. It is usual to illustrate this reduced form of the coenzyme by NADH$_2$ instead of NADH + H$^+$, a practice that will be followed in this book.

Redox Potential

Oxidation represents a release of energy from the substrate molecule. This energy is eventually either lost as heat or transferred to energy carriers. The ease with which the substrate molecule gives up its electrons is expressed by the redox potential of that molecule. This *redox potential* is given in volts and is either positive or negative. A substance more negative will reduce a reaction less negative and oxidize one more negative. However, this potential varies with pH, with the concentrations of substrate and products, and with temperature. At increased pH values, the potential decreases. In the plant cell, this potential is measured by redox dyes or by microelectrodes, both of which are usually inserted into the cell by micromanipulators, although sometimes the dyes can be taken up by the cell through its membranes.

Respiratory Chain

There are several pathways of respiration known to occur in plants at the present time. The two most common are fermentation and aerobic respiration, with the two most common types of aerobic respiration being the hexose-monophosphate-shunt and the combination of the glycolysis plus the Krebs' cycle. In all cases, these consist of a series of chemical reactions, some of which release hydrogen and accompanying electrons from the intermediate molecules and these electrons are used in the reduction of energy carriers with the hydrogen ending up combining with oxygen to form water. This activity which starts with the removal of a hydrogen and two electrons from the molecules, and ends with the formation of ATP and water is known as the *respiratory chain*. Although uncertainty exists about the intermediate coenzymes, in fact, it is possible that there are several alternative coenzymes that are used, one probable respiratory chain pathway is given in Fig. 8.3.

---

substrate + NAD $\longrightarrow$ NADH$_2$ + oxidized substrate
NADH$_2$ + flavoprotein $\longrightarrow$ flavoprotein H$_2$ + NAD
flavoprotein H$_2$ + cytochrome b $\longrightarrow$ flavoprotein + reduced cytochrome b
reduced cytochrome b + cytochrome c $\longrightarrow$ reduced cytochrome c + cytochrome b
reduced cytochrome c + cytochrome oxidase + $\frac{1}{2}$O$_2$ $\longrightarrow$ H$_2$O + cytochrome oxidase.

In essence, what is given above for the respiratory chain is only partially complete since the electrons removed from the substrate do not end up in the water. What happens is that somewhere along the chain, perhaps near the reduction of cytochrome b, the electrons and hydrogens separate. Coenzyme Q may remove these electrons. The two electrons plus ADP plus inorganic phosphate form 3 molecules of ATP, as given below:

$$3 \text{ ADP} + 3 \text{ Pi} + 2 \text{ electrons} \longrightarrow 3 \text{ ATP} \qquad \Delta F = 36 \text{ Kcal}$$

Fig. 8.3   Respiratory chain in plant cells.

The respiratory chain is much like the old time bucket brigade. No doubt, most of you readers are too young to remember such cooperate groups as a bucket brigade, but such a group consisted of a line of people, with the line usually extending from a well or creek to a house that was on fire. The person on the well end would fill a bucket with water and pass it on to the next man, and he would pass it to the next, etc., until the last man got the bucket and he would throw the water on the fire. As soon as a man passed a bucket on, he would reach back to the man before him and accept another bucket. In this way, many buckets of water could be thrown on the fire without anyone having to travel far. If we liken this to the respiratory chain, then each man would be replaced with a different electron acceptor, such as NAD, or flavoprotein or a cytochrome. The hydrogens, along with the accompanying electrons would be the buckets. The enzymes would accept the hydrogen and electrons, pass them on to the next, and reach back for another and so forth. The beginning of the line would be the substrate and at the end the hydrogen would be passed off to oxygen to form water. However, somewhere along the line the hydrogen and electrons would separate, almost as if someone had spilled the bucket.

Each enzyme is reduced as it receives electrons from the enzyme in back of it, and is oxidized as it passes the electrons off to the enzyme in front of it.

For every 2 protons and 2 electrons, removed from the substrate, there are 3 ATP molecules formed. In these ATP molecules, about 65% of the energy of the electrons is stored, with about 35% being lost as heat. In any respect, each ATP molecule stores about 12 kcal of energy and transfers this energy to endergonic chemical reactions.

The next question that might be asked is what chemical molecules act as a substrate for this ATP formation, and how are they formed. To answer this requires a discussion of the metabolic pathways associated with respiration.

PATHWAYS OF RESPIRATION

Hexose-monophosphate-shunt

One common pathway for respiration especially in older plants, is the *hexose-monophosphate-shunt pathway*. There is some indication this may be the principal pathway in plants generally, but is less active in young plants. The pathway is illustrated in Fig. 8.4. This type of respiration is both oxidative and aerobic so it will occur only in the presence of oxygen. The reduced energy carriers, $NADPH_2$, can either transfer their energy directly to endergonic reactions, or can participate in respiratory chain activity, generating ATP molecules. Therefore, a maximum of 6 ATP molecules can be synthesized for each glucose-6-phosphate molecule oxidized. Remember the glucose-6-phosphate got its energy originally from photosynthesis.

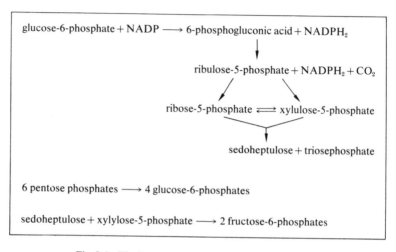

Fig. 8.4   The hexose-monophosphate-shunt pathway.

Glycolysis

Another common pathway by which glucose is broken down during both aerobic respiration and fermentation, is by the glycolytic pathway or *glycolysis*. This pathway is outlined in Fig. 8.5. Note the similarity of this pathway to the reversal of the photosynthetic carbon dioxide fixation pathway. Perhaps the main difference being that glycolysis occurs in the cytoplasm, whereas photosynthesis occurs in the chloroplast. All of the enzymes of glycolysis are soluble enzymes, suspended in the cytoplasm, with the exception of hexokinase which appears to be bound to membranes. If glycolysis starts with glucose, 2 ATP molecules are required, whereas 4 are formed. Therefore, some energy is released and transferred to ATP during glycolysis, but very little. Glycolysis will occur in either the presence or absence of oxygen. It begins with a sugar or sugar phosphate and ends with the formation of 2 molecules of pyruvic acid. Remember for each sugar molecule oxidized, 2 pyruvic acid molecules are formed. How much carbon dioxide is lost during glycolysis?

Fermentation

After the pyruvic acid has been formed via glycolysis, it is further broken down, but now the mechanism of its breakdown depends upon whether oxygen is absent or present. If it is absent, pyruvic acid is broken down to form ethanol, a process known as fermentation. The fermentation pathway is illustrated in Fig. 8.6. Since it is anaerobic, it releases very little of the energy originally stored in the sugar molecule, with much of this original energy remaining in the ethanol. Therefore, it is not very efficient, which is one reason a cell cannot tolerate such a system for long. The sugars are used up too rapidly for the energy released, and eventually the cell will deplete its energy reserves and die. Of course, another reason fermentation cannot long occur in most cells, is the toxicity of the alcohol produced. Yes, alcohol is toxic, and few if any cells can tolerate a concentration as high as 10%. As strange as it may seem, perhaps most plant cells possess the enzymes needed for fermentation to occur. Perhaps, fermentation is an anachronism. A relic of the past which represents the first mechanism for respiration but one which has largely been surpassed with more efficient methods. At times, fermentation may be valuable to most cells, for it will allow the cell to survive limited periods when the cell may be exposed to an anaerobic environment.

On the other hand, there are a few plant cells which survive with fermentation as their normal mechanism of respiration. Cells of roots of submerged plants where the roots are growing in mud at the bottom of ponds is an example. The center of large fruits, storage roots and tubers, various organs of succulent plants, woody stems and seeds, spores, etc. that have coats impermeable to gases are other potential examples.

Krebs' Cycle

Whenever oxygen is present the Pasteur effect is observed, or in other words, in the presence of oxygen, the pyruvic acid formed via glycolysis is not broken down to form alcohol so fermentation does not occur. Instead, the pyruvic acid molecules are broken all the way down to form carbon dioxide and water with a much greater release in energy. This breakdown is via a mechanism known as the Krebs' cycle and is illustrated in Fig. 8.7.

In contrast to the other pathways of respiration previously discussed, the Krebs' cycle occurs only in the mitochondrion of the cell. This is true in plants as well as animals, since this mechanism functions in both. More will be said about the mitochondrion later.

In Fig. 8.7, arrows can be seen to lead to 3 carbon dioxide molecules and to 10 hydrogens. These arrows represent locations where these are formed. The carbon dioxide are formed via decarboxylation reactions and the hydrogens leave via the respiratory chain, and end up in water molecules. Since 3 carbon dioxides are formed and since 2 pyruvic acid

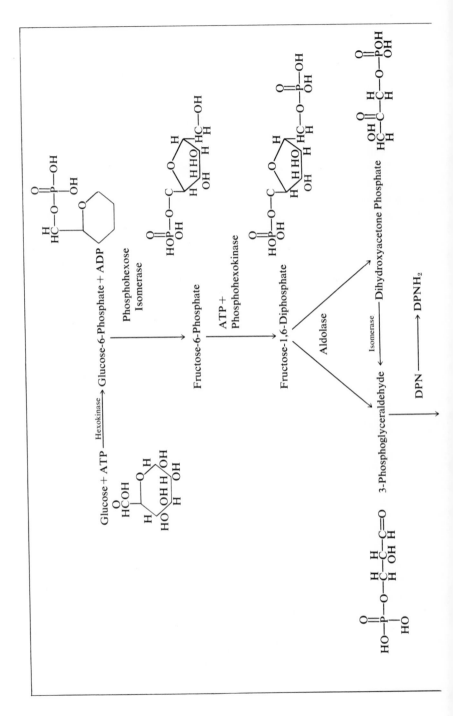

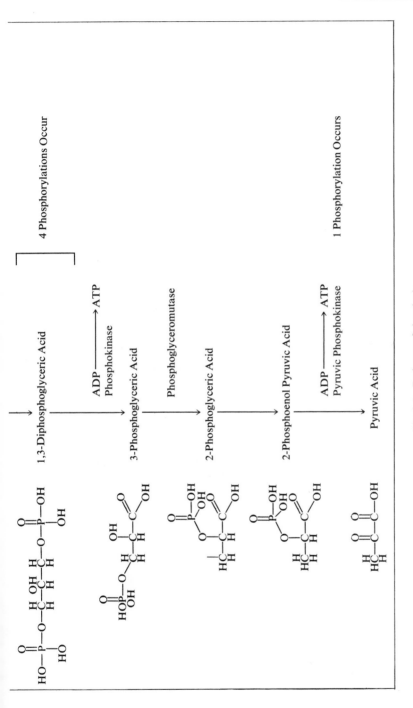

Fig. 8.5 Pathway of glycolysis.

$$\text{sugar} \longrightarrow 2\,C_2H_5OH + 2CO_2 \qquad \Delta F = -72 \text{ Kcal}$$
glycolysis

$$\text{sugar} \longrightarrow 2 \text{ pyruvic acid}$$

$$\text{pyruvic acid} \longrightarrow \text{acetaldehyde} + CO_2$$

$$NADPH_2 + \text{acetaldehyde} \longrightarrow \text{ethanol} + NADP$$

Fig. 8.6   Fermentation pathway in plants.

molecules are formed for each sugar molecule, this accounts for all of the carbons in the original sugar molecule.

The overall equation for aerobic respiration via glycolysis plus the Krebs' cycle is given below:

$$C_6H_{12}O_6 + 6O_2 \longrightarrow 6CO_2 + 6H_2O \qquad \Delta F = -688 \text{ kcal}$$

Note the much greater amount of energy released via aerobic respiration as compared with fermentation, namely, about ten times as much. This means that with a given amount of sugar, ten times as much energy would be released via aerobic respiration as through fermentation. Or for a given amount of energy released, ten times the amount of sugar would be

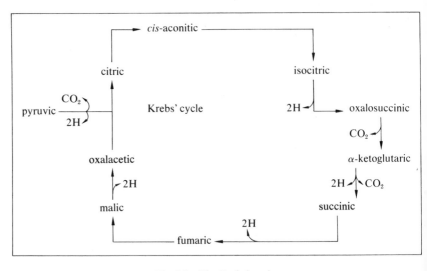

Fig. 8.7   The Krebs' cycle.

required via fermentation as through aerobic respiration. Perhaps these values, plus the absence of toxic products, will indicate the value of the preference for aerobic respiration in plant cells.

As in fermentation, not all of the energy released is conserved in ATP molecules. Some is lost as heat. However, about 65% is conserved in the 38 ATP molecules formed.

Oxidative Phosphorylation

The formation of ATP during respiration is known as *oxidative phosphorylation*. This is often studied by determining the P/O ratio, which is nothing more than the microatoms of phosphorus esterified per microatoms of oxygen consumed. Such a value averages about 3, indicating, as previously stated, that for every 2 hydrogens transferred to 1 oxygen to form water, 3 ATP molecules are synthesized.

PLANT MITOCHONDRIA

Although all respiration does not occur in the mitochondrion, much of it does, and the mitochondrion is looked upon as the powerhouse of the plant and animal cell. Therefore, let us look at the structure of this important organelle.

Distribution

Mitochondria are found in all eukaryotic cells, where they are located in the cytoplasm, suspended in the ground substance between the plasma membrane and the vacuole. In certain cells, they may be few in number, such as in the generative cell of pollen, whereas in other cells they may number over one thousand per cell. No good studies have been made to determine accurately the number of mitochondria in each plant cell, and no doubt this number will vary from one cell to another. It would be interesting to know what effect such variations may have on the rate of respiration and therefore upon the rate of metabolism per cell generally.

Shape

An electron micrograph of a plant cell will reveal an unusual uniformity of size and shape of the mitochondria in each cell. They are smaller than plastids but larger than sphaerosomes. However, when observed under the light microscope, these mitochondria do not appear to be either uniform in size or in shape. Instead they may appear spherical, elongated, dumbbell-shaped, or in a multitude of other shapes. Also, they change their shape constantly so that it is not possible to designate any one shape as indicative of the shape of mitochondrion in the living cell. Much deformity of their shapes occurs in the living cell, as the mitochondria interact with other cell organelles, especially during protoplasmic streaming, with such interactions resulting in continuous morphological changes in the mito-chondria. However, for simplicity, let us consider these organelles to be spherical or ellipsoidal in shape, a shape that is commonly observed in electron micrographs.

Structure

The mitochondrion is limited by a double membrane, the mitochondrial membrane. This membrane is differentially permeable as can be demon-strated by placing mitochondria in solutions that differ in water potential. In a solution of high water potential, they shrink, whereas in a solution of low potential, they expand in size.

The outer membrane of the mitochondrial double membrane is smooth and regular. The inner membrane is much invaginated, projecting out into the center of the mitochondrion, with such projections being long, and numerous, as seen in Fig. 8.8. These projections, or *cristae* as they are called, are similar to those found in animal cells except they are not as evenly spaced throughout the center of mitochondrion as in animal cells. They are not as parallel in distribution as in animal cells, and they even sometimes join to form a ring with adjacent cristae. The surface of these cristae are covered with projections, projections whose functions are unknown. These mitochondrial membranes contain the enzymes for

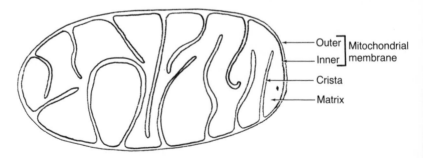

Fig. 8.8   A drawing of a plant mitochondrion – the powerhouse of the cell – showing its membranous structure.

Krebs' cycle activity, for respiratory chain activity, and for the synthesis of lipids and amino acids.

Chemically, the mitochondria consist primarily of water, proteins, and lipids. In fact, it appears that about 30% of the dry weight of the mitochondrion is made up of lipids, essentially all of which is phospholipids. In addition, there are numerous enzymes found in plant mitochondria. Three cytochrome b, two cytochrome c, as well as numerous other enzymes have been found there.

Formation

The mitochondria contain DNA and therefore should be capable of self-replication. Indeed, they do seem to divide, especially by fission, to form new mitochondria, and it appears that they do not arise except from pre-existing mitochondria. There are reasons to believe that mitochondria originally were free living cells, which entered other cells and took up a symbiotic relationship. If so, this must have been early in evolutionary history, since such mitochondria exist in all eukaryotic cells and therefore should have been present in these cells early in their evolutionary history. Earlier perhaps than the chloroplasts, since chloroplasts are found only in the plant cells and should have therefore appeared later.

It is possible that someday mitochondria will be cultured outside of the living cell. Indeed, today they are often extracted from the cell and their activities studied *in vitro*. However, duplication of these mitochondria *in vitro* has not been achieved, nor has their extended culture. The mitochondria become inactivated within a few hours after their removal from the cell.

Mitochondria are found in both plant and animal cells. However, plant mitochondria do differ from animal mitochondria. Plant mitochondria contain the respiratory chain enzymes in greater concentration. Plant mitochondria are less tightly coupled to phosphorylation because they have lower ADP:oxygen ratios. The plant mitochondrial membrane is more permeable than those of animals. Perhaps they differ in other ways too. It was previously pointed out how they also differ in the appearance of their cristae.

Function

In the living plant cell, the normal substrate, or food material, for the mitochondrion is pyruvic acid. Its products include ATP, amino acids, and fatty acids, as well as other materials. Since much of the energy carriers formed within the plant, and more specifically the ATP formed during oxidative phosphorylations, arise from the mitochondria, this organelle is sometimes referred to as the powerhouse of the cell. Its function primarily is to furnish the ATP needed as a source of energy for other metabolic activities. However, the synthesis of other materials, especially fats and amino acids, also represent important contributions the mitochondrion makes to the metabolism of the cell.

RESPIRATION RATE

Respiration is the basis for a dilemma with the plant. It must occur in all living cells at all times to supply the energy carriers needed for endergonic synthesis that are vital to the survival of the cell. On the other hand, respiration does use stored food materials and, once these are depleted, respiration even destroys more important cell chemicals, such as proteins. Therefore, it would be desirable that the rate of respiration be no higher than that needed. The rate of respiration should be rapid enough to supply the energy needed for metabolism, but no higher or excessive dry matter will be lost. The yield of the plant depends not only upon the rate of photosynthesis but also upon the rate of respiration, since as far as products and reactants are concerned, respiration is the reverse of photosynthesis, and uses dry matter rather than produces it. What is the minimum rate necessary for optimum development? That is a good question, and one that has not, to my knowledge, been answered, but should be.

The actual rate of respiration within the plant depends upon the number of living cells in each plant and the rate of respiration per cell. The latter depends upon the number of mitochondria per cell, the rate of respiration in each mitochondrion, and the fraction of the respiration rate attributed to extra-mitochondrial respiration. The rate in each mitochondrion depends upon a number of factors including enzyme activity, supply of pyruvic acid, and the supply of ADP and oxygen. It is impossible to measure all of these factors in each plant, even if single-celled plants are used for such determinations.

Measurement

It is possible to measure the respiration rate of the intact plant, or of tissue slices, and to correlate this rate with controlled factors of the environment, or with plant characteristics. Such measurements are usually made by enclosing the plant, or other sample, in a sealed container and determining the rate of oxygen uptake or of carbon dioxide evolution. Of course, this must be done in the dark, unless non-chlorophyllous tissues are used, to eliminate photosynthesis which would give erroneous results. Why?

Age

An observation that is commonly made is that the rate of respiration is related to the age or stage of development of the sample. Sometimes such results can be correlated with the amount of cell wall material as related to the amount of cytoplasm. As a cell increases in age or development, at least with most plant cells, there is an increase in the amount of cell wall per cell even to the extent that the cell wall may weigh more than the dry weight of all other cell constituents. Therefore, if the rate of respiration is reported on a dry weight basis, the rate would decrease with age just because the dry weight increase was due primarily to cell wall material. Respiration rate per cell may not change over the same period of time. One must be aware of the basis used to report respiration rates.

As the cell increases in size, there may be an increase in the total number of mitochondria which would result in an increased rate of respiration per cell. However, more studies need to be made on this subject. There does not seem to be any change in the mitochondria with age, but there may be a change in their number.

With increase in age or development, there is usually a change in the mechanism of respiration. It has often been reported that glycolysis and the Krebs' cycle predominates in seedlings, but as the plant matures or becomes senescent, the hexose-monophosphate-shunt mechanism predominates.

Temperature

Since all metabolic activities are affected by temperature, it would be expected that the rate of respiration is also. Its optimum temperature is quite high (30–40°), and would be higher were it not for protein denaturation. As the temperature is lowered, a two- to four-fold decrease in respiration rate is observed for each 10°C decrease. This decrease is due to a decreased rate of diffusion of oxygen and carbon dioxide, and a decrease in the kinetic energy of the molecules, which reduces their reaction rate. In cold weather, therefore, the rate of respiration would decrease, and in hot weather it would increase.

It is often desirable to decrease the rate of respiration of stored food produce. This can be done by storage at low temperatures—but at temperatures above freezing—although it must be considered that as the temperature is decreased, stored starch is converted to sugar. Have you ever eaten potatoes that were undesirably sweet? This is the result of prolonged storage at temperatures that were too low.

At temperatures of between 10 and 0°C, tropical plants exhibit a marked decrease in respiration rate. This appears to be associated with chilling injury because this is not reversible.

Some plant tissues can, and do, survive temperatures much below freezing (−50°C) without permanent injury. Even at these low temperatures, respiration is occurring, but at a very low and often imperceptible rate. Certain plant species—those of alpine or arctic habitats—always are found at temperatures below 30°C. They have their optimum temperature at lower values (10–20°C). Certain plant species are found where temperatures often exceed 40°C—desert species—and have their optimum temperature above 30°C. Therefore, the optimum temperature for the most rapid rate of respiration is related to the temperature range of the plant's natural habitat.

In the field, temperature fluctuates over each 24 hour period, so plants are not exposed to a constant temperature for more than a few minutes. Therefore, temperature does change the rate of respiration from hour to hour, with the rate being lower at night and reaching a high in the afternoon. When studying the rate of respiration in the laboratory, the plant tissues are maintained at a constant temperature to eliminate temperature variations as a factor which would alter the respiration rate determinations.

Water

Generally, the greater the water content of the cells, the higher the rate of respiration. This is particularly evident with seeds or spores or with pollen grains. The respiration rate of seeds is so low as to be often undetectable. However, when seeds are planted and watered, the seed takes up water and by the time its water content is about 30%, the respiration rate increases considerably, as seen in Fig. 8.9.

Oxygen

As one might expect, the oxygen content is also important as a factor influencing the rate of respiration. This is particularly noticeable when the

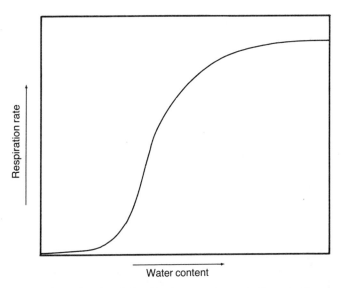

Water content

Fig. 8.9   A graph illustrating the relationships between the respiration rate of seeds and their water content.

content decreases below 10%. The normal oxygen partial pressure of the atmosphere is about 20%, and at values above 10% there is little increase in respiration rate, but at lower values a definite decrease occurs. One might expect, therefore, that oxygen content is rarely a variable except for roots in water-saturated soil or roots submerged in ponds, etc. Also, this could be a factor in large fruits or in tree trunks and similar locations.

As an aid to assuring that a favorable oxygen concentration is present to cells within the structure of multicellular plants, the plant often has several structural modifications. Some of these are common to many plants, such as lenticels on tree trunks, large branches and fruits, inter-cellular spaces among cells generally, and stomates in the leaves. Some, particularly plants growing with their roots in the mud at the bottom of ponds have specialized structural modifications, such as the pneumato-phores or breathing pores on the roots of swamp cypress or the large intercellular cavities in the leaves and stems of many cattail and ranun-clus species.

Carbon Dioxide

Since carbon dioxide is given off as a product, one might expect, just from chemical equilibria considerations, that excessive amounts of carbon dioxide may inhibit or reduce the rate of respiration. Indeed it does if the concentration exceeds about 10%. However, such concentrations of carbon dioxide are seldom encountered so probably this is not an important factor in the natural environment.

Light

Light is a factor affecting the rate of respiration. Of course, respiration is not a photochemical reaction and therefore is not directly affected by light but light is responsible for stomate opening and closing and is responsible for sugar synthesis via photosynthesis. It also causes an increase in temperature which increases the rate of respiration. Indeed, light may not play a direct role in this respiration, but it does have a tremendous influence. In fact, it also may be a factor in abscission of leaves and in determining what limbs the farmer should prune from his trees. If the leaves are near the base of a tree where they no longer receive direct sunlight, they are not able to produce photosynthate faster than it is used in respiration so the leaves will abscise (fall off) or will be pruned off since these leaves are now a liability to the tree rather than an asset.

Contamination

Atmospheric contamination is another factor that affects respiration. This is becoming increasingly important because of the increased occurrence of smog. Whether such contamination increases or decreases the rate will depend upon which contamination chemical is present. It is difficult to make predictions without experimentation, or determinations.

Stimulation

Another factor, one that would hardly be considered if its effects had not been frequently demonstrated, would be mechanical stimulation. Just moving the leaf, may cause an increase in the rate of respiration, and mechanical injury does likewise. It is necessary to remember this when cutting plant samples for respiration studies.

It is of prime importance that as much photosynthate as possible be conserved, not only to the farmer, but also to all biological entities that depend upon this as their source of energy and carbon. Therefore, any substrate used for plant respiration above that essential for the normal activities of the plant represents a waste. Anything that can be done to reduce this loss is of great value to all of the biological world, and especially to mankind. A challenge of the future will be to find better ways of controlling this loss than we now have.

THE CLIMACTERIC

One very interesting respiration phenomenon associated with the stage of development is that known as the *climacteric*. It occurs especially in ripening fruits, but has also been reported in flower petals and other plant tissues. An example of this climacteric can be demonstrated with apple fruits. As the fruit reaches its maximum size, but is still green, the rate of respiration declines to a low value. Shortly before ripening, there is a tremendous increase in the respiration rate and this rise, known as the *climacteric rise*, continues until it reaches the *climacteric peak*, which is its maximum rate. At this time, the apple fruit is just ripe. Then the rate gradually declines and this decline will continue until the fruit rots. To date, the cause of this rise is unknown, although it does seem to be associated with a tremendous increase in ethylene production. Ethylene is known to be a plant hormone. Whether it causes the climacteric or is a result of it, is not known.

HEAT

During metabolism, energy stored as chemical energy is degraded to heat and such degradation is the driving force for living activities. However, although energy is transferred and transformed within the cell, all energy eventually ends up as heat. This heat is lost from the plant, by conduction, convection, or radiation, and represents wasted energy. However, this and other sources of heat do control the temperature to which all plant cells are subjected, and temperature is an important factor in living activity. Therefore, it is appropriate that the subject of the temperature effects on plants be now discussed.

Temperature affects every metabolic activity of the plant cell. Therefore, one might expect that with increased temperature the only change would be a similar increase of all metabolic activities and at lower temperatures a similar decrease. However, this is not so, since temperature alters each metabolic activity at a different rate and therefore, as the temperature changes, each activity is altered to a different extent.

Temperature changes have a greater effect on chemical activities than on physical activities. The temperature coefficient (Q_{10}) for chemical reactions will vary from 2 to 4. That for physical activities will be about 1.2.

Therefore, physical changes over the range of biological temperatures are little affected, whereas chemical activity is greatly affected.

Plant Temperature

The temperature of the plant is near that of its environment. Its temperature may vary a few degrees from that of its environment, being lower at night and higher during the daytime, but these variations are small.

Optimum Temperature

In spite of the differential variations in reaction rates within the cell in response to temperature changes, it is possible to speak of the optimum temperatures for plants. *Optimum temperature* would refer to the ideal or best temperature for a given characteristic, but would vary with the characteristic discussed.

Alpine Plants

Generally, for normal development of the plant, arctic and alpine species have optimum temperatures near 10°C. At lower or higher temperatures, growth is decreased. This seems like a rather low temperature but it is the temperature to which the plants are often exposed during the daytime in their growing season, and apparently evolution has adapted them for such conditions. In fact, many boreal shade plants, species that normally grow in the shade in the cold climates, have an optimum temperature near freezing, and some species of plants can be found growing under and through snow.

Temperate-region Plants

Plants species found in the temperate regions have optimum temperatures near 25–30°C, the temperature to which they are often exposed during the daytime in their growing season. Similar optima are also observed in plant species found in the tropical regions of the world. Such regions are noted for their constant temperatures, with air temperature varying only slightly from 25 to 30°C from hour to hour, day to day, or season to season. Since both temperate and tropical plants have the same optimum temperature requirements, one might ask why tropical plants cannot survive in temperate regions. The answer lies not in their optimum temperature requirements, but in the extremes in temperature to which they are tolerant. Tropical plants will not tolerate great extremes in temperature while such extremes are tolerable to many temperate climate species.

Variations in Optima

The optimum temperature requirement must be considered to be variable not only from one species to the next, but also from one stage of development to the next, from one organ to the next, and even from one metabolic activity to the next. For example, cotton plants shortly after seed germination have an optimum temperature for root elongation that is near 33°C, whereas a few days later the optimum is about 27°C. Another even more striking example is with the tomato which reportedly has an optimum temperature requirement by the seedling of 30°C, but that of the mature plant is about 13°C. In this case, the difference appears to be that in the mature plant the distance for the translocation of food to the growing points is greater, and this requires greater transport through the phloem and this transport has an optimum temperature near 13°C. On the other hand, meristematic activity itself has an optimum near 30°C.

The optimum temperature for flowering is greater than that for vegetative growth and that for vegetative growth is greater than that for seed germination, generally. Therefore, plants do not grow so well under constant temperature conditions as under changing conditions. This, of course, corresponds with normally observed changes in their natural environment, since seeds germinate early in the year when the temperatures are cool, vegetative growth occurs when the temperatures are normally a bit warmer, and by the time the plant flowers, the temperature is usually quite warm.

A noticeable difference in the optimum temperature for different metabolic activities within a given plant or cell is that of photosynthesis and respiration. The optimum temperature for photosynthesis is lower than that for respiration. This is important for the well-being of the plant since as the temperature increases the rate of respiration will increase faster than that of photosynthesis, so if the temperature remains high, it would be possible for the plant to use up its stored food materials and die, since such food materials would not be replenished fast enough.

Chilling Injury

As the temperature is lowered below that of the optimum, and heat is lost, changes take place within the plant. Generally, metabolic activities slow down, but not all at the same rate. Eventually such differential metabolic rates result in such changes in the amounts of metabolites produced that the plant is injured. Such injury, at temperatures above freezing, is referred to as *chilling injury*. Chilling injury is especially noticeable in tropical species, as previously indicated, since their range of tolerance to low temperatures is particularly narrow. Such injury may occur at temperatures as high as 10°C, and the range from 0 to 10°C is the temperature range over which chilling injury usually occurs. Have you ever noticed how some species of plants are killed in the Fall even before temperatures get down to freezing?

Although the range for chilling injury is normally between 0 and 10°C, there are exceptions. Melons, sorghum, and date palm plants stop growing at temperatures as high as 15 to 18°C, whereas peas, wheat, and rye continue to grow until temperatures reach −2 to −5°C.

Chilling injury increases with the duration of exposure. Exposing a plant to chilling temperature for one night may not injure it permanently, but continued exposure to the same temperature, or exposure to that temperature for several nights, will result in its death.

Chilling injury is often associated with the following changes: Fruits on the plant fail to ripen after exposure to chilling temperatures. Instead, they remain green until they rot. Respiration rate often increases when the plant is exposed to chilling temperatures, but such increase is only temporary since the rate later decreases, as the tissues and their cells die. Protoplasmic streaming has also been observed to slow down and eventually stop. Of course, wilting may be evident and any growth will cease.

As evidence that chilling injury may be caused by upsets in metabolism due to different Q_{10} values of the various metabolic activities, one might consider the example reported for cosmos plants. Cosmos stops growing at a temperature below 20°C, unless they are supplied with the vitamin thiamine. At optimum temperatures, thiamine has no effect on this species, so apparently at low temperature the synthesis of thiamine by the plant is decreased to such an extent that this vitamin becomes deficient which results in injury to the plant. At higher temperatures, cosmos plants synthesize this vitamin in adequate amounts for their needs, but not at lower temperatures. In other species, other metabolites may become deficient at lower temperatures, and as the temperature continues to decrease, more than one metabolite may be deficient.

Differential rates of metabolism may not result in a deficiency of a metabolite, but may result in such an accumulation of any one that this may be toxic to the plant. Nevertheless, most injury to plants at temperatures below optimum, but above freezing, appear to be caused by metabolic upsets which result either in a deficiency of an important metabolite or in the accumulation of some chemical to the point where it becomes toxic. Perhaps the chemical causing the trouble varies from one species to another and can only be determined by investigation.

Frost Injury

As the temperature is lowered, it eventually reaches the point where frost injury becomes evident. Just what this temperature is at which frost injury occurs varies from one plant to another, but is somewhere below 0°C. As previously indicated, pea plants may not be injured until the temperature becomes lower than -5°C, and pears can tolerate Winter temperatures to -7°C. No doubt, leaves of our northern conifers can also tolerate temperatures below freezing without frost injury occurring. However, irrespective of the species or of the conditions under which it is growing, as the temperature continues to fall, eventually it becomes so low that frost injury does occur.

As the temperature is lowered below freezing, ice may begin to form in the intercellular spaces of the plant. The water in these intercellular spaces is nearly pure water, with few solutes, and therefore has a freezing point higher than the water in the cells. *Intercellular ice formation* is particularly evident if the temperature is lowered slowly, as occurs in nature. With intercellular ice formation, the water potential of the intercellular spaces increases. This causes water to move out of the cell into the intercellular space, reducing the water content of the cell and thereby its osmotic pressure, which decreases its freezing point. Therefore, intracellular ice formation is prevented or at least delayed by such activity. Intercellular ice formation is rarely injurious to the plant unless the cells become so desiccated that they die from the lack of water.

However, as the temperature continues to decrease, eventually, the point is reached at which ice forms within the cells, a condition known as *intracellular ice formation*. Intracellular ice formation is lethal and does result in the death of the cell, although we do not know the mechanism involved. It may cause mechanical disorganization of cell organelles or membranes but freezing is also known to denature lipoproteins, which could account for the injury.

It is worthy of note that plant cells can be exposed to a very low temperature if the rate of freezing and thawing are rigidly controlled. In fact, some plant structures can be stored for long periods of time at very low temperatures.

Hardening

Needless to say, in certain climates it would be of advantage to the plant, and to man if crop plants were concerned, to have the freezing point of the cell contents lowered as Winter approaches. This is exactly what happens. If a plant, such as alfalfa, is subjected to low temperatures in Summer, the temperature at which injury occurs will be much higher than if it is subjected to the same treatment in the Fall. The reason being, that as the temperature slowly decreases in the Fall of the year, changes take place within the cells of plants which make these cells more resistant to injury at low temperatures. This condition is known as *hardening*.

We do not know just what changes occur within the cell that increase their resistance to low temperatures, as hardening occurs. There is, however, a good correlation between that plants state of hardening and its resistance to low temperatures, to high temperatures, and to drought injury. Therefore, one might suspect that osmotic relations are involved. Indeed, in some cases, the osmotic pressure of the cells is increased, due to a decrease in their water content or to an increase in their solutes, especially sugars and amino acids. Starches may be converted to sugars to increase the sugar content and proteins may be hydrolyzed to amino acids. The cells of hardened plants are smaller with thicker walls, and there is also some evidence of an increase in the number of sulfhydryl groups of proteins associated with hardening. Also, hardening may protect certain lipoproteins against denaturation at low temperature. No doubt, the mechanism of hardening varies from one species to the next.

Plants that become so well hardened that they survive the Winter, become dehardened as the temperatures warm up in the Spring, so that by Summer, all evidence of hardening to low temperatures has gone. However, hardening may be induced during the growing season, by drought or by high temperatures.

Winter-kill

Another type of injury often seen in climates where the temperatures during the Winter are liable to vary significantly is that known as *Winter-kill*. This injury is caused by the water freezing in the soil with the water potential of the soil becoming high. Then a warm breeze may develop causing evergreen plants to lose water much more rapidly than they can take it up and therefore, Winter-kill is due to desiccation. This is common in the plains states.

HIGH TEMPERATURE INJURIES

So much for low temperature injury to plants. If the temperature is increased above that of optimum, more heat enters the plant and again this results in an upset of metabolism, due to differences in the Q_{10} values of the various metabolic activities. Such injury is especially notable at temperatures above 30–40°C, but may occur even at lower values in some species, as we shall see shortly.

Local Injury

Some of this injury is local, occurring in only isolated areas of an individual plant. Stem girdling is sometimes observed where the plant may be injured just where its stem or trunk comes in contact with the soil surface. Since the temperature of the soil surface, when the soil is exposed to the sun, is often higher than the air, this could account for such girdling. Desert perennial plants seem to be particularly resistant to this injury and such injury does not occur below temperatures of about 70°C on desert plants, but other species often exhibit such injury at temperatures above 45°C.

Sunscald is another type of injury that is local. It can often be observed on fruits or woody stems that are exposed to the sun. The more foliage on a tree, the less likelihood of sunscald. It is caused by the rapid changes in temperature of the exposed plant structure as the sun shines on it.

General Injury

Of course, much of the injury at temperatures above optimum is much more general, affecting the plant generally.

Metabolic Disturbances

Injury at temperatures above optimum may be caused by three mechanisms. First, metabolic disturbances may be induced, second, desiccation may occur as the rate of water loss exceeds that of water uptake, and third, at the excessively high temperatures, those over about 50°C, direct thermal effects are induced which results in a coagulation or denaturation of proteins, denaturing the proteins and enzymes of the cell and causing its death.

As a cell is exposed to increasing heat, certain changes can be observed within the cell. First, protoplasmic streaming decreases as the protoplasm becomes more viscous. Photosynthesis is also greatly reduced resulting in a decrease in the synthesis of carbohydrates. Respiration rate continues to increase, using more rapidly the supply of carbohydrates present. Then membrane structure becomes altered with a great change in membrane permeability. No doubt other metabolic changes are occurring resulting in deficiencies or toxicities of metabolites. Finally, the proteins become denatured with the subsequent inactivation of the enzymes essential for survival of the cell.

An example of metabolic disturbances may be that previously considered, namely, the relative rates of photosynthesis and respiration. Or we might consider the pea plant. Pea plants have an optimum temperature near 20°C. At 35°C, the plant becomes chlorotic—turns yellow—and dies. However, this injury can be alleviated by adding the purine, adenine, to the plant. This indicates that at the optimum temperature, pea plants synthesize sufficient adenine for their needs, but at higher temperatures, adenine becomes deficient. *Lemna minor* behaves similarly, whereas subterranean clover requires certain amino acids and *Arabidopsis* sp. requires biotin or cytidine. Therefore, it would appear that what metabolite becomes deficient depends upon the species. It might also be expected that at even higher temperatures, other metabolites may become deficient or some chemicals may become toxic. Much work remains to be done on this subject to reveal all the answers we would like to have.

Since heat injury to cells is associated with a tremendous increase in permeability of the plasma membrane, an often-used-criterion for determining heat resistance is to determine the temperature at which permeability increases by measuring the electrical conductivity of the medium in which the cells are suspended.

Hardiness

Heat hardiness is one means by which a plant can survive periods when the environmental temperature is above optimum. However, this is not the only method of such protection. Increasing the transpiration rate will reduce the temperature of the leaf somewhat, and also the leaves of many plants alter position during times of high temperatures such that they are no longer perpendicular to the sunlight, and therefore, less infrared radiation is absorbed by the leaves, so the leaves remain cooler than they might otherwise be. Unlike many animals, plants cannot escape from hot weather but they do have certain adaptations that allow them to tolerate it, at least within limits. Nevertheless, temperatures above optimum are not conducive to the best growth and development.

Protein Denaturation

Protein denaturation has been studied for many years. Unlike other causes of injury at higher temperatures, it is irreversible. By that I mean, if a plant wilts, this can be overcome by supplying water to it. If a metabolite upset develops, this can be overcome by supplying the metabolite that is deficient. However, if denaturation sets in, this cannot be reversed, even by lowering the temperature. It is a permanent type of injury which, if extensive enough, will lead to the death of the cell. Proteins differ considerably in the temperatures at which they are denatured. Certain blue-green algae live naturally in hot springs where the temperature is over 75°C, and their proteins are not denatured. Desert plants also have proteins that are more resistant to denaturation. However, as previously indicated most plants sustain such injury at much lower temperatures. This does not, however, indicate that all proteins of each individual plant are denatured at the same temperature. Individual proteins differ considerably in this characteristic. However, if a plant is injured at a given temperature, and this injury is due to protein denaturation, this indicates that at least one kind of protein is denatured at the injurious temperature.

THERMOPERIODICITY

Plants are not accustomed to living in an environment of constant temperature, with the exception of those of the tropical rainforest. Through evolution, temperate and arctic or alpine species have become adapted to variation in diurnal temperatures and therefore exhibit *thermoperiodicity*. In other words, they grow best when the temperatures fluctuate from night to day. This is true of many stages of development. Seed germination is best when the temperature fluctuates, hardiness also is better developed under these conditions, as is fruit and flower formation. Perhaps some of these observations may be due to the fact that the translocation of organic materials is most rapid at lower temperatures, whereas cell division and enlargement are more rapid at higher temperatures.

Plants can become hardened to heat, or can become more heat resistant, just as they can become hardened to low temperatures. Such hardening is due to changes that make the proteins more resistant to denaturation. As previously indicated, this hardening is often correlative, with plants becoming hardened to heat being more resistant to desiccation and low temperature injury. This is true of many species, but in some, these hardening processes seem to be unrelated, and a plant hardened to low temperature may not exhibit heat resistance until being conditioned by exposure to warmer temperatures.

As might be expected, the heat resistance of a plant is directly related to the natural environment of the plant. Those living in normally hot climates, have more heat resistance than alpine or arctic species. Heat resistance is also related to the stage of development of the plant cell. Cells that are mature and dormant have greater heat resistance than young cells, or of plant tissues that are actively growing. This is true of cold resistance too.

In this chapter, the subject of energy release has been considered, and we have learned that this energy release is associated with the formation of energy carriers, and the loss of heat. In subsequent chapters, we shall learn how these carriers transfer their energy to other chemical reactions which result in the synthesis of the many chemical molecules found in the living plant and its cells.

Chapter Nine | Amino Acid Synthesis and Metabolism

THE MINERAL NUTRIENT NITROGEN

As shown in Chapter 5, nitrogen is one of the essential elements for all plants. In fact, it is a macronutrient, indicating that it is required in relatively large amounts by the plant. Indeed, it is required in larger amounts than is any of the other mineral nutrients, and comprises about 1–4% of the dry weight of the plant. Within the plant, nitrogen is found as a constituent of amino acids, chlorophyll, proteins, purines, nucleic acids, alkaloids, vitamins, and other important organic compounds. With the exceptions of water, and the carbohydrates, nitrogen-containing compounds form the bulk of the plant constituents.

Characteristics

As a chemical element, nitrogen has some interesting characteristics. It is not found naturally as a significant constituent of the rocks of the earth, and is rare in the lithosphere, with the exception of the Chilean nitrate deposits and a few smaller widespread concentrations. Still it comprises about 78% by volume of the atmosphere, and therefore terrestrial plants are literally bathed in it. In spite of this apparent abundance of nitrogen in the atmosphere, it is the mineral nutrient most often needed as a fertilizer for cultivated plants. How can a plant which is bathed in nitrogen, become deficient and even die from this deficiency?

Nitrogen Cycle

Nitrogen exists in various forms on the earth and these forms are inter-convertible. Their interconversion comprises what is known as the nitrogen cycle of nature, and is illustrated in Fig. 9.1. We can see, from this illustration, that 78% of the atmospheric gases are composed of

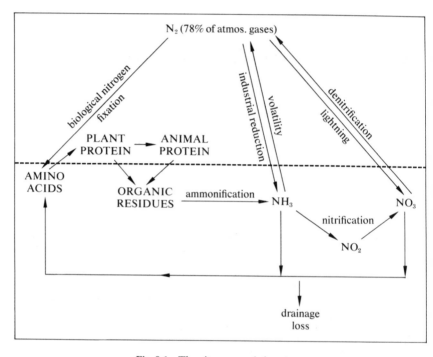

Fig. 9.1   The nitrogen cycle in nature.

nitrogen, and this is largely present as the inert nitrogen gas (N_2). How-ever, there are several equilibria established between this "pool" and other stages of the cycle. Nitrogen is being removed from the atmosphere mainly by three processes: electrical fixation (lightning), industrial reduc-tion, and biological nitrogen fixation. It then circulates throughout the living and non-living carriers, and is finally returned to the atmosphere as nitrogen gas by the processes of oxidation and denitrification. The magnitude of this N pool can be seen by the estimate that 3×10^8 lb of N exist above an acre of ground and only 1/60,000 as much in an acre of soil.

Electrical fixation methods involve the use of an electrical discharge to convert the nitrogen gas into a form that can be used by the plant. The most primitive of these electrical fixation methods is lightning. Whenever a flash of lightning is produced, a small amount of nitrogen gas is converted into ammonia or nitrate. This is then dissolved in the moisture of the atmosphere, and enters into the soil with the precipitation. The amount of available nitrogen which enters the soil from this source is about 3–8 lb for each acre of soil per year. When this is compared to the 200–300 lb of nitrogen fertilizers which are sometimes applied to an acre of soil under cultivation, it does not appear to be important, but it is a very important source of nitrogen for non-cultivated plants.

Within the present century, man has proceeded to duplicate nature by using electrical currents to convert nitrogen gas of the atmosphere into available forms of nitrogen. At the present time, this is the source of some of our commercial nitrogenous fertilizers, and the importance of this process is increasing annually. Its greatest obstacle is the large amounts of electricity required for the process.

Biological nitrogen fixation involves a very few species of plants that have the unique ability to use nitrogen gas and incorporate it into their organic nitrogen compounds. Some of these, such as the bacterium, *Azotobacter* sp., and the cyanophyta, *Nostoc* and *Anabaena* sp., use energy derived from the oxidation of organic matter or from photosynthesis, and independently fix nitrogen. Others, such as the *Rhizobium* bacterium function symbiotically, being located on the roots of legume plants, where nodules are formed around them. The *Rhizobium* sp. use energy obtained by oxidizing organic compounds supplied by the legume host. When these bacteria cells die, their organic nitrogen components become available and furnish the host with a great deal of usable nitrogen. An estimate of the amount of usable nitrogen added to a soil from these various sources is given in Table 9.1.

Table 9.1   Sources of nitrogen added to each acre each year.

Source	Amount (lb/acre/yr)
Lightning + rain + irrigation	5–7
Symbiotic fixation	12
Non-symbiotic fixation	6
Fertilizers	0–several hundred

Soil Nitrogen

Within the soil, most of the nitrogen, other than nitrogen gas, is present in the form of proteins and other nitrogenous compounds which represent the remains of dead plant and animal cells. In fact, 90–98% of the available nitrogen may be in such a form. Due to the action of ammonifying bacteria, this nitrogen is released in the form of ammonia as these bacteria use the proteins as a source of energy for their respiration. In acid or water-saturated soils, such as we often find in swamps, bogs, or river or lake bottoms, the ammonia may accumulate to significant concentrations. In aerated soils, such as most of the soils with which we are familiar, such accumulation of ammonia does not occur. Due to the action of nitrifying bacteria, the ammonia is rapidly oxidized to nitrite and finally nitrate. Nitrite is quite unstable so does not accumulate, but nitrate is stable and does accumulate in the soil to significant concentrations. However, nitrate is not adsorbed to the soil colloids, but is very soluble in water so it is easily leached out of the soil with the gravitational water. In fact, it has been estimated that approximately 30% of the soil nitrate is removed by crops and 70% is lost by leaching or erosion. This loss, plus the high nitrogen requirement of plants, is responsible for giving nitrogen the reputation of being the fertilizer that is most added to soils to improve plant growth.

AVAILABLE FORMS OF NITROGEN

Since nitrogen gas can be used by only a few species of plants, we do not consider it to be available to plants generally. Some organic nitrogen compounds can serve as a source of available nitrogen. These include the amino acids and urea, but not the proteins and other high molecular weight compounds. Therefore, although 90–98% of the soil nitrogen exists as organic nitrogen compounds, only about 2% of this is immediately available to the plants, and therefore does not serve as an important direct source of available nitrogen for the plant. Ammonia, ammonium salts, nitrites, and nitrates are readily available but of the four, only ammonia, ammonium, and nitrate are important sources of available nitrogen to the plant.

Uptake as Ammonia

Although the nitrogen which is available to plants is generally taken up as nitrates, there are certain conditions under which this is not so. The oxidation of ammonium to nitrate requires the presence of certain microorganisms, as well as an ample supply of oxygen. Whenever these requirements are met, even though fertilizers may be applied as liquid ammonia or as ammonium salts, the oxidative process occurs so rapidly that much of the available nitrogen enters the plant roots as nitrates rather than as ammonium ions. However, if these microorganisms are not present or ample oxygen is lacking, the situation changes. At the present time, there is an increased use of soil fumigants to destroy pathogenic agents prior to sowing. While these fumigants do their assigned tasks very well, they are unable to differentiate between the harmful pathogens and the beneficial microorganisms. As a result, the microorganisms responsible for the oxidation of ammonium to nitrates are also destroyed. If this happens, and is followed by fertilization with ammonium salts, then much of the available nitrogen would be taken up by plants as ammonium.

If oxygen is deficient in a soil, as is frequently the case with puddled or water-saturated soils, then conditions will not be favorable for oxidation, but will be favorable for reduction. As a result, the available nitrogen in these soils will be present in their reduced form, and will be taken up as ammonium ions or possibly as ammonia. Many soils throughout the world are permanently or periodically saturated with water. Especially is this true in swampy or boggy areas, but at certain periods of the year this condition is more widespread than is generally realized.

If the available nitrogen is taken up in the reduced form, this would appear to be especially valuable to the plant because ammonium does not require the energy-consuming process of reduction within the plant that is required if nitrates are taken up. Indeed, it is sometimes observed that plants supplied with ammonium exhibit greater vegetative growth and require less carbohydrates than those supplied with nitrates. However, it must be remembered that ammonia or ammonium salts are toxic to plants even when present within the plant in rather low concentrations. Also, plant cultures supplied with ammonium require more oxygen to be available to the roots than those supplied with nitrates. This may be due to the release of oxygen by the roots as nitrates are reduced. Therefore, any attempt to compare the benefits derived from furnishing available nitrogen in the form of ammonium salts or nitrates must take the above mentioned facts into consideration.

When the soils are swampy or boggy or generally water saturated, ammonium or ammonia is the form taken up by the plant, but in most soils, under natural conditions, available nitrogen enters the plant in the form of nitrates.

Uptake as Nitrate

Nitrates are taken up by the plants in the form of the nitrate anion (NO_3^-). The amount taken up is reduced by high phosphate concentration, or by high concentrations of ammonium salts or chlorides in the soil. Also, oxygen must be present. The mechanism of uptake appears to be similar to that for anion uptake generally, as discussed in Chapter 5.

Nitrate uptake usually occurs very rapidly, resulting in an increase in the pH value of the nutrient medium, if the medium is not buffered, and an accumulation of nitrates within the root cells. The extent of accumulation depends upon the extent of uptake of the nitrate, and the rate of nitrate assimilation. Some species of plants are known to be nitrate accumulators, species such as stinging nettle, elderberry, burdock, and Canada thistle, whereas in other species the nitrate concentration is always low, species such as milkweed, dandelion, yarrow, dogwood, and spirea and other shrubs. Some species accumulate nitrates to such an extent that nitrate poisoning of cattle occurs when these plants are foraged. Such poisoning occurs when the nitrate concentration exceeds about 0–6% of the dry weight of the plant.

Nitrate Accumulation

If one attempts to study the accumulation of nitrate within a plant, one soon finds that the extent of accumulation varies from one part of the plant to another. We know that nitrates enter primarily through the roots of higher plants, and yet nitrogen is found as a constituent of all of the plant cells. Therefore, it must be translocated throughout the plant. Many tests have been made to locate nitrates in various plant tissues. These tests reveal the presence of nitrates in the roots, but may or may not detect it in other plant parts. We explain this apparent anomaly by saying that the nitrate may be rapidly reduced to ammonium or some closely related form in the roots. Plant cells cannot use nitrates directly as such. It is entirely possible to visualize a plant which contains large concentrations of nitrates, and yet one that is starving due to the lack of suitable nitrogen. Indeed, there appear to be a few species that are inherently unable to reduce nitrates, and therefore accumulate nitrates without ever using them. In these cases, only the nitrogen absorbed as ammonium or organic nitrogen would be of benefit to these plants.

Other species accumulate nitrates to varying degrees depending upon the rates of absorption and reduction. In apple and other deciduous trees, asparagus, narcissus, and various perennials, the nitrates are rapidly reduced in the root cells. Therefore, they do not accumulate in other cells although such accumulation would be possible if the nitrates were available. In other species, such as tomato, tobacco, raspberry, cucurbita, and cereals, the reduction takes place more slowly in the roots and therefore, nitrates can be and are translocated to other cells where they accumulate. This is well to keep in mind when using the diphenylamine or related field or laboratory tests for nitrates in plant tissue. Just because the test does not show nitrates present does not indicate nitrogen deficiency, but may merely indicate that the nitrates were reduced before they reached the tissue sampled.

NITRATE ASSIMILATION

As stated previously, nitrate cannot be used directly by the plant to satisfy its nitrogen requirements, but first it must be assimilated. Such an assimilation results in its reduction to form ammonia or ammonium ion, which is then incorporated into the organic compounds of the plant. Such a reduction requires a great deal of energy, and the reaction series can be summarized as follows:

$$NO_3 + H_2O + 2H^+ \longrightarrow NH_4^+ + 2O_2 \qquad \Delta F = +83 \text{ kcal/mol}$$

The energy requirement for nitrate assimilation is met either through respiration or the light reaction of photosynthesis.

Nitrate Reduction

The reduction of nitrate to ammonia or ammonium ion is referred to as *nitrate assimilation*. It involves a number of chemical reactions and a number of enzymes. The first reaction is often referred to as nitrate reduction and is catalyzed by the enzyme known as nitrate reductase. This enzyme is an inducible enzyme, not being present in the cell when nitrate is absent but its formation is induced by the presence of the substrate — nitrate ions. This enzyme contains the micronutrient molybdenum. If molybdenum is deficient, nitrates accumulate within the plant, sometimes reaching such a concentration that burning the plant causes a mild explosion, since nitrates can be explosive. Even though the nitrate concentration of these plants may be very high, they show symptoms of nitrogen deficiency by turning chlorotic with the yellowing appearing first on the oldest leaves but spreading rapidly over the entire plant. Nitrate reductase is a soluble enzyme, not being attached to any of the cell structures. The nitrate reduction reaction is illustrated as follows:

$$NO_3 + NADH_2 \longrightarrow NO_2 + NAD + H_2O$$

As can be seen above, the energy is supplied as the $NADH_2$ nucleotide.

Nitrite Reduction

After the nitrite has been formed, it in turn is reduced via a reaction known as nitrite reduction. This involves an enzyme known as nitrite reductase, an enzyme that contains copper and iron micronutrients. This reaction derives its energy from ATP and $NADPH_2$, and is suppressed by ammonia. Therefore, ammonia accumulation decreases nitrate assimilation by a mechanism known as feedback and prevents a buildup of ammonia to the extent that ammonia would become toxic. Feedback is a good, and often used, mechanism for the regulation of metabolism within the plant.

Ammonia Formation

The next reaction or series of reactions involved in nitrate assimilation are not well known. The nitric oxide, believed to be formed during nitrite reduction, is somehow converted to hydroxylamine (NH_2OH).

The reduction of hydroxylamine requires an enzyme named hydroxylamine reductase, an enzyme that requires the micronutrient manganese. Its energy source is $NADPH_2$ and the reaction forms ammonia or ammonium. Actually which of these two forms of reduced nitrogen is formed is unknown, but we shall assume it is ammonia.

ENTRANCE OF AMMONIA INTO ORGANIC COMPOUNDS

In all, the assimilation of 1 mole of nitrate forms 1 mole of ammonia and requires eight moles of hydrogen which, as indicated above, represents the expenditure of a great deal of energy. However, nitrate assimilation does release some oxygen, which can be used by the cells for respiration.

As stated earlier, ammonia cannot accumulate within the plant cell in high concentrations since it will kill the cell. Such accumulation is normally prevented by feedback control of nitrate assimilation and by the rapid incorporation of ammonia into the organic compounds of the cell.

Amino Acid Formation

The next question we wish to consider is how does the ammonia get into the organic nitrogen compounds of the plant. Research has shown us two phenomena of special interest in regards to this problem. First, the nitrogen appears to enter into combination with organic compounds by forming amine (–NH$_2$) groups. Second, the formation of organic nitrogen compounds occurs at the expense of carbohydrates. In fact, the carbohydrate concentration of the plant is reduced when available nitrogen is present, since the carbohydrates are used for the synthesis of organic nitrogen compounds instead of being accumulated. These facts indicate that reduced nitrogen reacts with carbohydrate derivatives to form amino acids. The carbohydrate derivatives appear to be any α-keto organic acid, but especially α-ketoglutaric acid. An α-keto acid is any organic acid which has a keto (C=O) group attached to the first carbon atom next to the carboxyl (COOH) group.

Actually, it is theoretically possible for ammonia to react with any α-keto acid in the plant, but the evidence for any significant synthesis involving other than α-ketoglutaric acid and pyruvic acid is lacking, and it appears that essentially all of the nitrogen enters organic compounds by combining with α-ketoglutaric acid. This reaction is illustrated below:

$$
\begin{array}{l}
\text{COOH} \\
\text{C=O} \\
\text{HCH} \\
\text{HCH} \\
\text{COOH} \\
\text{\scriptsize α-ketoglutaric} \\
\text{\scriptsize acid}
\end{array}
+ NH_3 + NADH_2 \longrightarrow
\begin{array}{l}
\text{COOH} \\
\text{HCNH}_2 \\
\text{HCH} \\
\text{HCH} \\
\text{COOH} \\
\text{\scriptsize glutamic acid}
\end{array}
+ H_2O + NAD
$$

It can be seen that this is an energy requiring reaction, which requires reduced NAD as the energy carrier. This NADH$_2$ may be formed during respiration or during the light reaction of photosynthesis. Those cells that cannot use ammonia as a source of nitrogen probably lack the enzyme involved in this reaction, an enzyme called glutamic dehydrogenase. This enzyme is probably present in most plant cells, but not in animal cells. Therefore, animals are dependent upon plants for their source of nitrogen, which is usually supplied to them as amino acids.

TRANSAMINATION

Glutamic acid is the first organic nitrogen compound formed in plants. However, we know that there are more amino acids in plants than glutamic acid. Therefore, our next concern should be, how is the nitrogen from glutamic acid used to form other amino acids in plants, and later we will learn how it is used to form the other organic nitrogen compounds.

Other than glutamic acid, and possibly some alanine, all amino acids are formed by transferring the nitrogen from glutamic acid to an α-keto-acid, a mechanism known as *transamination*. This mechanism is illustrated below:

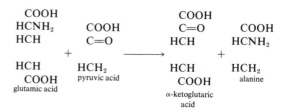

Any amino acid can react with any α-keto acid to give a new amino acid and a new α-keto acid, through this mechanism of transamination. The enzymes involved are called transaminases, with a different transaminase required for each amino-keto acid substrate. Transaminases are widely distributed in plants occurring in most plant cells, associated with the mitochondria, the site of respiration, close to the energy carriers. The prosthetic group of at least many of these enzymes, is vitamin B_6, also known as pyridoxyl phosphate.

It would be possible then to have as many amino acids formed as there are α-keto acids within the plant. Indeed, many amino acids are known, and more are being discovered each year. Over a hundred have been reported, which are many more than are found in animals. A list of the plant amino acids normally found in proteins is given in Table 9.2. Other amino acids will be considered in Chapter 13. Of them, aspartic acid, glutamic

Table 9.2   Amino acids used in protein synthesis.

Name	Structure
Alanine	CH_3—CHCOOH NH_2
Arginine	H_2N \\ CNHCH$_2$CH$_2$CH$_2$—CHCOOH / NH_2 HN
Asparagine	NH_2OCCH_2—CHCOOH NH_2
Aspartic acid	$HOOCCH_2$—CHCOOH NH_2
Cysteine	$HSCH_2$—CHCOOH NH_2

Table 9.2 (*Continued*)

Name	Structure
Glutamic acid	HOOCCH$_2$CH$_2$—CHCOOH 　　　　　　　　NH$_2$
Glutamine	NH$_2$OCCH$_2$CH$_2$—CHCOOH 　　　　　　　　NH$_2$
Glycine	H—CHCOOH 　　　NH$_2$
Histidine	CH$_2$—CHCOOH 　　　　NH$_2$
Isoleucine	CH$_3$CH$_2$ 　　　＼CH—CHCOOH 　／　　　NH$_2$ CH$_3$
Leucine	CH$_3$ 　　＼CHCH$_2$—CHCOOH 　／　　　　　NH$_2$ CH$_3$
Lysine	H$_2$NCH$_2$CH$_2$CH$_2$CH$_2$—CHCOOH 　　　　　　　　　　　　NH$_2$
Methionine	CH$_3$SCH$_2$CH$_2$—CHCOOH 　　　　　　　　NH$_2$
Phenylalanine	CH$_2$—CHCOOH 　　　　NH$_2$
Proline	—COOH
Serine	HOCH$_2$—CHCOOH 　　　　　NH$_2$
Threonine	CH$_3$CHOH—CHCOOH 　　　　　　　NH$_2$
Tryptophane	CH$_2$—CHCOOH 　　　　NH$_2$
Tyrosine	HO⟨　⟩CH$_2$—CHCOOH 　　　　　　　NH$_2$
Valine	CH$_3$ 　　＼CH—CHCOOH 　／　　　NH$_2$ CH$_3$

acid, alanine, and serine are present in greatest abundance. Not all of the amino acids are found in each plant species.

STRUCTURE OF THE AMINO ACIDS

All amino acids have certain structural similarities. Each has a carboxyl group, with an amino group on the adjacent carbon. However, as can be seen in Table 9.2, many amino acids have more than one carboxyl group and some have more than one amino group. When the amino acid has two carboxyl groups, we call it a dicarboxyl amino acid. When it has two amino groups, we refer to it as a basic amino acid. *Amides* are amino acids with a $CONH_2$ group. Two very common and widespread amides are the amides of glutamic acid and aspartic acid, amides called specifically glutamine and asparagin respectively. These amides are formed by removing an hydroxyl group from the second carboxyl group and replacing it with an amino group. The mechanism for glutamine synthesis is illustrated below:

$$
\begin{array}{l}
\text{COOH} \\
\text{HCNH}_2 \\
\text{HCH} \quad + \text{ATP} + \text{NH}_3 \longrightarrow \\
\text{HCH} \\
\text{COOH} \\
\text{glutamic acid}
\end{array}
\qquad
\begin{array}{l}
\text{COOH} \\
\text{HCNH}_2 \\
\text{HCH} \quad + \text{ADP} + \text{H}_2\text{O} \\
\text{HCH} \\
\text{CONH}_2 \\
\text{glutamine}
\end{array}
$$

Magnesium or cobalt are required for this reaction and the enzyme is called glutamine synthetase.

SULFUR AS A MINERAL NUTRIENT

Further consideration of the structure of the amino acids reveals that some of these amino acids contain sulfur in the reduced form. Since these sulfur-containing amino acids, especially methionine and cysteine, are of great importance to the plant, it will be worthwhile now to consider how sulfur is used by the plant and how it gets into the amino acids that contain it.

Available Sulfur

Elemental sulfur cannot be used by most plant species, but sulfur is one of the essential elements, and therefore is required. In the environment, sulfur may be present in several forms. Sulfur dioxide (SO_2) gas is found in the atmosphere with its concentration increasing near the larger cities and near industrial areas. It is often given off in the smoke of industrial plants, and can be used by plants. It is absorbed by the leaves, reduced and serves as an important source of sulfur if sulfur is deficient in the soil. The sulfur content of a plant has been observed to increase with the nearness of its habitat to industrial areas.

In the soil, sulfur may be present largely as a constituent of several organic compounds which may or may not be absorbed by the plant. However, the greatest supply of sulfur is derived from the divalent sulfate anion ($SO_4^=$) present in the soil solution. This sulfate is either formed from the organic sulfur compounds or from gypsum ($CaSO_4$). Since the sulfate ion is but weakly adsorbed by the soil colloids, it enters the plant from the soil solution wherein it is dissolved. Being so dissolved, it is also easily leached from the soil. The soil solutions in arid regions, such as the Western states, have a high concentration of sulfate dissolved in them. In fact, sulfate is often the anion present in greatest amount, along with chloride, in these soils.

Sulfite is unstable in contact with the air, being easily oxidized to sulfate, and is not accumulated in the soil in even moderate amounts. However, when it is present in the soil, it is absorbed, at a rate which increases with decreased pH, and may become toxic even at low concentrations. Thiosulfate is also unstable in the soil, and is not accumulated.

Sulfur Content of Plants

Sulfur is found in plants in rather high concentration, especially in meristematic cells. The total sulfur content is between 0.01% and over 1.5% on a dry weight basis. Much of this is located in the leaves, but the seeds may contain a great deal. In fact, sulfur is often surprisingly uniformly distributed among all plant organs. The following table, Table 9.3, gives some examples of sulfur distribution within the plant and the total sulfur content of various organs.

If it were not for the fact that available sulfur is present in gypsum and in phosphate fertilizers added to the soil, and in the atmosphere as sulfur dioxide, sulfur deficiencies would be reported much more frequently in crop plants than they are. Actually, crops take up about as much sulfur as phosphorus during the growing season (3–20 lb per ton of crop produced); some taking up less (cereals) and some much more (turnips, cabbage). When plants are deficient in sulfur, they are stunted both in shoot and in root growth, their leaves are pale green due to a lower concentration of chlorophyll, and they accumulate excessive amounts of starch but less

Table 9.3   Total sulfur content of plants* (% sulfur on a dry weight basis).

Species	Leaves	Stems	Roots	Fruit or seed
Corn	0.23–0.25	0.05–0.17	0.03–0.28	0.004–0.30
Oats	0.14–0.17	–	–	0.02–0.29
Sunflower	0.43–0.66	0.09–0.17	0.34	0.02
Carnation	0.08	0.14	0.06	–
Red clover	0.04–0.05	0.04–0.06	–	0.04
Horsechestnut	0.08	–	–	0.02–0.03

*From *Handbook of Biological Data*, W. S. Spector (Ed.), 1956. W. B. Saunders Co., Philadelphia, Pennsylvania. © Federation of American Societies for Experimental Biology. (Reprinted with permission.)

sucrose and reducing sugars. Since sulfur is necessary for nitrate reduction and for protein synthesis, nitrates accumulate in large amounts, as do ammonia, amides, and amino acids, with sulfur deficiency.

When plants accumulate an excess of sulfur, which they may do if growing on gypsum soils or saline soils, their leaves become prematurely senescent and abscise (drop off).

THE ENTRANCE OF SULFATE INTO THE PLANT

Sulfate is dissolved in the soil solution and enters the plant through the roots. Its rate of uptake varies greatly, being rapid if the plant is deficient in sulfur, and much slower when the plant has adequate amounts. During the daytime, the rate of entrance is much greater than at night. This is attributed to a greater rate of movement of water through the plant when the rate of transpiration is high and sulfate is being rapidly translocated from the roots as solute in the transpiration stream. Any condition, such as cloudy weather, a dry soil, or an increased relative humidity, which will decrease the rate of transpiration will reduce sulfate uptake.

Within the plant, sulfur is translocated as sulfate which accumulates, largely in the cell vacuole, until the sulfur is incorporated into the organic compounds. One can determine this extent of accumulation by analyzing the plant for sulfate sulfur. Normally, the sulfate will represent about 20–65% of the total sulfur, but as sulfur becomes less available to the plant, the total sulfate:total organic sulfur ratio decreases to a lower value. If large amounts of sulfate are made available, the ratio may reach values of 5–10, before injuries result. This indicates that there must be a limiting factor in the mechanism of sulfate assimilation and incorporation into the amino acids. Excess sulfate accumulates as sulfate rather than greatly increased organic sulfur.

SULFATE ASSIMILATION

The ability to reduce sulfate and to incorporate it into their organic compounds is characteristic of plants but not of animals. Animals must be furnished their sulfur in the form of organic sulfur compounds and cannot use inorganic sulfur. Supplying them with certain sulfur-containing amino acids will often suffice.

Sulfate reduction, is an energy-requiring process, with rather large amounts of energy being required. This energy is supplied largely by ATP. The common mechanism of reduction is outlined as follows:

$$\underset{\text{sulfate}}{SO_4} \xrightarrow[\text{enzymes}]{ATP} \underset{\text{sulfite}}{SO_3} \xrightarrow[\text{enzymes}]{} \underset{\text{sulfide}}{S^=} \xrightarrow[\substack{\text{serine} \\ \text{sulfhydrase}}]{\text{serine}} L\text{-cysteine}$$

This series of reactions can be further broken down to give the following sequence which represents our current ideas of the common mechanism of sulfate reduction:

$$sulfate\,(SO_4^=) + ATP \longrightarrow adenosinephosphosulfate\,(APS) + 2Pi$$
$$APS + ATP \longrightarrow phosphoadenosine\ phosphosulfate\,(PAPS) + ADP$$
$$PAPS + C(SH)_2 \longrightarrow phosphoadenosine\ phosphate\,(PAP) + SO_3 + CSS$$
$$SO_3 + 3\,NADPH_2 \longrightarrow S^= + NADP + 3H_2O$$
$$S^= + serine \longrightarrow cysteine$$

As you can see, there is much similarity between nitrate and sulfate. Both are reduced before they are used by the plant. The reduction of both requires a great deal of energy supplied either through respiration or photosynthesis, and both appear in organic compounds first as constituents of amino acids.

Cysteine is the first organic compound into which sulfur is incorporated. It is a 3-carbon amino acid with a carboxyl group at one end of the molecule and a sulfhydryl group at the other end, as seen below:

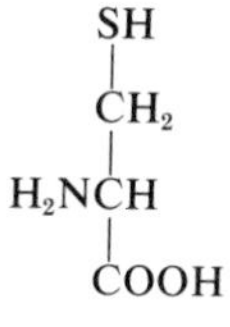

cysteine 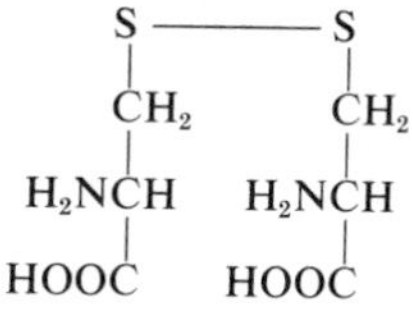   cystine 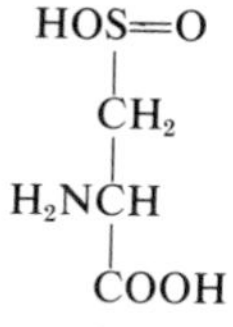   cysteic acid

It is ubiquitous in plant cells, occurring both as the free molecule and as a constituent of other organic compounds. If one supplies a plant with radioactive sulfate, the radioactive sulfur can be found in cysteine in a short period of time. However, cystine and cysteic acid will also be labelled. Cystine and cysteic acid are easily formed by the oxidation of cysteine, both *in vivo* and *in vitro*. Therefore, if we extract the amino acids from the plant, and find cystine and cysteic acid among them, this is no evidence that these existed within the plant, since they might have been formed during the extraction process. Also, because of this possibility, we cannot quantitatively determine the concentration of cysteine, cystine or cysteic acid within the living plant. Cystine is also readily oxidized to cysteic acid.

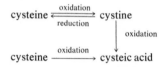

Cysteine serves as a substrate for many metabolic reactions, the result of which is the incorporation of sulfur into other organic compounds within the plant (Fig. 9.2).

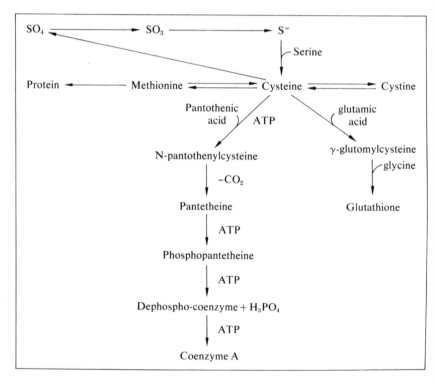

Fig. 9.2   An outline of some pathways of sulfur metabolism in plants.

TRANSULFURATION

The sulfur-containing amino acids, including cysteine, are able to transfer their sulfur to other amino acids. Homocysteine is the principal product of this *transulfuration* reaction, and is formed as follows:

cysteine + homoserine ester $\longrightarrow$ cystathionine + ester group
cystathionine $\longrightarrow$ pyruvic acid + homocysteine

Homocysteine can then be used to synthesize methionine as follows:

homocysteine $\longrightarrow$ homocystine
homocystine + methyl donor $\longrightarrow$ methionine

Methionine

Methionine is an amino acid that is rarely found as a free molecule in normal plant cells. However, it may be extracted from all plant tissues that have been subjected to treatment which results in the breakdown of proteins, such as by boiling plant material in 6N HCl which hydrolyzes the proteins. Such treatment also oxidizes some of the methionine to methionine sulfoxide and to methionine sulfone.

Methionine is very important to the plant because it is a constituent of plant proteins and is probably the chief source of methyl groups used in the synthesis of many methyl-containing metabolites. Actually there are probably three methyl donors in plants, but all three are derived from methionine as shown below:

methionine + ATP $\longrightarrow$ adenosylmethionine + 3Pi
methionine + adenosylmethionine $\longrightarrow$ methylmethionine + adenosylhomocysteine
methionine + methyl group $\longrightarrow$ dimethylpropiothetin

The mechanism through which these methyl donors transfer their terminal methyl groups to other compounds is designated as *transmethylation*. An example of transmethylation is given below:

methionine + ATP $\rightleftharpoons$ *S*-adenosylmethionine + 3Pi
S-adenosylmethionine $\rightleftharpoons$ homocysteine + CH$_3$- + adenosyl group
homocysteine + serine $\longrightarrow$ cysteine + an unknown

Actually *S*-adenosylmethionine serves as the direct source of methyl groups, in this example.

Methionine can also function as an amino acid in transamination reactions, and the product formed can supply methyl groups and is converted to α-ketobutyric acid:

$$\text{methionine} + \alpha\text{-keto acid} \xrightarrow{\text{transaminase}} \alpha\text{-keto-}\gamma\text{-methylthiobutyric acid}$$

$$\downarrow \text{transmethylation}$$

$$\alpha\text{-keto butyric acid} + CH_3-$$

Methionine represents an example of a phenomenon often encountered in metabolism studies. Its presence, in the free form, is suspected but little evidence of its accumulation is found. Therefore its presence in minute concentrations is assumed, and it forms other compounds nearly as rapidly as it is produced.

GLUTATHIONE

Cysteine will also react with the amino acids glutamic acid and glycine to form a tripeptide, reduced glutathione (GSH), as follows:

$$\text{cysteine} + \text{glutamic acid} \xrightarrow[\text{Mg}^{++}]{\text{ATP}} \gamma\text{-glutamylcysteine}$$

$$\gamma\text{-glutamylcysteine} + \text{glycine} \xrightarrow[\text{Mg}^{++}]{\text{ATP}} \text{glutathione}$$

$$\underset{\substack{NH_2}}{HOOC-CH-}CH_2-CH_2-\overset{\displaystyle O}{\overset{\|}{C}}-NH-\underset{\substack{CH_2 \\ SH}}{CH}-\overset{\displaystyle O}{\overset{\|}{C}}-NH-CH_2-COOH$$

glutathione

Glutathione is ubiquitous in plants and has been studied very intensively, in the past. It readily participates in oxidation-reduction reactions, although the cell is able to keep the glutathione in the reduced form most of the time so oxidized glutathione (GS—SG) does not accumulate.

$$2GSH \underset{\text{reduction}}{\overset{\text{oxidation}}{\rightleftharpoons}} GS-SG + 2H$$

Oxidation may be brought about non-enzymatically by proteins, etc. containing —S—S— bonds by oxygen in the presence of heavy metals, by peroxides, quinones, or by iron porphyrin compounds, or by certain enzymes.

Glutathione accounts for a proportion of the sulfhydryl groups that are so important in the physiological activities of the plant. However, it is very reactive and experiences a rapid turnover. At one time, glutathione was considered to function in the respiratory chain of respiration. This role has since been attributed to the cytochromes. However, glutathione is believed to participate in protein synthesis, and to participate in triose phosphate oxidation, and it may cause the reversible formation of sulfhydryl and —S—S— bonds between protein molecules

$$(NADPH_2 + RSSR \xrightarrow[\text{(proteins, etc.)}]{GSH} 2RSH + NADP)$$

Much of the organic sulfur in plants is present in glutathione, cysteine, or in the other amino acids that are part of the protein molecules. However, small amounts of sulfur are found in a large number of other organic compounds, some of which are found in, and essential to, all cells, and others which have no known essential function and are considered metabolic by-products. In the former group are coenzyme A, thiamine, biotin, thioctic (lipoic) acid, and cytochrome c. The latter includes the glucosides sinigrin and sinalbin and their hydrolytic products, the mustard oils, vinyl and albylisothiocyanate, allyl, methyl and vinyl sulfides, and mercaptans, which are found in onions and mustards especially, and dithiol isobutyric acid which occurs in asparagus.

FUNCTIONS OF AMINO ACIDS

So far, we have learned how amino acids function as a source of methyl groups, how they function in forming sulfur to sulfur bonds by oxidation and reduction, and how sulfur groups can be transferred through transulfuration. Let us now consider some other functions of the amino acids in plants.

Buffers

Amino acids are buffers which help to maintain a favorable pH value within the plant cell. This they do since they contain both acid and basic groups. The carboxyl group is a proton donor, and the amino group a proton acceptor. These groups allow the amino acid to maintain a constant hydrogen ion or proton concentration in the medium surrounding it. The amino acids function in this capacity both as the free amino acid and as constituents of polypeptides and proteins.

Polarity

Since the amino acid molecule does have both acid and base groups, it is a polar molecule. This polarity disappears at a certain pH value of the medium, a value which is somewhat specific for each amino acid. When the pH is favorable for eliminating the polarity of the amino acid, we say that the amino acid is at its isoelectric point. Due to these properties of the amino acids, proteins also have isoelectric points, and protein functions are very much influenced by the pH of the medium. More will be said about this in the next chapter.

Detoxication

Amino acids function in plants also by removing from the cell the ammonia that is present. This is particularly a function associated with amide formation, and the concentration of amides increases as ammonia concentration of the cell increases. This is one mechanism by which the cell is protected from ammonia toxicity.

Source of Food

Amino acids can serve as a source of carbon and energy. When carbohydrates become deficient in the plant, amino acids are deaminated, releasing the ammonia and the organic acid from which the amino acid was originally formed. The organic acid can then enter the Krebs' cycle, to be broken down to release energy through respiration.

Protection

There is also the possibility that some of the amino acids serve to protect the plant against pathogens that could damage the plant. As previously stated, there are some hundred amino acids found in plants. Of these, only twenty-two are found in proteins. The others may have additional functions, some of which include protection, and some of which are unknown.

Syntheses

Amino acids also function in the synthesis of other organic compounds. Their role in protein synthesis has already been mentioned. They also form amines, purines and pyrimidines, alkaloids, vitamins, enzymes, and others. Amine formation results merely by the removal of the carboxyl group from an amino acid, as shown below:

$$H_3C\diagdown$$
$$CH-CH_2-CH-COOH \longrightarrow$$
$$NH_2$$
$$H_3C\diagup$$

$$H_3C\diagdown$$
$$CH-CH_2-CH_2-NH_2 + CO_2$$
$$H_3C\diagup$$

leucine   isopentylamine

Various amines are found in plants in small amounts. These are frequently volatile and add aroma to flowers, etc. Some amines are also used for the synthesis of the important phospholipids. Widely distributed amines include isopentylamine, methylamine and trimethylamine.

The amino acids are synthesized, at least largely, in the mitochondria Fig. 9.3. From there they go into the mesoplasm where they move by diffusion, mass movement, and protoplasmic streaming to various parts of the cell, to be used in metabolism or stored in pools. For instance, at the ribosomes they are used in protein synthesis, and in the vacuole they are

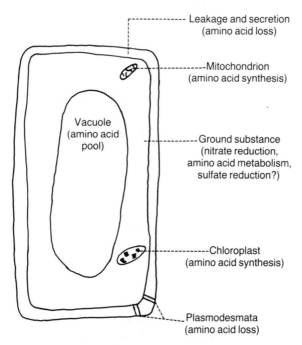

Fig. 9.3   Possible locations within the cell of amino acid synthesis and loss.

stored. They also move out of the cell by leakage or secretion or through plasmodesmata and are transported throughout the plant. Long distance translocation occurs as the amino acids move through the phloem tissues. In fact, such amino acid transport represents the means by which most nitrogen moves, in the organic form, throughout the plant.

Let us now go on to the next chapter where we will learn how the amino acids are used for protein synthesis.

Chapter Ten

Plant Proteins

In the last chapter, the synthesis and functions of the amino acids were considered. One of the functions was the use of the acids in protein synthesis. Although not the only one, this function is vital to the cell. To learn how amino acids are used for protein synthesis, it is necessary to review the structure and function of the polynucleotides, the nucleic acids.

THE NUCLEIC ACIDS

Nucleic acids are synthesized primarily in the nucleus, but some are also synthesized in both the mitochondria and the chloroplasts. The two nucleic acids are ribonucleic acids, abbreviated RNA, and deoxyribonucleic acids, abbreviated DNA, the "ribo-" and "deoxyribo-" referring to the type of sugar present in the molecule. Other differences will be discussed later in this chapter.

DNA

Some DNA is found in both the chloroplasts and mitochondria of plant cells, but it is much more abundant in their nuclei. There it forms part of the chromosomes or of the chromatin material, depending on whether the observation is made during nuclear division or during interphase. Just how many molecules of DNA make up one chromosome is not known, but chromosomes are composed of DNA and protein molecules.

DNA molecules are large, stable molecules, the largest of all cell molecules. They are composed of two, long, unbranched chains of

DNA-nucleotides, with these nucleotides joined together through their phosphate groups. Each chain is joined to its complementary chain to form a double-chained molecule, the DNA molecule. Within the cell, this DNA molecule is present in the form of a helix, a configuration that looks like a coiled spring. To understand how the two chains are joined together and how the helical form is maintained requires a consideration of the structure of the DNA-nucleotides and their orientation within the DNA molecule.

There are four types of *DNA-nucleotides*, as seen in Fig. 10.1. Each is made up of three molecules, one each of phosphoric acid, deoxyribose sugar, and of a nitrogen base (purine or pyrimidine). The only four species of nitrogen bases found in the nucleotides of DNA are *adenine, cytosine, guanine,* and *thymine.* Adenine and guanine are purines, and cytosine and thymine are pyrimidines. Although the phosphoric acid and the sugar are the same in each DNA-nucleotide, any one of four possible types of nitrogen bases could be present. Hence, four types of DNA-nucleotides are used in the formation of each DNA molecule, and each DNA molecule can differ by the sequences of each nucleotide in each chain, and by the total number of nucleotides in each molecule. At the present time, it is not known where within the cell the nucleotides are actually formed, but they are found at the site of DNA synthesis.

DNA synthesis, which is considered in detail in Chapter 16, occurs in the mitochondria and chloroplasts, and during interphase in the nucleus, but only in the presence of intact DNA molecules.

To form the double-chained molecule the two complementary chains of the DNA molecule are held together by bonds formed between the purine and pyrimidine molecules of adjacent nucleotide chains. Hydrogen bonds certainly play a role in this bonding. In any respect, only certain combinations of purines and pyrimidines can join. Adenine joins only to thymine and cytosine only to guanine. This pairing is referred to as *base-pairing.* As a result of this base-pairing, once one chain of the DNA molecule is synthesized, the structure of the other chain has been determined. (Why?)

RNA

The function of the DNA molecule is to synthesize RNA molecules, and all RNA is synthesized in the presence of DNA. Therefore, RNA synthesis occurs only in the nucleus, mitochondria and chloroplasts of the cell but this continues in all living plant cells and at all times.

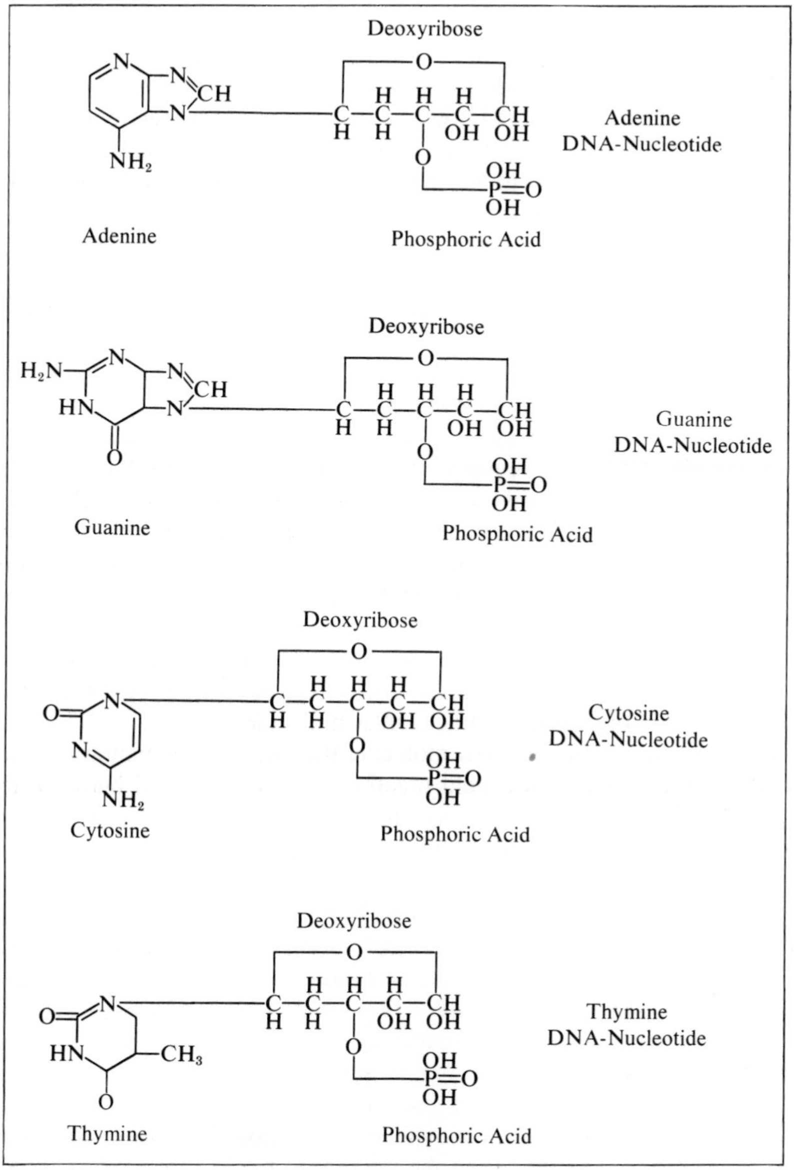

Fig. 10.1   Structure of the DNA-nucleotides.

Two methods can be used to classify RNA: first, according to its location, as nuclear, cytoplasmic, mitochondrial, chloroplastic or even leaf, stem, etc. RNA, second, and most commonly, according to its function within the cell and in this case it is known as either messenger, ribosomal, transfer, or storage RNA.

All RNA molecules are similar in being single-chained, polynucleotides as contrasted with DNA which are double-chained. Each RNA molecule also consists of four types of *RNA-nucleotides*. These are listed in Fig. 10.2. It will be seen from this figure that RNA-nucleotides differ from those of DNA in two ways. First, the sugar is ribose, instead of deoxyribose; and second, the pyrimidine thymine is absent, its place being taken by *uracil*.

Synthesis of the nucleotides involves adding a nitrogen base to the sugar molecule to form a *nucleoside*, and then adding inorganic phosphate to form the nucleotide. Two more phosphates are then added to form the triphosphate form of the nucleotide, and this form is used in the nucleic acid synthesis, with pyrophosphate given off each time an RNA-nucleotide triphosphate is added to the chain.

Various RNA molecules differ in the number of nucleotides per molecule, and therefore in their molecular weight and length, and also in the sequence of nucleotides. The transfer RNA, abbreviated either *t*-RNA or *s*-RNA, molecules are the shortest and smallest; the messenger RNA, abbreviated *m*-RNA, molecules are the longest and largest. The ribosomal RNA, abbreviated *r*-RNA, are intermediate in size.

RNA synthesis involves the enzyme RNA polymerase, DNA, and the RNA-nucleotide triphosphates as illustrated below:

$$\text{RNA-nucleotide triphosphates} \xrightarrow[\text{DNA}]{\text{RNA polymerase}} \text{RNA} + \text{pyrophosphates}$$

The exact way in which the DNA molecule functions in this capacity is unknown, but the nucleotide sequence in the RNA molecule is determined by the base-pairing between one strand of the DNA molecule and the RNA molecule being formed. It is through this mechanism that DNA controls the sequence of the nucleotides in the RNA molecule, and therefore the properties of the RNA synthesized.

m-RNA is synthesized in the cell nucleus, at the surface of the DNA molecules of the chromosomes. After its formation, it moves to the cytoplasm, or other area where protein synthesis is to occur, functions in this protein synthesis, and is then destroyed. It is much shorter-lived than either *t*-RNA or *r*-RNA.

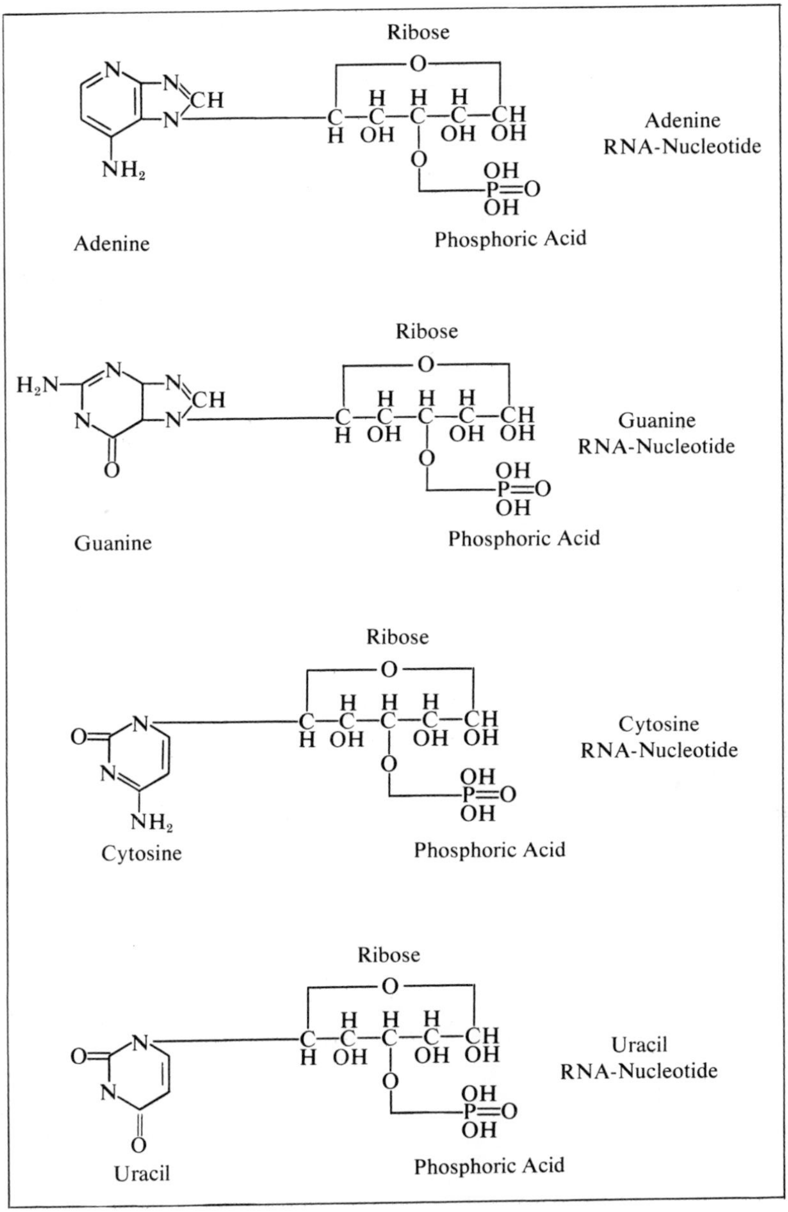

Fig. 10.2   Structure of the RNA-nucleotides.

r-RNA is the most abundant RNA in the cell, and is synthesized at the nucleoli. Nucleoli were discussed in Chapter 1 and the reader would do well to review nucleolar structure and function. It is interesting to speculate whether all ribosomal RNA is synthesized in the nucleoli, since r-RNA is also found in the mitochondria and chloroplasts. One may ask if these synthesize their own r-RNA or if they obtain it already formed from the nucleus. Actually, the r-RNA is also synthesized by the DNA, since part of the chromatin material is bound to the nucleoli and therefore it is better to say that the synthesis of r-RNA occurs at the surface of DNA molecules where the DNA molecules are bound to the nucleoli. In any respect, r-RNA is not often found as free molecules, but bound to proteins to form the ribosomes, also discussed in Chapter 1. Ribosomes are found in the nucleus, mitochondria, and chloroplasts, but most of them occur in the cytoplasm where they are either attached to the endoplasmic reticulum to form rough endoplasmic reticulum, or suspended in the ground substance of the cytoplasm. Those in the ground substance are larger than those in the mitochondria or chloroplasts.

Unattached ribosomes are found either free as individual ribosomes, or are joined together with numerous other ribosomes to form the *polysomes*. Polysomes are considered to be aggregates of ten to forty ribosomes in the process of protein synthesis, and are held together by one m-RNA molecule. Polysomes may be free, or may be attached to endoplasmic reticulum. Bound ribosomes are considered to be more active in the process of protein synthesis than are free ribosomes, and are easily disassembled by shock or by respiration inhibitors. All of the factors that determine the ratio of free to bound ribosomes in the cytoplasm, are not known.

The location of t-RNA synthesis is unknown, but since this is the smallest RNA molecule and is soluble, it is found in many parts of the cell. It is probably synthesized in the nucleus at the chromatin material or at other locations of DNA molecules, since all RNA is made by DNA.

As t-RNA molecules are the smallest RNA molecules, with molecular weights near 25,000, their nucleotide sequence has been studied more than has that of other RNA molecules. Such studies have revealed that t-RNA molecules may contain less than a hundred nucleotides in each molecule, but each of the more than twenty kinds of t-RNA molecules differs in its nucleotide sequence.

The function of t-RNA in protein synthesis is to transfer amino acid molecules to the site of protein synthesis and to participate in determining the sequence in which these amino acids occur in the polypeptide chain.

All proteins are made up of the same twenty different kinds of amino acids shown in Table 10.1, and each *t*-RNA is specific for one kind of amino acid. This justifies the statement that there must be at least twenty different kinds of *t*-RNA within the plant cell, each named according to the amino acid for which it is specific, as alanine *t*-RNA, glycine *t*-RNA, etc.

This amino acid must be activated before the *t*-RNA molecule can pick up the amino acid. Such activation occurs at the expense of ATP as shown below:

$$\text{amino acid} + \text{ATP} \xrightarrow{\text{amino acyl synthetase}} \text{amino acyl} + \text{PP}$$

An example of this reaction is:

$$\text{alanine} + \text{ATP} \xrightarrow{\text{alanine acyl synthetase}} \text{alanine-AMP} + \text{PP}$$

Since there are twenty amino acids involved in protein synthesis, there must be twenty different types of amino acid synthetase enzymes and twenty types of amino acyls, as seen in Table 10.1. The amino acyl

Table 10.1   Amino acids, their acyls and acyl *t*-RNAs required for protein synthesis in plant cells.

Amino acid	Amino acyl	Amino acyl *t*-RNA
alanine	alanine-AMP	alanine acyl alanine *t*-RNA
arginine	arginine-AMP	arginine acyl arginine *t*-RNA
asparagine	asparagine-AMP	asparagine acyl asparagine *t*-RNA
aspartic acid	aspartic acid-AMP	aspartic acid acyl aspartic acid *t*-RNA
cysteine	cysteine-AMP	cysteine acyl cysteine *t*-RNA
glutamic acid	glutamic acid-AMP	glutamic acid acyl glutamic acid *t*-RNA
glutamine	glutamine-AMP	glutamine acyl glutamine *t*-RNA
glycine	glycine-AMP	glycine acyl glycine *t*-RNA
histidine	histidine-AMP	histidine acyl histidine *t*-RNA
isoleucine	isoleucine-AMP	isoleucine acyl isoleucine *t*-RNA
leucine	leucine-AMP	leucine acyl leucine *t*-RNA
lysine	lysine-AMP	lysine acyl lysine *t*-RNA
methionine	methionine-AMP	methionine acyl methionine *t*-RNA
phenylalanine	phenylalanine-AMP	phenylalanine acyl phenylalanine *t*-RNA
proline	proline-AMP	proline acyl proline *t*-RNA
serine	serine-AMP	serine acyl serine *t*-RNA
threonine	threonine-AMP	threonine acyl threonine *t*-RNA
tryptophane	tryptophane-AMP	tryptophane acyl tryptophane *t*-RNA
tyrosine	tyrosine-AMP	tyrosine acyl tyrosine *t*-RNA
valine	valine-AMP	valine acyl valine *t*-RNA

molecules are available for attachment to the t-RNA molecules to form amino acyl RNA as indicated below:

$$\text{amino acyl} + t\text{-RNA} \xrightarrow{\text{ATP}} \text{amino acyl } t\text{-RNA} + 2\text{ADP}$$

An example of this reaction would be:

$$\text{alanine-AMP} + \text{alanine } t\text{-RNA} \xrightarrow{\text{ATP}} \text{alanine acyl } t\text{-RNA} + 2\text{ADP}$$

This activity results in the formation of a t-RNA attached to an amino acid.

It is not possible, at this time, to reveal what differences in sequence exist among them as the nucleotide sequence of all of these t-RNAs have not yet been determined. It seems that the t-RNA chain is folded back near one end of the chain and that, near the folded area, there is a sequence of three nucleotides. This sequence that appears to determine where the amino acyl t-RNA becomes attached to the m-RNA molecule during polypeptide synthesis, is called the *anticodon*. The triplet of this anticodon would be specific for each type of amino acid attached to the t-RNA molecule, but the amino acyl t-RNA does not, in itself, determine the position of the amino acid in the polypeptide chain.

PROTEIN SYNTHESIS

Actually, it is the m-RNA that is the important factor in determining the sequence of amino acids in the newly-formed polypeptide chain. This m-RNA is synthesized at the surface of the DNA molecule of the chromatin material, and its nucleotide sequence is extremely important in determining both the sequence of amino acids in the polypeptide chain and the length of the chain. The sequence of the nucleotides in the m-RNA molecule is determined by base-pairing with the DNA molecule. Each three nucleotides of this m-RNA molecule is a triplet or *codon* which represents the site of attachment of the anticodon of the t-RNA molecule. After the m-RNA is synthesized, it moves out of the nucleus into the cytoplasm where it attracts ribosomes to form the polysome. These ribosomes move along one side of the m-RNA molecule, and, as they do so, the amino acyl t-RNA molecules join on the m-RNA molecule by base-pairing between the anticodon and the codon. As this union is made, the amino acids of adjacent amino acyl t-RNAs join and the t-RNAs separate from both the messenger RNA and the amino acid, and move out to pick up another amino acid. The amino acids remain attached together and to the last *t*-RNAs, and by continually adding to this chain of amino acids, a polypeptide molecule is formed. Since the only amino acyl *t*-RNAs that can attach at a given location on the *m*-RNA chain are those with anticodons that can base-pair with the codon at that location, the *m*-RNA determines the sequence of amino acids in the polypeptide chain. Eventually, as the chain becomes long enough, it separates to form the free polypeptide molecule. Protein synthesis is illustrated in Fig. 10.3.

RNA, unlike DNA, is not stable but is often broken down within the cell. In fact, RNA is stored in cells of seeds, in order to be broken down as the seed germinates, the nucleotides being used for the synthesis of more RNA within the embryo. The sequence of the breakdown of the RNA molecule appears to be as follows:

$$\text{RNA} \xrightarrow{\text{RNA depolymerase}} \text{nucleotides}$$

$$\text{nucleotide} \longrightarrow \text{nucleoside} + \text{P}$$

$$\text{nucleoside} \longrightarrow \text{ribose} + \text{nitrogen base}$$

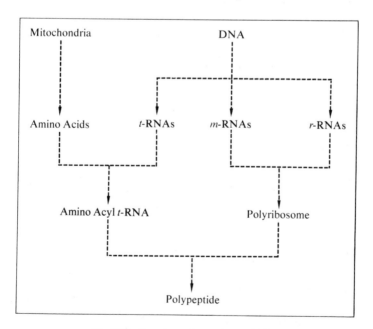

Fig. 10.3   Summary of protein synthesis.

This breakdown need not go all the way, but can stop at any stage. Messenger RNA is particularly unstable and is rapidly broken down. In fact, much of that synthesized within the nucleus probably never reaches the cytoplasm. On the other hand, both r-RNA and t-RNA are more stable and are quite long-lived.

After the polypeptide chain has been synthesized, it will either serve as the protein molecule or, more often, will unite with other polypeptide chains to form the protein molecule, frequently through the formation of disufide bonds between adjacent chains.

The newly-formed protein molecule assumes a certain spacial configuration; a configuration that is specific for each type of protein, and that is essential for its biological function.

It has been estimated that each plant cell contains about one protein molecule for every three million water molecules, or about one billion protein molecules. These billion protein molecules represent about 2000 different types of protein, which exist within any one plant cell, with an average of about 500,000 molecules of each type. These estimates are only approximations but give some concept of the numerous different kinds of proteins present within each cell. Although the kinds of proteins show some similarity from cell to cell, they show differences too. There is probably at least one kind of protein that is specific for each plant species and that is found only in that species. However, many kinds of proteins are more widely distributed.

PROTEIN CLASSIFICATION

Since the number of types of proteins is so large, it would be desirable to have a good system by which these proteins could be classified. Many such systems exist, but no one of them is ideal. Each is designed for a different purpose, and each is satisfactory for that purpose as long as its limitations are known.

Location

One system is based on the location of the protein within the plant organ. There are leaf proteins, root proteins, fruit proteins, flower proteins, pollen proteins, stem proteins, etc. For purposes of specifying where the protein is found, this system is good, but it must be realized that there are thousands of different kinds of proteins in each of these classes.

Proteins can also be classified by considering the location of the protein within the cell, so that there are nuclear proteins, chloroplast proteins, mitochondrial proteins, and cytoplasmic proteins. This method is satisfactory to specify the cell organelle that contains the protein, but again it must be realized that there are many different kinds of proteins within each cell organelle.

Solubility

One system of protein classification that has been used for many years, and that will continue to be used due to its convenience, is based upon the solubility of proteins in different solvents (Table 10.2). The four main protein types in this system are the albumins, globulins, glutelins, and prolamins. The difficulty with this method is that some of the types have no sharp lines of demarcation but their solubilities integrate or overlap from one solvent to another. Also, there are only four classes, whereas it is known that many thousands of proteins exist.

Table 10.2   Classification of proteins based upon the solubility of the protein in various solvents. (+ means soluble, − means insoluble.)

Protein class	water	Solubility in: dilute neutral salt solution	dilute alkali or acid	70% ethanol
Albumins	+	+		
Globulins	−	+		
Glutelins	−	−	+	
Prolamins	−	−	−	+

Protein Structure

Perhaps the best system of protein classification would be one based on the structure of the protein molecule, since each protein has a definite chemical and physical structure, which is further subdivided into primary, secondary, and tertiary structures. Such a method is only in its infancy since much remains to be discovered about the structure of most of the proteins found in plant cells. In fact, the complete chemical structure and spacial configuration is not accurately known for any. However, methods of determining this structure are now available, and all that is needed is the motivation and time necessary for a complete analysis. Usually, in a laboratory equipped to do this work, at least a year of hard work is required to elucidate the primary structure of a protein, and this applies only if the protein can be isolated in its absolutely pure form. It is currently very difficult or even impossible, to obtain pure protein from most plant cells. In fact, there is no good criteria to reveal when an extracted protein is pure, so if pure protein cannot be obtained, its structure cannot be elucidated. Much work remains before the complete structure of many plant proteins is known.

As indicated above, each protein has a primary, secondary, and tertiary structure which must be understood if a structural system of classification is to be used. The *primary structure* involves the amino acid composition of the protein, the sequence of these amino acids in each polypeptide chain of the protein, the constituents of the protein that are not amino acids, such as heavy metals, etc., and the size and molecular weight of the protein molecule. One study contributing to our knowledge of the primary structure of plant protein has been carried out using an allergenic protein from the pollen of ragweed, a species of plant particularly noted for its cause of hayfever. The amino acid composition of this protein is given in Table 10.3. The molecular weight of this protein was found to be about

Table 10.3 The amino acid composition of a protein that is responsible for the hayfever caused by ragweed pollen. (Data of King, *et al.*, (1964). From *Biochem.* **3**: 458–468. © 1964 by The American Chemical Society. Used with permission of the copyright owners.)

Amino acid	No. of amino acid residues per protein molecule
Lysine	18
Histidine	6
Aspartic acid	49
Arginine	16
Threonine	17
Serine	26
Glutamic acid	25
Proline	15
Glycine	37
Alanine	31
Cysteine, incl. cystine?	7
Valine	24
Methionine	7
Isoleucine	20
Leucine	21
Tyrosine	4
Phenylalanine	12
Tryptophan	6

38,000, and leucine was determined to be one of the terminal amino acids of the polypeptide chain. As complete as these data are, they still do not reveal the sequence of the amino acids in the chain, nor do they reveal the other terminal amino acid.

To the best of our knowledge, most proteins have the same *secondary structure*, that of an α-helix. The α-helix is a structure that looks like a coiled spring, with the coil going in the direction to the right. If you were looking at the structure with the coils going away from you, they would be coiled in a clockwise direction.

As detected by X-ray techniques, the protein molecule is usually folded and bent in various ways, giving the *tertiary structure*. This is in addition to the α-helical structure.

The tertiary structure has more importance than is at first apparent since it is associated with function. In fact, when the protein becomes denatured, this involves a change in the tertiary structure of the molecule. This tertiary structure is usually maintained by hydrogen bonds and by sulfur to sulfur bonds, the latter being known as disulfide bonds. If these hydrogen or disulfide bonds are broken, the tertiary structure of the molecule changes as does its biological activity.

It appears, however, that secondary and tertiary structure of the protein molecule are determined by the primary structure. As a result of its primary structure, each protein molecule assumes definite and specific secondary and tertiary configurations, and these represent the most thermodynamically probable structure of the molecule in view of its primary structure.

Yes, a classification system based upon the primary, secondary, and tertiary structure of the protein molecule would be very desirable, but the time when such a system becomes *the* system of protein classification is a long way off.

Function

Another system of protein classification that is very often used is one based upon the function of the protein within the plant or its cell. There are reserve proteins that are found primarily in the cells of the cotyledons and endosperm of seeds and in fruits, but are not limited to these locations. These are the proteins that are stored in the cell—proteins that show little or no biological acitivity. There are structural proteins that form part of the membranes of the cells, and there are enzymatic proteins that are constituents of the many cell enzymes. Reserve proteins are a distinct class, but it is not always possible to differentiate between structural and enzymatic proteins. It is certainly possible and likely that some proteins can and do function both as structural and enzyme proteins at the same time.

Reserve proteins have been studied most extensively and are a distinct group with properties that differ from structural and enzymatic proteins. Seeds store a great many of these reserve proteins, those with fleshy cotyledons, such as beans, having higher protein contents (about 25% protein) than those which store their protein in the endosperm, such as cereals (about 8% protein).

Early in the development of these seed storage tissues—the endosperm and cotyledons—their cells become so filled with starch and aleurone grains that other cell organelles become difficult to find. Aleurone grains originate from vesicles, which themselves arise from Golgi bodies. So much reserve protein is laid down in these vesicles that aleurone grains are little more than large grains of protein surrounded by a membrane.

Evidently, the proteins that serve as reserve proteins are synthesized within the aleurone grain, since all of the RNA necessary for protein synthesis are found therein. One interesting question that seems to need an answer is, what is the source of *m*-RNA in these grains?

Many kinds of protein are stored in aleurone grains. In the cotyledons of dicots, the reserve protein appears to be primarily globulin, but many types of globulins may be present and each type is not specific for a given plant species. In the cotyledons of many legumes, the reserve protein has been described as being primarily the two globulins, vicilin and legumin, which are probably mixtures of globulins. Both are very widely distributed among the legumes, and both disappear rapidly following germination, not to appear again as globulins, but to produce amino acids that later appear in the structural and enzymatic proteins in the embryo or developing seedling.

The reserve protein of cereal grains are not globulins. Several globulins have been isolated from cereal grains but they are present in very small amounts, and not in the endosperm. The reserve protein of cereal grains are prolamins and glutelins, which make up about 70% of the total protein. The most often studied of these include gliadin of wheat, hordein of barley, and zein of corn. However, these are not pure proteins either but represent mixtures of prolamins and glutelins.

The aleurone grains swell and cavities develop within them during seed germination. Then these grains break up into fragments and finally disappear altogether. In fact, the aleurone grains seem to disappear faster than the globulin or other reserve protein.

Although much remains to be learned about reserve proteins, they have been studied more extensively than have the structural and enzyme proteins of plants. Perhaps this is due to the economic importance of the seed proteins and to the fact that they are more easily extracted. As difficult as it is to obtain pure reserve proteins of seeds, it is much more difficult to obtain pure structural and enzymatic proteins.

Leaf proteins, particularly those found in the chloroplasts, have been studied often due to their association with photosynthesis. In the chlorophyll-containing cells of plants, about one third to one half of the protein is located in the chloroplasts, with very little in the nucleus and perhaps none or only traces in the vacuole. The chloroplast proteins are different from those of the other cell components, and contain about 70% of all leaf protein sulfur in the form of sulfur-containing amino acids. Chloroplast proteins are located in the membrane, grana, and stroma.

One of the most studied of the leaf proteins is that known as fraction I protein. This is a soluble, globular protein of the stroma and has associated with it the enzyme needed for the addition of carbon dioxide to ribulose diphosphate. It is widely distributed in the leaves of many plant species.

It has been amply demonstrated that the protein, chlorophyll, and carotenoid of the chloroplast vary dependently. All three increase or decrease together and to about the same extent. This implies that chloroplast pigments are bound to the chloroplast proteins, and such binding may be essential for pigment stability.

PROTEINS AS FOOD

There has recently been an upsurge of interest in plant leaf proteins as a possible source of protein for human nutrition. Since plants represent a much more efficient source of nutrients than animals and since the leaves of plants are quite high in protein content, they have been viewed as a good potential source of food for the rapidly-increasing world population. However, there appear to be three main problems associated with the use of leaves as a source of protein. First is the problem of making the protein available and digestible. Second is the high fiber content of the leaf, and third, there are toxic and allergenic substances in many plant leaves. Nevertheless, this source of protein looks promising for the future, when a significant amount of our food proteins may come directly from plant leaves. Unfortunately, such proteins are deficient in the amino acid methionine which will have to be added as a supplement.

PROTEIN BREAKDOWN

In all living cells, there is a continuous turnover of protein molecules. Some are being synthesized while others are being broken down, so that each protein molecule has a *turnover time* — a time interval from synthesis to degradation.

Proteolytic Enzymes

Protein breakdown is brought about by *proteolytic enzymes*. These enzymes are of two types. The *proteases* break the polypeptide chain into smaller units or peptides, which consist of several amino acids. Then the *peptidases* go to work on the peptides and break them down into their constituent amino acids. So, the proteolytic enzymes break the proteins down into their amino acids. It is interesting to note that these proteolytic enzymes are proteins themselves and yet are able to resist their own action.

The concentration of any one species of protein within the cell or plant is dependent upon the relative rates of synthesis and breakdown. During growth, there is a net increase in structural and enzymatic proteins as synthesis exceeds degradation. As the cell or plant matures, the two rates become equal, or nearly so, and the concentrations of the proteins approach a constant value.

However, there are times in the development of the cell or plant, when net protein breakdown occurs, resulting in a decrease in protein concentration. Such net breakdown occurs in the storage cells of the seed during seed germination and during the early stages of seedling development. Here the reserve proteins are broken down into their amino acids, and these amino acids move from the cotyledon or endosperm cells to the embryo or developing seedling where they are used in the synthesis of structural or enzymatic proteins. Net breakdown also occurs in the Spring of the year as growth is resumed in perennial plant species. It occurs in other storage tissues of plants, such as in roots and tubers, when the protein of these are needed to support growth. It occurs during senescence in all cells, and it occurs during periods of drought and during frost hardening or heat hardening in plant cells. Net protein breakdown also occurs in many plant parts when these parts are removed from the plant, such as when leaves or flowers are removed, and it increases when the plant or cells are grown in extended periods of darkness.

Nitrogen Deficiency

When nitrogen is deficient, there is a net protein breakdown, since, under this condition, there is insufficient nitrogen present for the synthesis of new amino acids. This explanation would not be valid if protein breakdown merely resulted in the release of the amino acid constituent of the protein. However, the amino acids formed during protein breakdown are further broken down to form glutamic acid and ammonia. This ammonia may be used in amide synthesis and the glutamic acid may be broken down, via the Krebs' cycle (page 161), into carbon dioxide and water. Thus, during protein breakdown, there is a net loss of amino acids.

Diurnal Variations

The protein content of leaves varies with a number of factors that control either synthesis, breakdown, or both. For instance, the protein content of leaves has been shown to be higher at evening than in the morning. This has been attributed to an increase in breakdown over synthesis during the night, with increase in synthesis over breakdown during the day. This is due to photosynthesis which supplies more carbohydrates for protein synthesis during the daytime.

Temperature

Exposure to high temperatures also decreases the protein content of leaves. Thus, the protein content of leaves will decrease during the heat of the day, only to increase rapidly toward evening. This has been attributed to a greater acceleration of protein breakdown at the higher temperatures, than to protein synthesis.

Age

The age of the leaf is also a factor to consider. The protein content of young leaves is greater than that of old leaves. This is due to the greater rate of protein synthesis in the young leaves, and one should be sure to use leaves of the same age when sampling a plant for protein analysis. At a given time, a plant will often contain leaves of various ages.

The age or stage of development factor is also important when considering the storage of nitrogen within the plant. Since hydrolysis of proteins predominates in older and in senescent leaves, as the leaves age, nitrogenous substances become solubilized and are translocated to other parts of the plant. Some go to the younger leaves that may be developing, but much will go to the seeds in the case of annual plants, or to the roots, stems, or bulbs of biennial and perennial plants. It has often been reported that leaves on trees lose a considerable amount of nitrogen just before abscission in the Fall of the year, and that much of this is translocated to the permanent parts of the plant where it is stored. This stored nitrogen can then be converted back into the amino acids and used for protein synthesis the following Spring during early growth of new leaves.

Generally, the protein content of meristematic cells is very high. This may be due in part to their high rate of metabolism, to their small vacuole, as well as to their thin cell walls. As the cell ages, these factors change, reducing the protein content. Young plant tissues, tissues such as those of seeds, flowers, fruits, etc., have a high protein content. As these age, the protein contents decrease, and for those parts that will abscise, protein is largely broken down and the amino acids translocated to more permanent parts of the plant prior to abscission.

In this chapter, the synthesis, storage, and breakdown of protein have been considered. In the next chapter, the subject of the synthesis, structure, and function of that very important class of plant proteins known as enzymes will be studied.

Chapter Eleven

Plant Enzymes

In the previous chapter, one method of protein classification was discussed which classified proteins either as reserve, structural, or enzymatic proteins. This chapter will be devoted to those proteins that function as enzymes or as part of the enzyme molecule.

THE STRUCTURE OF ENZYMES

Metabolism is the mechanism of life. And yet, it too has its controls. Certain of these controlling agents are known as enzymes, and these enzymes determine which metabolic reactions will occur and which will not. Each chemical reaction has its enzyme.

Enzymes have some general properties that are worth considering. All contain proteins. Some of these consist of just one polypeptide chain, but most consist of two or more polypeptides. Some enzymes are made up of only protein molecules whereas others have other chemical elements or molecules attached to the protein to form the complete enzyme. The protein is called the *apoenzyme*, and the attached molecule, the *coenzyme* or *prosthetic group*. A given enzyme may consist then of just the apoenzyme, or of an apoenzyme plus a coenzyme. If it consists of just the apoenzyme, then the protein determines both the identity of the substrate molecule upon which the enzyme will act, and the type of action it will catalyze. Examples of enzymes which appear to be composed of just a protein molecule include the enzymes: urease, amylase, papain, and protease. These are all hydrolytic enzymes which function to break down the substrate molecule by adding water molecules to its components.

If the enzyme consists of both an apoenzyme and a prosthetic group, as most enzymes do, then the apoenzyme determines the substrate and the prosthetic group determines the activity. Prosthetic groups in these enzymes are either organic molecules or inorganic ions called *activators*. Enzyme activity is greatly influenced by, and often even entirely dependent upon, the presence of certain enzyme activators. Such activators include inorganic salts of alkali or alkaline earth metals. The concentration of these need not be great, but the effects will be. How these enzymes are activated is unknown, but perhaps such activators function at the union between the enzyme and the substrate, allowing the two to make contact, whereas such contact would be impossible in the absence of the activator. Of course, this is one of the roles of the mineral nutrients in plant cells, and one reason why their deficiency results in injury to the plant. More than fifteen different inorganic elements are known to function as activators of plant enzymes. Some of these are listed in Table. 11.1. In fact, of the mineral nutrients discussed in Chapter 5, only boron and nitrogen are not known to serve as activators. Examples of enzymes that consist of both a protein and a prosthetic group, include dehydrogenase, oxidase, peptidase, and phosphatase. More will be said about the coenzymes later.

ISOENZYMES

It is possible for the protein structure of the enzyme to vary considerably without altering the activity of the enzyme. Thus there are a number of enzymes that use the same substrate and perform the same function, but whose molecular structures or configurations differ. If these enzymes are all within the same plant species, they are called *isoenzymes*. However, sometimes differences exist among enzymes performing the same function in different species of plants, and such enzymes are called *heteroenzymes*. The difference between isoenzymes and heteroenzymes then, is whether they are found in the same species or in different species. Both denote enzymes that catalyze the same reactions, but differ structurally or in other physical properties.

Isoenzyme formation involves alterations in that portion of the protein molecule that is not directly associated with enzyme function. In many cases, this difference is due to differences in protein configuration. Such configuration may be changed by such factors as any change in the sequence of amino acids in the protein molecule, forming amides from amino acids after they are part of the chain, altering configuration of the

Table 11.1   Inorganic elements known to be activators of plant enzymes. (Reprinted from Schrenk and Frazier. *Plant Food Review*, Vol. 10, No 3, Fall 1964 — a publication of the National Plant Food Institute, Washington, D.C.)

Element	Relative amount needed (parts per million of element in nutrient solution)	Role in plant
Potassium (K)	200	Activates an enzyme in breakdown of sugar (an early step in the process of respiration).
Calcium (Ca)	120	An important part of the middle section of plant cell walls. Important in maintaining cell membranes.
Phosphorus (P)	63	Involved in the system plants use in utilizing energy from such food reserves as carbohydrates and fats. Also contained in enzyme cofactors.
Sulfur (S)	32	A component of the sulfhydryl group ($-SH$) which is essential to many enzymes. Contained in cofactors for the energy release of an important intermediate in respiration, namely pyruvic acid.
Magnesium (Mg)	24	Required for the action of many enzymes in respiration, photosynthesis, and other aspects of metabolism. Also a constituent of chlorophyll.
Iron (Fe)	5.6	Constituent of iron-containing substances, as cytochromes concerned with electron transport in respiration, and ferrodoxin which participates in photosynthesis.
Manganese (Mn)	0.60	Necessary for the formation of oxygen from water in photosynthesis; the oxygen is given off as a by-product.
Zinc (Zn)	0.070	A part of several important enzymes (dehydrogenases) that catalyze oxidation reactions in plant respiration where hydrogen is removed from organic compounds.
Copper (Cu)	0.020	Concerned with an oxygen consuming enzyme of hydrogen transport in plant respiration, and in the reduction of nitrate ions to ammonium ions.
Molybdenum (Mo)	0.010	Is in the reductase enzyme that converts nitrate ion to nitrite ion. Is also concerned with nitrogen fixation by free-living bacteria, e.g. azotobacter, and by those in root nodules, e.g., rhizobia.
Cobalt (Co)	0.001	Plays a role in nitrogen fixation by bacterial action in legume root nodules.

protein with small molecules, polymerization of the protein molecule, and by changing the way in which the protein chain is folded. It is well known that most of the area of the enzyme molecule is not directly involved in catalytic activity, that catalytic activity is limited to a very small portion of the molecule.

Perhaps isoenzymes represent one way in which the plant is able to adapt to changing environments. If one enzyme is blocked, its iso-enzyme takes over the function, thus insuring the continuity of meta-bolism so important for the survival of the individual.

MECHANISM OF ENZYME FUNCTION

In Chapter 7, it was shown that the rate of a chemical reaction can be speeded up by either increasing the temperature — but such an increase would require temperatures much above that tolerable to the plant cell — or by lowering the activation energy requirement. Enzymes do the latter. They speed up the rate of chemical reactions by lowering the activation energy requirement of the reaction.

Many of the chemical reactions which occur in the plant cell could also occur outside of the plant cell in the absence of their specific enzymes as long as the substrate contained all of the chemicals needed for the reaction. However, such reactions would occur very slowly, in fact, much too slowly to be of any use to the living cell. For instance, if sucrose is mixed with water, and placed in a sealed container on the laboratory bench, the sucrose molecules will hydrolyze and form glucose and fructose, but it would take more than a hundred years before any great amount of the monosaccharides was present. This is much too slow for the living plant cell. This cell brings about the same reaction in a matter of minutes or even seconds. This tremendous speed-up in the reaction rate is due to the presence of an enzyme within the cell, and this is the func-tion of plant enzymes, namely, to speed up the rate of chemical reactions within the cell so that such reactions occur rapidly enough to support life.

Enzyme-substrate Interaction

Even today, the means by which this activation energy requirement is lowered is not completely understood. In some way, the enzyme attaches itself to the substrate molecule or molecules, molecules which usually are smaller than the enzyme molecule itself. This attachment occurs for only a very short period of time. When the attachment is broken, the product is released from the enzyme, as seen below, where E is the enzyme, S is the substrate, and P is the product:

$$E + S \longrightarrow ES \longrightarrow E + P$$

Perhaps this attachment is accomplished by the lock and key mechanism which can be illustrated by referring to a technique used in criminalistics. A bullet fired from a gun can be used to identify the gun that fired it. If the microscopic ridges on the bullet correspond with similar microscopic grooves on the inside of the barrel, this shows that the bullet fits perfectly in the gun and was fired from it. Similarly, if grooves in the enzyme molecule correspond with ridges on the substrate molecule, the substrate molecule will attach itself to the enzyme. If these grooves and ridges do not correspond, such attachment cannot be made.

Enzyme Involvement

The enzyme does participate in the reaction, but it is released after the reaction is complete, and is released in its original form, now being available to become involved in another similar role. In short, the enzyme participates in the reactions, but is not permanently altered as a result of this participation. It also does not initiate a reaction that is not already proceeding, and it does not change the equilibrium constant of the reaction, which means that after the reaction is complete the ratio of substrate to product molecules is the same in an enzyme-catalyzed reaction as in the same reaction that is allowed to reach equilibrium without a catalyst. It does catalyze the reversible reaction in both directions, with the direction of net activity determined by the law of mass action.

Mass Action

A knowledge of the *law of mass action* is very important for understanding chemical reactions and the cation exchange phenomenon discussed in Chapter 2. It states that the rate at which a chemical reaction occurs is proportional to the molar concentration of those chemicals being used in the reaction. A simple example of this can be shown with the following reaction:

$$A \longrightarrow B$$

If the concentration of substrate A is increased, the rate of the reaction will be increased. This law also can be applied to other activities. For instance, in cation exchange reactions, the more sodium ions in the soil solution, the greater their percentage adsorbed to the soil colloid.

Rate of Reaction

Some indication of the rate of increase of the reaction rate by an enzyme is shown using sucrose hydrolysis as an example. The hydrolysis of sucrose by HCl and without a catalyst has an activation energy requirement of 26,000 cal, but hydrolysis of the sucrose by malt invertase, an enzyme, has an activation energy requirement of only 13,000 cal. Thus, the enzyme reduces the activation energy requirement to one half of its original value. This is enough to increase the reaction rate about 500,000 times or more. A tremendous increase in reaction rate.

Since the purification of enzymes is very difficult, the concentration of enzyme present in a plant or in its cells is measured as the activity of the enzyme, by determining the rate at which its catalyzed reaction occurs. The enzyme is very efficient. It has been suggested that one catalase enzyme molecule can cause the conversion of over one million molecules of hydrogen peroxide to oxygen and water per minute, and one mole of the enzyme can catalyze up to three million moles of the substrate. Imagine how fast a man would be working if he held and released three million objects each minute. However, these values are for the pure enzyme in optimum environment. On the other hand, it has been calculated that in a living cell, on the average, one enzyme molecule catalyzes the conversion of 1000 substrate molecules, to form their products, per minute. Of course, this value will vary greatly from one enzyme to another, but perhaps this indicates the rapidity with which enzyme-catalyzed reactions do occur within the cell.

Enzyme Complexes

Sometimes enzymes function independently of other enzymes, but often they function in complexes with others, much like an assembly line in a factory. Such complexes consist of a number of different kinds of enzymes, each of which receives the product of one enzyme and converts it into a product that can be used as the substrate by another. In such complexes, the enzymes are close together spacially, so that such interchange of products can occur very rapidly. In such cases, when the scientist tries to remove and isolate one specific enzyme, the function of the enzyme is destroyed, and therefore one cannot tell if the enzyme is actually functioning within the cell or not. To isolate such complexes without altering their function is very difficult and in many cases is impossible. It is desirable, and often necessary, to isolate an enzyme or group of enzymes from the plant and study them *in vitro*. When this is done, it must be realized that such extraction does alter the enzyme and the medium used *in vitro* is no doubt different from the natural environment of the enzyme within the cell. Therefore, it is difficult to transfer our knowledge of enzyme activity *in vitro* to that *in vivo*.

Factors Altering Enzyme Activity

Many enzymes continue to function within the cell after the death of the plant. Of course, enzymes vary as to how long they can remain active under such conditions but death does not necessarily involve the cessation of all enzyme activity. This makes death hard to define, precisely. Of course, many enzymes can be removed by destroying the plant cell, and extracting the enzymes from the cell. Under such conditions, many enzymes will continue to function outside the cell in a flask, at least for some time, usually several hours. However, the removal of the enzyme from the cell is not an easy process since there is high acidity in the vacuoles that destroys many enzymes unless the extraction medium is strongly buffered. Also, certain chemicals, such as the tannins found within the vacuole destroy the enzymes when the cell is damaged, during the extraction process. Then too, enzymes are very sensitive to many environment factors, as will be seen shortly. They are denatured by high temperatures, certainly by those above about 50°, and once out of the cell appear to be denatured by temperatures much lower, such as at room temperature. Enzymes are altered by pH extremes, and may be destroyed during extraction unless precautions are taken to prevent such pH extremes from occurring. The extraction medium is often buffered for this protection. Excesses of heavy metals, such as silver, lead, mercury, etc. also denatures the enzyme molecules, so such materials must be absent from the medium.

Although this is not often a factor during the extraction process, certain types of radiation, such as ultraviolet, cosmic, and gamma radiation also denature the enzyme molecule. This is one reason why such radiation can be used to destroy bacteria and other microorganisms, and is useful for sterilizing the environment.

Enzyme Size

The enzymes, being composed of at least proteins, are very large molecules with high molecular weights. Usually, their molecular weight is greater than 70,000 and values as high as 300,000 are not uncommon. Due to this large size, the enzyme molecule falls within the range of the colloidal particle, so such enzymes are colloids and behave as such. They show cation exchange properties, since they are negatively charged colloids, and impart viscosity to their colloidal systems since they are hydrophillic. Also, their diffusion rates are slow, so travel by diffusion over considerable distance is not accomplished. Some of them do move about in the cell, but such movement does not involve great distances and is often accomplished by some form of protoplasmic streaming which supplements diffusion.

Specificity

Enzymes are specific in their activity. A given protein can act upon only substances having one molecular pattern, such as lipase can only act upon fats, and protease can act only upon proteins. Each protein can also only perform one type of activity. For instance, dehydrogenases can only catalyze reactions whereby hydrogens are removed from the substrate, and oxidases can only catalyze oxidation reactions. Therefore, there is a different enzyme for each type of reaction and for each type of substrate used in the reaction. Since there are thousands of such reactions within a cell, it is no wonder why such a large number of different kinds of enzymes must be present.

Another example of enzyme specificity, one that is interesting and valuable, but sometimes dreaded, is that of participating in immunological reactions. When I first went to UCLA as a graduate student, I was surprised to find the plant physiologists growing rabbits in their laboratory. As I later discovered, these rabbits were used to detect the presence of specific proteins through their immunological reactions. Such proteins, including those of enzymes, when introduced into an animal, cause the animal to form antibodies that are released into the blood stream. If the protein is again injected into the rabbit, a reaction occurs between the protein and the antibody. This reaction is very specific and can be caused only by the protein that incited antibody formation originally. Hayfever is a good example of an immunological reaction, since the protein from the pollen of ragweed or other allergenic species, gets into the victim's blood stream, causing antibody formation, and then when a later exposure occurs, the antibody and protein interact to give the symptoms of hayfever. Immunological tests are very good tests to use to detect and identify proteins and only trace amounts of the protein are required for these tests.

ENZYME NOMENCLATURE

Each of the many kinds of enzymes within the cell has a name. This name is given to it as its function within the cell becomes known, and is derived from two considerations. The enzyme may be named after the substrate it catalyzes, examples of which are cellulase, an enzyme that hydrolyzes cellulose, or sucrase, the enzyme that hydrolyzes sucrose. Also lipase, is an enzyme that catalyzes lipids. As is evident, the name of the enzyme is derived from the stem of the substrate with "-ase" added as a suffix.

Some enzymes are named not for their substrate, but for the type of reaction they catalyze, with examples being dehydrogenase or oxidase. However, more often the enzyme is named both for its substrate and its action. For example, succinic dehydrogenase is an enzyme that acts upon succinic acid and removes hydrogens from the molecule. Ascorbic acid oxidase is an enzyme that catalyzes the oxidation of ascorbic acid. (How would isoenzymes be named?)

ENZYME INHIBITION

Enzymes can be inhibited by certain organic chemicals. Such inhibition may be either competitive or non-competitive. An example of *competitive inhibition* is the action of malonic acid. Malonic acid will compete with succinic acid for position on the enzyme in the Krebs' cycle reaction whereby succinic acid is usually converted to fumaric acid and inhibit this reaction. Such inhibition, as is true of competitive inhibition generally, can be reversed by adding more of the normal substrate. *Non-competitive inhibition* occurs when a stable combination occurs between the enzyme and the molecule acting as the inhibitor. Carbon monoxide, azide, iodoacetate, and fluoride act in this manner, as non-competitive inhibitors. Inhibitors of enzyme activity are often used by the plant physiologist to study enzyme-catalyzed reactions that occur within the cell. Some of these inhibitors are specific — they inhibit only one reaction — but most are general, inhibiting many types of reactions or substrates within the plant cell. Natural protein inhibitors often exist in the living plant cell. One such inhibitor is the powerful trypsin inhibitor found in soybeans. This has been shown to be a protein, but enzyme inhibitors of other chemical compositions are also present in cells. Such natural inhibitors may function during differentiation in the plant.

The physical environment has a great effect on plant enzymes whether these enzymes are in the cell or outside.

Temperature is important in relation to chemical reactions generally, and enzyme-catalyzed reactions are no exception. The Q_{10} of these reactions usually will vary from two to four, but variations also occur from one type of reaction to another. This results in metabolic disturbances being associated with temperature changes, as discussed in Chapter 8.

As previously stated, pH is a very important factor of the environment that may damage enzymes when they are being extracted from the cell. In fact, pH is a factor of importance to enzymes at any time and will affect enzymes in several ways. First, an enzyme has a narrow pH range over which it is most stable. At higher or lower values, the enzyme is denatured. The pH may also alter the substrate, making it more or less usable by the enzyme. This is particularly true when the substrate is an acid or a base, or when it contains acid or basic groups. Enzymes often act only on the acid or base ion or on the undissociated molecule, but not on more than one of these. An example of this is shown in Fig. 11.1, which indicates the inhibition of a weak acid, transcinnamic acid on pollen germination. Note that inhibition occurs at low pH values but decreases as the pH value increases. The difficulty with such studies is to separate the effects of the pH on the enzyme activity and on membrane permeability.

pH

Since the enzyme is composed of numerous amino acid molecules, and since amino acids are amphoteric, proteins are also *amphoteric*. This means that the protein or enzyme molecule contains both acid (carboxyl) and basic (amino) groups. As a result of this, at a certain pH value, the enzyme molecule will exactly have the acid and base groups neutralized so that it will be a neutral molecule electrically. This is the pH value that is known as the *isoelectric point*. Each type of protein has a specific isoelectric point but the pH value at this point will vary from one type of enzyme to another. At this value, the enzyme is most easily denatured. Not far from it, is the optimum pH value, the enzyme is most active so for optimum enzyme activity, the pH value of the cell would need to be maintained at the optimum value of the enzyme. Unfortunately, the enzymes differ from one to another so there is no one pH value that is optimum for all enzymes. Perhaps few or any enzymes are as active in the cell as they could be, due to adverse pH value.

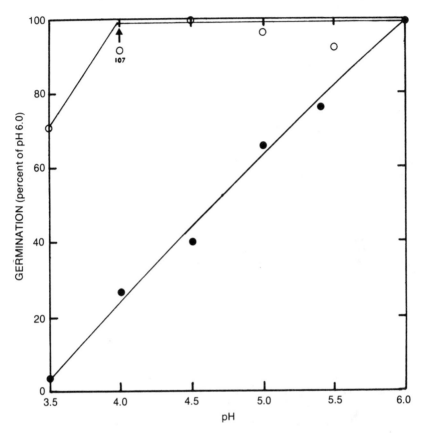

Fig. 11.1   Pollen germination percentage plotted against the pH of the germination medium. The black circles are values obtained in the presence of a weak acid (transcinnamic acid) and the white circles are in its absence. (From Goss, *American J. Botany*, 1968, **55** (1): 73–77.)

Hydration

Another interesting property of enzymes is their hydration requirements. Normally the water content of the cell is high, so the enzymes are fully hydrated. However, as the water content of such tissues as those of mature seeds, spores, pollen grains, etc. dry out, eventually enzyme activity decreases until it is often undetectable in the dry tissues. Then if the seed, etc. is placed in a moist environment, as is the case when the seed is planted, within a few hours enzyme activity again becomes very great. (*See* Fig. 8.9.) There is a tremendous increase in enzyme activity with such hydration. This is assumed to be due to the orientation of the water molecules around and within the enzyme protein molecule in such a way that the protein molecule becomes hydrated and its configuration changed back into that needed for enzyme activity. As indicated earlier, such a change in protein configuration can cause a considerable change in its enzyme activity without any chemical change being made. The magnitude of enzyme activity changes, with hydration, varies from one type of enzyme to another and perhaps the location of the enzyme within the cell may be important. However, more study must be done for a complete understanding of enzyme hydration.

Enzyme Concentration

The rate of an enzyme-catalyzed reaction is also influenced, as might be expected, by the enzyme concentration. The more enzyme molecules at the reaction site, the more rapid the rate of the reaction. Doubling of the reaction rate by doubling the enzyme concentration is observed at most enzyme concentrations.

Substrate and Product Concentration

Since enzymes do not change the equilibrium constant of the reaction, one might expect that changing either the substrate or product concentration of a reaction would alter the rate of the reaction, and indeed they do. Increasing the substrate concentration increases the reaction rate and increasing the product concentration decreases the rate. However, in an enzyme-catalyzed reaction, the rate limiting step, the slowest part of the reaction, is assumed to be the breakdown of the enzyme-substrate complex to release the enzyme plus the product. This explains the observation that one can increase the rate of such a reaction by increasing the substrate concentration only up to a point, beyond which, increases in concentration have no stimulation or inhibition on the rate. At this time, the enzyme is saturated with substrate and the rate can be increased only by adding additional enzyme molecules.

Redox Potential

Enzymes are very much affected by the redox potential of the medium. This is because many of them have reduced sulfur (–SH) groups that must remain in their reduced form for the enzyme to function. Such enzymes differ greatly in the ease with which these groups can be oxidized and therefore the ease with which they can be inactivated. In any case, any chemical that alters this relationship will alter enzyme activity. Chemicals such as iodoacetate, p-chloromercuribenzoic acid and heavy metals such as divalent copper and mercury are known to do this, but in some cases, the purified enzyme is inactive even in distilled water unless a reducing chemical is present.

ENZYME LOCATION WITHIN THE CELL

Enzymes are found in all living cells, and found there very abundantly. Few of them are located within the cell wall or in the intercellular spaces, but most are in the protoplast. Some are generally distributed throughout the protoplast, and some are located at specific sites, such as in the mitochondria or in the nucleus. Some enzymes are found in cells of many species or of many cells within the same individual, whereas others are found in only a few cells or in cells of only a few species. Some are present in the cells of the individual throughout its life, whereas others are present only during certain stages of the development of the individual. More will be said about both the spacial- and time-distribution of enzymes in the chapter on metabolic regulation.

ENZYME LOSS

Plant cells lose enzymes to the external medium. Just how this loss occurs is not definitely known, but perhaps both secretion and leakage are involved. Such enzymes are amylase, peroxidase, acid phosphatase, indoleacetic acid oxidase, invertase, RNAase and DNAase, as well as other large organic chemical molecules have been reported to be lost to the external environment by such plant structures as tissue cultures, pollen grains, and plant roots. Insectivorous plants have been known for many years to secrete hydrolyzing enzymes into their environment and these enzymes digest insects that have been trapped, releasing the amino acids from the insect proteins which are then taken up by the plant to supply or at least supplement its nitrogen requirements. The pitcher plant, as seen in Fig. 11.2, is well noted for this. Its leaves are shaped like a vase and are always partially filled with water. Due to a slick surface on the inside of the leaf, any insect that ventures into the leaf loses its footing

Fig. 11.2 Drawing of a pitcher plant showing five leaves at the base. These leaves are united at their margins to form a cup in which water collects.

and slips into the water. There it drowns, and the enzymes secreted by the leaf go to work on it.

Fungal cells are also well known for the hydrolyzing enzymes they give off to the environment. These enzymes then digest food materials of the substrate and the soluble products are taken up by the fungal cells to supply their food requirements. One interesting problem in plant physiology is how the cell can control the loss of enzymes so that those lost are not predominantly those that must be retained by the cell for its metabolism. What selective mechanism determines which enzymes will leave the cell and which will remain?

TYPES OF ENZYME-CATALYZED REACTIONS

How many types of chemical reactions occur within the living plant cell is not known, but certainly the number must run into the thousands. These reactions may however, for aid in our understanding of them, be grouped together into six main groups, namely, amination-deamination, carboxylation-decarboxylation, hydrolysis-condensation, methylation-demethylation, oxidation-reduction, and phosphorylation-dephosphorylation reactions, as illustrated in Table 11.2. To understand these reactions

Table 11.2   Principal groups of biochemical reactions that occur in plants.

Group	What happens
Amination-deamination	NH_2 added or removed
Carboxylation-decarboxylation	Addition or loss of CO_2
Hydrolysis-condensation	Addition or loss of a water molecule
Methylation-transmethylation-demethylation	Addition or loss of CH_3
Oxidation-reduction	Removal or addition of electrons, usually accompanied by loss or addition of hydrogen and addition or loss of oxygen
Phosphorylation-dephosphorylation	Addition or loss of PO_4

is to understand much of biochemistry. Therefore, let us consider these reactions individually and some of their characteristic enzymes.

Amination-deamination

Amination-deamination reactions are associated with the addition of ammonia to a keto acid, to an amino acid during amide formation, and during the breakdown of amino acids at which time the amino group is released as ammonia, either from amides to form the corresponding amino acid, or from amino acids to form the corresponding organic acid.

A common amination-deamination reaction in plant cells is that by which glutamic acid is deaminated to form α-ketoglutaric acid plus ammonia or vice versa. This is an easily-reversible reaction and accounts both for the entrance of ammonia into glutamic acid and for the loss of ammonia from the amino acids. The enzyme that catalyzes this reaction is glutamic dehydrogenase, which is really not a deaminase. It may seem strange that a dehydrogenase would be catalyzing a deamination reaction and technically it does not. What happens is that the dehydrogenase removes two hydrogens from the glutamic acid molecule (oxidizes it) after which this molecule rapidly and non-enzymatically loses its amino group. This reaction then first involves an oxidation reaction, followed by automatic deamination.

Perhaps the best studied of the amination-deamination reactions is that associated with transamination. As shown in Chapter 9, transamination is the reaction whereby the amino group of an amino acid is exchanged with a keto group of a keto acid, forming a new amino acid and a new keto acid. This is the way essentially all of the amino acids are formed, except for the glutamic dehydrogenase reaction.

Transamination is catalyzed by an enzyme known as *transaminase*.

Actually there is a different transaminase for each amino acid formed from transamination, which means that more than twenty transaminases exist within the plant cell. These transaminases are named by listing the amino acid and then the keto acid formed, and adding transaminase, as for example the enzyme-catalyzing transamination between glutamic acid and aspartic acid is known as glutamic-aspartic transaminase. In most cases, the amino acid used is glutamic acid, and α-ketoglutaric acid is released in the reaction. All transamination reactions are easily reversible.

The transaminase enzyme consists of a protein to which is bound a coenzyme. In all cases, the coenzyme is pyridoxal phosphate, but the protein moiety will differ with each transaminase. The structure of pyridoxal phosphate is given in Fig. 11.3.

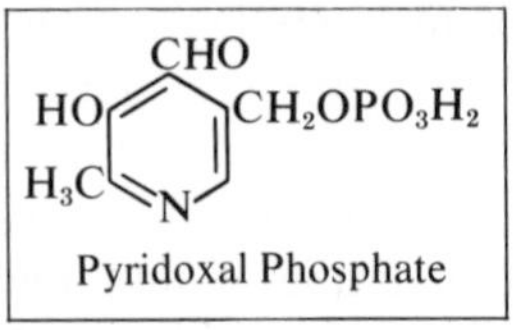

Pyridoxal Phosphate

Fig. 11.3   The structure of pyridoxal phosphate, the coenzyme of transaminases.

Carboxylation-decarboxylation

Carboxylation reactions were discussed in Chapter 6, and some decarboxylation reaction in Chapter 10. These reactions involve either an addition of carbon dioxide to a molecule to form a carboxyl (–COOH) group, or the removal of the carboxyl group to form carbon dioxide respectively. The enzymes involved are known as *carboxylases* or *decarboxylases*. In decarboxylases, the coenzyme is biotin, whose structure is shown in Fig. 11.4. or thiamine pyrophosphate (vitamin B_1). These reactions are reversible enough that they can be used either for carboxylation or decarboxylation reactions.

Biotin

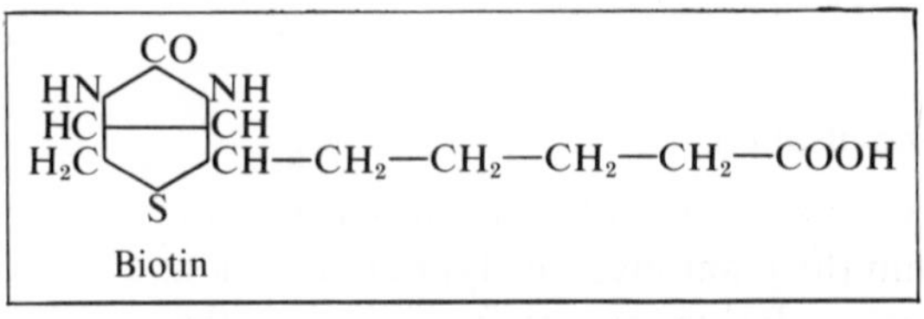

Fig. 11.4   The structure of biotin, a coenzyme of carboxylases.

Biotin is believed to be synthesized from pimelic acid via desthiobiotin. The site of such synthesis within the cell, and many of the intermediate steps of its synthesis, are unknown.

Reactions that are predominantly carboxylation reactions include those of Fig. 6.4. Among these are included some of the most essential reactions of the cell's metabolism, so that a continuation of carboxylation reactions is essential at all times. During fruit ripening, a tremendous increase in the rate of carboxylase activity has been reported, but the reason for this is unknown.

Perhaps the best known carboxylation reaction is that catalyzed by ribulose diphosphate carboxylase, whereby, during the dark reaction of photosynthesis, carbon dioxide is irreversibly added to ribulose diphosphate with the resulting production of two molecules of phosphoglyceric acid. This enzyme contains the soluble fraction I protein, a protein that is widely distributed in the leaves of many plant species. It is found only in the leaf chloroplasts, as a soluble protein of the stroma.

Decarboxylation reactions include those associated with respiration as well as with the decarboxylation of amino acids to form amines. Perhaps the best known is the irreversible reaction involving the decarboxylation of pyruvic acid to form carbon dioxide and acetyl CoA, a reaction which does occur in the mitochondria even though it is not technically part of the Krebs' cycle. The coenzyme for this reaction is not biotin, but is thiamine pyrophosphate, whose structure is given in Fig. 11.5. Thiamine pyrophosphate is synthesized by adding the pyrophosphate ester of 2-methyl-4-amino-5-hydroxymethyl-pyrimidine to the phosphate ester of 4-methyl-5-(B-hydroxyethyl)-thiazole.

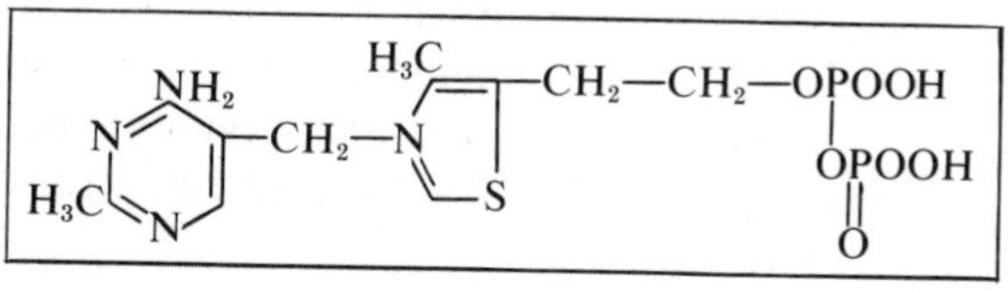

Fig. 11.5   The structure of thiamine pyrophosphate.

Hydrolysis-condensation

The synthesis and degradation of polymers, such as proteins, starch, fats, etc., within the plant involves types of reactions known as *hydrolysis* and *condensation*. Polymers are formed from their building blocks by condensation, and broken down to release these building blocks by hydrolysis. For example, proteins are formed from amino acids by condensation, whereby a water molecule is removed between two amino acid molecules, and the protein is broken down to release these amino acid residues of the protein chain, an action known as hydrolysis. However, certainly condensation reactions, and occasionally hydrolysis reactions, are summary reactions of activities that involve several different enzymes and often phosphorylation-dephosphorylation reactions.

Three groups of *hydrolytic* enzymes that are very widely distributed and that have been well studied, are the proteolytic enzymes that hydrolyze proteins, the amylases that hydrolyze starch, and the lipases that hydrolyze fats. These are all proteins that function with no coenzyme; proteins with a large molecular weight. And yet, they are not enzymes *per se*, but classes of enzymes. For instance proteolytic enzymes include enzymes such as papain, bromelin and ficin, and amylases include α-amylase and β-amylase.

Methylation-demethylation

Methylation-demethylation reactions are common in plant cells. These are reactions by which methyl (CH_3-) groups are added to or released from a molecule. In most cases, it appears as if this involves primarily transmethylation reactions, as discussed in Chapter 9, but certainly a molecule must be methylated before transmethylation can occur. This methylation may come from formic acid although, if so, the formic acid concentration must be low since it is usually undetectable in most plant structures. One exception, of course, is the leaves of stinging nettle, where formic acid is the cause of irritation when this leaf is touched.

Redox

Oxidation-reduction, sometimes referred to as *redox*, reactions involve the transfer of electrons from a chemical that is oxidized to one that is reduced. Thus oxidation reactions and reduction reactions occur simultaneously. The electron transfer usually accompanies hydrogen transfer, whereby two hydrogens are either added to or removed from an organic compound. Accompanying this hydrogen addition or removal is the addition or removal of two electrons, one accompanying each hydrogen.

The enzymes involved in these reactions are the *oxidases*, the *dehydrogenases*, and the *cytochromes*, and are often intimately involved in cell respiration, where they comprise the enzymes of the respiratory chain.

These enzymes all contain a micronutrient as part of their coenzyme molecule, and this micronutrient is usually either iron or copper.

One well-known dehydrogenase, not directly involved in respiration, is *glutamic dehydrogenase*, an enzyme which catalyzes the reaction earlier mentioned which results in the reversible reaction between α-ketoglutaric acid and glutamic acid. This enzyme is very widely distributed in plants, and is perhaps even omnipresent, where it is located in the mitochondria of the cell. In pea plants, at least seven isoenzymes of glutamic dehydrogenase have been found.

The redox enzyme *cytochrome c* has been studied in some detail, both in plants and in animals. Its molecular weight is near 13,000, and its apoenzyme consists of a single polypeptide chain, made up of 104–108 amino acids. To this protein is bound an iron-containing heme molecule, whose structure is seen in Fig. 11.6. Note its similarity to chlorophyll.

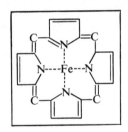

Fig. 11.6   The structure of a heme molecule, the prosthetic group of cytochrome.

Cytochromes are red in color instead of green like chlorophyll, and are found incorporated into the inner mitochondrial membranes instead of in the chloroplasts. They are found in either their oxidized or reduced forms, and these forms differ markedly in their absorption spectra, which allows the form of the enzyme to be easily determined in plant cells. Cytochromes also differ from many enzymes in that they do not use normal organic compounds as substrate, but instead receive the electrons and hydrogen from other enzymes. These other enzymes may be dehydrogenases, flavoproteins, or other cytochromes.

Perhaps the redox enzyme whose activity is most obvious is the enzyme known as *tyrosinase* or *polyphenol oxidase*. This is a copper-containing oxidase that oxidizes phenolic compounds in the presence of oxygen. If an apple fruit is cut in two, the cut surface is white or nearly so. However, a few minutes later, this cut surface turns brown, as seen in Fig. 11.7. This color change is brought about largely by the oxidation

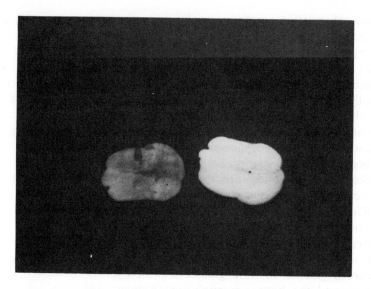

Fig. 11.7 Illustration of tyrosinase activity in cut potato tuber. Tuber half on the left was cut and exposed to the atmosphere for one hour. That on the right was cut and washed in running water for one hour.

of phenolic compounds by this oxidase when the cells are exposed to the higher oxygen content of the atmosphere. (Why is the oxygen content within the intact fruit lower?) Such darkening can be prevented by immediately submerging the cut surface under water. This activity is not limited to apples, but can be observed with many other fruits and vegetables, as well as with other plant tissues.

Other enzymes of the redox type include catalase, cytochrome oxidase, cytochrome *b*, flavoproteins, the pyridine nucleotides, such as NAD and NADP, ubiquinone coenzyme Q, peroxidase, ascorbic acid oxidase, and a large variety of dehydrogenases, such as succinic dehydrogenase, malic dehydrogenase, etc.

Phosphorylation-dephosphorylation

Since all chemical reactions either require or release energy, and since this energy must be supplied in a usable form and often captured to be reused in other reactions, there must be a mechanism by which this can be accomplished. This mechanism often involves the synthesis or transfer of high energy phosphate groups and is of the *phosphorylation-dephosphorylation* type.

Phosphorylation reactions involve, directly or indirectly, ATP as a source of energy. Sometimes this ATP is used directly, as in the phosphorylation of glucose, to give glucose-6-phosphate, but at other times it is used indirectly to synthesize other triphosphate nucleotides which then function in phosphorylation reactions. These nucleotides are shown structurally in Fig. 10.2. Of course, you already know that the ATP is synthesized in plants either in the chloroplasts during photosynthesis, or during respiration primarily in the mitochondria. However, even today, it is not known exactly the mechanism by which the ATP is formed from ADP, although of course energy and inorganic phosphate are required. Other triphosphate nucleotides are formed at the expense of ATP by transferring energy and inorganic phosphate to their nucleosides.

Kinases are enzymes that catalyze the transfer of phosphate from one of the triphosphate nucleotides, such as ATP, to a molecule to phosphorylate it. Therefore, kinases are phosphorylation enzymes. Perhaps the best known is *hexokinase*, the enzyme that transfers the phosphate group to glucose to form glucose-6-phosphate, used during both glycolysis and the hexose-monophosphate-shunt. Kinases also exist, in plants, which can add phosphate in a similar manner to other sugars, such as fructose, mannose, etc.

In sugar metabolism, the intermediates are sometimes nucleotide sugars, such as UDP-glucose, GDP-glucose, ADP-glucose, or TDP-glucose. These are formed by a reaction similar to that given below:

$$\text{UTP} + \text{glucose-6-phosphate} \xrightarrow{\text{pyrophosphorylase}} \text{UDP-glucose} + \text{pyrophosphate}$$

It can be seen that in these reactions, the enzyme involved is a *pyrophosphorylase*. Many such enzymes are known in plants.

Of course, adding a phosphate group to a molecule is only a temporary situation and eventually this phosphate group is lost, usually as inorganic phosphate. Enzymes which catalyze this loss are known as phosphatases, and are no doubt present in all living plant cells. They catalyze the reaction whereby the sugar phosphate is dephosphorylated to form the free sugar molecule and inorganic phosphate. The inorganic phosphate can then be reused in the synthesis of new nucleotide phosphates. An example of a phosphatase reaction is given below:

$$\text{sedoheptulose-1,7-diphosphate} \xrightarrow{\text{phosphatase}} \text{sedoheptulose-1,7-diphosphate}$$

In this chapter, the nature of plant enzymes has been discovered. Since these do control metabolism, this knowledge will be valuable in the next chapter where the control of metabolism is discussed.

Chapter	The Regulation of
Twelve	Metabolism

In many respects, the regulation of plant metabolism is the very essence of plant physiology. Any change in the appearance or function of the plant or of its cells must be preceded by a change in the regulation of its metabolism.

METABOLIC REGULATION DEFINED

What is "metabolic regulation"? As discussed in a previous chapter, metabolism is the synthesis and degradation of organic compounds within the living plant or cell. Therefore, the regulation of metabolism would refer to those processes by which the rate of such syntheses and degradations is controlled. Just as the speed at which an automobile travels is regulated by such factors as the slope of the road, the direction and velocity of the wind, the horsepower rating of the engine, the amount of friction associated with the moving parts, and other factors, the rate at which metabolism occurs within the cell is regulated by many factors too, as seen in Table 12.1. In this chapter, these factors which regulate metabolism will be discussed and it will be shown how they are integrated to give that strange but amazing phenomenon known as life.

Its Continuity

Life is the sum total of all of the metabolic activities found within the living cell. It has no contemporary beginning, since there is no beginning of metabolic regulation within the cell. Instead, metabolic regulation continues during cell division and on into the new cell. It is continuous

Table 12.1   Factors that affect the regulation of metabolism in plant cells.

A. Environmental
   1. Temperature
   2. Light
   3. Chemical composition of the medium

B. Internal
   1. Substrate concentration at the reaction site
      (a) Membrane permeability
      (b) Distribution of substrate within the cell
      (c) Rate of substrate synthesis
   2. Enzyme distribution
   3. Enzyme synthesis
      (a) Constitutive enzymes
      (b) Induced enzymes (enzyme induction and repression)
   4. Enzyme inhibition
      (a) Negative feedback
      (b) End-product inhibition
      (c) Competitive inhibition, by substrate or product

from parent to offspring. Similarly, there is no end to it either, at least as long as the cell remains alive. In fact, some metabolic regulation continues even within the cell for some time after the cell is dead.

Organization

The living plant or one of its cells is a very complex and fascinating machine. A machine is something that can do work, and the living plant cell certainly can do that. To do such work requires both a continuous expenditure of energy and a control mechanism. It is not possible to understand the living plant cell by just conducting a complete chemical analysis of its constituents any more than it is possible to find out how an automobile functions by grinding up its parts and making a chemical analysis of them. Life is much too complex to understand in this manner. The living cell is indeed more than a collection of chemical molecules; rather it is an organization of chemical molecules. This is a tremendous difference, a difference that corresponds to the difference between the living and the dead cell. Such complex organization requires a continuous expenditure of energy, since it works in opposition to the second law of thermodynamics and represents very low entropy, but just as important is the need for a functional control mechanism. Its metabolism must be regulated to maintain this organization.

REGULATION BY TEMPERATURE

The rate of each of the metabolic reactions is controlled by the prevailing temperature, but since the Q_{10} is not the same for each reaction, it is not difficult to see that metabolism is very much regulated by the temperature of the cell. And, as was discussed in Chapter 7, changes in temperature which result in temperature above or below optimum will result in metabolic upsets, since reaction rates of some metabolic activities are altered more than those of others, causing deficiencies or excesses of certain metabolites. Therefore, the metabolic activities in a cell at one temperature are not the same as in the same cell at another temperature. Since temperature variations are common from hour to hour, no doubt the metabolism of the cell is regulated accordingly and therefore also varies from hour to hour. Temperature is definitely a regulator of metabolism, and an important one.

Floral Initiation

One of the results of temperature alterations of metabolic regulation can be seen in the effects of low temperature treatment on flowering. As a result of exposure of the plant to low temperatures, flowers are formed on some species whereas in the absence of such treatment, the plant continues its vegetative development indefinitely. Such temperature treatment may either cause the plant to form flowers, or may increase the number of flowers formed, or both. Causing the plant to form flower buds is known as *floral initiation*. Floral initiation is a result of changes in the metabolism of the plant—changes known as floral induction. In other words, *floral induction* is the sum total of metabolic alterations in the plant that will result eventually in floral initiation. Floral induction may be caused by temperature or by other treatments.

Vernalization

Vernalization is the induction or increase of flowering by exposure to low temperatures. It is characteristic of most biennial, and some perennial, plant species and also of some of the well-known cereal crops, such as Winter wheat. Not all grasses require vernalization for floral induction but this phenomenon has been studied perhaps most intensely in the cereals. In these species, vernalization treatment can be applied to the seeds or to the seedlings. Winter wheat is planted in the Fall of the year and the seeds germinate forming seedlings before Winter sets in.

During the Winter, the seedlings are exposed to low temperature, and these low temperatures regulate the seedling metabolism in some unknown way, which will allow floral initiation to occur the following Spring.

Vernalization treatment can occur in the seeds of cereals, and has been studied most extensively there. However, for vernalization treatment to bring about vernalization, certain conditions must prevail. First, the seeds must take up water before they can be vernalized. As a result of such uptake, the water content of the seeds must exceed about 50% of their dry weight, but these seeds need not have maximum water content, and prior-germination is not necessary. Second, vernalization cannot begin for 10–24 hours after the increased water content is reached. This can be interpreted to mean that the increased water content hydrates certain enzymes necessary for the formation of required metabolites. Third, vernalization is an aerobic process with oxygen being required. In the absence of an adequate oxygen content of the atmosphere, vernalization does not occur, irrespective of the length of time of exposure to low temperatures.

Vernalization itself requires both exposure to low temperatures and time of exposure. It does not appear to be a result of an instantaneous change in metabolism upon exposure to favorable temperature, but is a result of such exposure over an extended period of time. The temperature required for vernalization is low but above the point at which frost injury occurs. With cereals, the optimum temperature is about 1–6°C, but the range of temperatures that will cause vernalization is larger.

The duration of exposure to low temperatures for vernalization is rather long, having a magnitude of 1–3 months. Usually, the longer the period of exposure, the more flowers that will eventually form on the plants that develop as seen in Fig. 12.1.

Vernalization can be stopped by either gently drying the seeds or by exposing them to mild temperature of about 15°C. It is common to vernalize cereal seeds, dry the seeds, and store them for subsequent planting. Such drying stops vernalization but does not delete it. In fact, the seeds can be allowed to take up water, exposed to a limited duration of low temperature treatment, dried, allowed to again take up water, and exposure to low temperature continued. The initial stage of vernalization is not erased by drying, so the second exposure need only be of such duration as to bring the total exposure to the same duration as would have been required if the treatment had been continuous.

Using cereal seeds, studies have been carried out to determine what part of the seed must be exposed to the low temperature treatment for vernalization to occur. For such studies, seed parts are excised and cul-

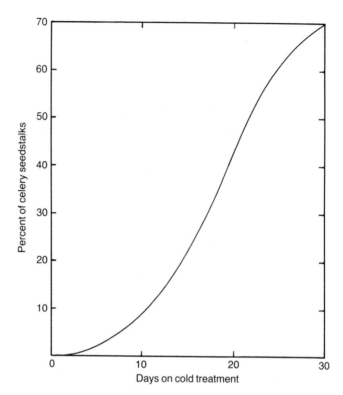

Fig. 12.1   Illustration of the effects of cold temperature treatment on the extent of flowering.

tured in a flask with subsequent low temperature treatment. When this is done, it is found that the embryo itself must receive the treatment. However, such treatments are without effect unless carbohydrates are available to the embryo during vernalization. Why carbohydrates are required is not known.

Of course, cereals are not the only plants that require vernalization for flowering. Most biennial plants and some perennial plants do also. However, there is a difference among species in their requirements for vernalization. In some, the seeds can be vernalized, whereas with other species the seedling or mature plant must be so treated. In these the root must be intact, but vernalizing of only the root will not induce flowering. Perhaps the root is required as a source of carbohydrates, since carbohydrates are known to be stored there. Actually, the part of these plants that must be cooled is the apical bud. This bud will eventually form the flowering stock and the flowers. Evidence that this apical bud is the site of vernalization comes not only from *in vitro* studies, where only the apical bud is so treated, but also from grafting studies where a vernalized apical bud is grafted onto a non-vernalized plant, causing the non-vernalized plant to flower.

An interesting aspect of vernalization is observed in those plants where a scion from an annual plant can be grafted onto a rootstock of the biennial plant of the same species, causing the biennial plant to flower without vernalization. This indicates that, as a result of vernalization, some chemical is produced (a hormone?) which is required for flowering, whereas in the annual plant this chemical is normally produced in adequate amounts, without vernalization, for flowering. What this hormone is, is not known.

Attempts have been made to determine what changes occur in the metabolic regulation of the plant during vernalization, but to little avail. Such studies have revealed that during the low temperature treatment there is an increased hydrolysis of carbohydrates, storage proteins, and lipids; a reduction in the auxin concentration; changes in the pH value of cell sap; a modification of enzyme activities; a change in vitamin B and vitamin C contents; and the isoelectric points of proteins become more acid with no change in their seriological properties, but such changes occur in plants that do not require vernalization too, when they are exposed to low temperature. Therefore, the changes listed above indicate that low temperatures alter the regulation of metabolism, but do not reveal which of the metabolic regulations are necessary for vernalization. When vernalization is complete, and the plants returned to warmer temperatures, the chemistry of both vernalized and non-vernalized plants appears to be identical. If vernalization does induce the formation or accumulation of a hormone, such must be present in concentrations that are so low as to be undetectable, or the hormone is so unstable that it cannot be detected by present methods of analysis. That RNA synthesis is essential during vernalization implicates this as a possibility.

Another indication of the effects of low temperature on metabolic regulation is the fact that some plants do not require vernalization for flower induction but form more flowers per plant if previously exposed to low temperatures, and often the number of flowers formed increases as the duration of exposure to low temperatures increases.

In many plant species, the need for vernalization can be replaced by the application of gibberellin to the plant. These gibberellins represent a group of plant hormones that have some very astonishing effects on certain plant activities – especially they stimulate growth. At the time this book was written, twenty-nine different gibberellins were known to exist in plants. At the time you are reading this, probably even more have been discovered. These molecules vary among themselves primarily by having either nineteen or twenty carbon atoms per molecule, and in the presence or absence of hydroxyl groups in the molecule. The present method of nomenclature names each type of gibberellin with an A, plus a characteristic number as a subscript. For instance, there is a gibberellin A_1, a gibberellin A_2, etc. up to and including a gibberellin A_{29}. One or another form of gibberellins is found in all higher plants and in many lower plants.

The plant synthesizes the gibberellins from acetyl CoA through mevalonic acid, by synthesis of a diterpenoid, named geranylgeranyl pyrophosphate, which forms kaurene and this forms gibberellin. There are some intermediate steps in this pathway, that are not mentioned, but this will give some idea of the major intermediates along the biogenetic pathway.

In the plant, the gibberellins are synthesized primarily in the young leaves of the apical bud, the root tips, and in developing seeds. From the site of synthesis they are transported throughout the plant through both the phloem and the xylem tissues.

By what mechanism the gibberellins can replace the vernalization requirement is unknown.

Bud Dormancy

Another phenomenon that is, in many respects, similar to vernalization and one that further implicates the alteration of metabolic regulation by low temperatures is the breaking of bud dormancy by exposure to low temperatures. Both the vegetative and the flowering buds of perennial plants become dormant the Summer or Fall prior to the time they open, a dormancy known as *Summer* or *Winter dormancy*. It now appears that such dormancy may be due to an increase in the ratio of the hormones abscissic acid to gibberellin. As long as this ratio is high, the buds stop growing and are said to be dormant. They will remain in this condition until the ratio is altered enough to allow growth to continue, at which time the buds are said to have broken bud dormancy.

Abscissic acid, abbreviated ABA, is a plant hormone that inhibits growth, rather than stimulates it. Chemically it is a 15-carbon organic acid, soluble in many organic solvents. As is true of other plant hormones, it occurs in plants in very low concentrations but even minute amounts can produce astonishing results. Among the physiological activities attributed to it include the prevention of seed germination while the seeds are still in the fruit, inhibition of flowering of long-day plants kept under short-day conditions, counteraction of gibberellin and auxin stimulation, induction of senescence, inhibition of shoot growth, delay of anthesis, induction of dormancies, and the stimulation of abscission.

Abscissic acid is very widely distributed in plants and may even be ubiquitous. It can be synthesized by tissues of all ages, and can be inactivated and metabolized in the same tissues. It is readily translocated from one part of the plant to another. It functions by counteracting the action of stimulating hormones so that an interaction occurs in plant tissues between the growth stimulators and growth inhibitors. The results of this interaction are expressed as growth or growth inhibition.

Abscissic acid is then a growth inhibitor. However, there is evidence that other growth inhibitors also exist in plant species and the next few years should reveal the nature of some of these.

Bud dormancy is important to the plant since the meristematic region is especially sensitive to low temperature injury when it is active in growth, but resistant to such injury when growth has ceased. Therefore, it is a means by which the buds can survive the low temperatures of Winter. However, bud dormancy is not present on tropical species. If it were, there would be no perennial tropical plants, because bud dormancy is broken, in the field, only by exposure to low temperatures of Winter, and temperatures low enough to break bud dormancy do not exist in the tropics. This is also the reason some species of fruit trees cannot be grown in warmer climates. For instance, when I was living in Southern California, I witnessed several Summers when peach trees failed to form flowers, fruits, or even many leaves because the Winter had not been cold enough for a long enough period of time to break the bud dormancy.

Yes, bud dormancy is broken naturally by exposure of the bud to low temperatures, but each species is specific as to how low the temperature must be and how long the buds must be exposed to it. For instance, pear trees require a temperature of 7°C for 1200 hours, whereas peach tree buds break dormancy at a similar temperature when exposed about 1000 hours. Also, each bud of the tree acts as an isolated system from the other buds on the same tree, and it is possible to break the dormancy of one bud, or a few buds, on a plant while the other buds still remain dormant. Therefore, if the temperature is a little too high or the duration is too short, only a few buds will be found to have broken dormancy next Spring.

Since bud dormancy appears to be caused by a high ratio of abscissic acid to gibberellin, bud dormancy can only be broken by treatments which alter this ratio favorably. A few years ago, I transferred some sumac shrubs to the greenhouse in the Fall. To the apical buds of some, I applied a few drops of gibberellic acid solution while others were left untreated. Those to which the hormone had been applied broke bud dormancy and formed leaves, even though it was still Winter, whereas the untreated remained dormant. This indicates, as one might expect, that the application of gibberellin will break dormancy by lowering the ratio of abscissic acid to gibberellin in the buds.

Bud dormancy can also be broken by dipping the buds in warm water or by the application of ethylene chlorohydrin to the buds. Low temperature, and these other treatments, alter the regulation of metabolism to restore the favorable hormone concentrations needed for subsequent growth and development of the buds.

Seed Dormancy

Another example of metabolic regulation by low temperatures is the breaking of seed dormancy. Some seeds are dormant because within their embryo or endosperm an inhibitor or deficiency exists which will not allow germination to occur. Such seeds are rendered germinable by storage in moist soil or peat moss for 2–3 months at a temperature near 4°C. In the field, such dormancy is broken naturally during the low temperatures of Winter as long as the temperatures that prevail are low enough and prevail for a long enough period of time. If either the temperature is not low enough or does not remain low enough for the required period of time, the dormancy will not be broken so the seed will not germinate. The value of this dormancy to species survival should be evident. Dormancy of this type is common in species such as apple, peach, juniper, linden, dogwood, iris, ash, rose, and pines, as well as many others. The mechanism of this dormancy is not known and perhaps may vary from one species to another. However, in some cases abscissic acid, is present and inhibits germination. What metabolic changes are brought on by the low temperature exposure that results in the breaking of this dormancy is not known. However, the low temperature treatment effects can be reversed by   her oxygen deficiency, excess water, or exposure to high temperatur This sounds very much like the condition associated with vernalization and bud dormancy.

The above examples should be adequate to show the importance of temperature in the regulation of metabolism within the plant, although many more examples could be sited. However, temperature is not the only environmental factor important in this respect, as seen in Table 12.1. Solar radiation is another important environmental factor which regulates metabolism.

REGULATION BY SOLAR RADIATION

If an absorption spectrum is determined for any organic chemical found within the plant cell, using wavelengths in the range from low ultraviolet to high infrared, it can be shown that many of these chemicals have an absorption spectrum; that each absorbs radiation at some wavelength within this range. Each absorbing molecule is either destroyed by this radiation it absorbs, which in itself may alter metabolism, (How?), or is activated by it. Such activation will result either in a loss of the energy by fluorescence or other means, which has no direct effect on metabolic regulation, or the energy is retained and increases the activation energy of the molecule allowing it to participate in chemical reaction in which it could not previously be active. One example of such alteration of metabolism by increasing the activation energy of the molecule would be the absorption of light by chlorophyll. This results in photosynthesis which changes the amount and kinds of chemicals present within the cell and therefore changes cell metabolism.

Another example of the consequences of the regulation of cell metabolism by light is the differences which exist between etiolated and light-grown seedlings, as seen in Fig. 7.3. Of course, this is the result of a number of photochemical activities, not just one.

Other examples of the effects of light exposure on metabolic regulation are those associated with the phytochrome pigment, and include such phenomena as opening of the hypocotyl hook, spore discharge by fungi, seed germination, chlorophyll synthesis, leaf expansion, Winter bud dormancy, and photoperiodic floral induction in plants. Phytochrome has been found to be attached to the cell membranes and may function in enzyme induction or repression (to be studied later).

Then also, there are phototropisms and the high-light intensity effects on plants. Surely, these are adequate to indicate the importance of light in regulating metabolism in plants and their cells.

REGULATION BY ENDOGENOUS CHEMICALS

The chemical composition of the medium in which the cell is bathed is also a very important factor relative to metabolic regulation. However, this cannot be separated from characteristics of the permeability of the membranes of the cell because the important factor is not the chemical composition of the medium outside the cell, but the chemicals available for metabolic reaction within the cell. Just because a certain chemical has a given concentration outside the cell does not imply that it has the same composition at the site of metabolism within the cell. Although an inhibitor, for instance, may be present in the external medium, it will have no effect on metabolism if the membranes are impermeable to it. In fact, rarely would these parameters be the same. Characteristics of membrane permeability will be considered in detail in Chapter 14, but remember that it is the concentration of the chemical at the site of its metabolism that is most important for the regulation of metabolic activity.

REGULATION BY SUBSTRATE CONCENTRATION

No doubt, the concentration of the substrate, at the site of the reaction, is a very important consideration in respect to metabolic regulation. However, not only can one not determine what this concentration will be by measuring its concentration in the external medium, but even determining its concentration within the cell may lead to erroneous conclusions because the cell is compartmentalized so that the substrate may be at one location within the cell, but its enzymes may be located elsewhere. This presents a big problem that must be evaluated when studying cell metabolism. How does one determine the actual concentration of the reactant at the site of its reaction? Nevertheless, the concentration of the reactant at this site is very important in metabolic regulation, and any condition that controls this will regulate metabolism.

The concentration of the substrate molecules at the site of reaction is controlled by its concentration in the external medium, the permeability of the plasma membrane to it, its rate of movement within the cell from the site of uptake to the site of utilization, which would include permeability of the organelle membranes involved, the rate at which it may be synthesized within the cell, and the rate at which it would be used in the reaction. Any of these factors would regulate metabolism by altering substrate concentration. However, once the substrate molecule has arrived at the site of its utilization, the rate at which it is utilized will depend also upon other factors.

EQUILIBRIA

Thermodynamic Equilibrium

Metabolism within the plant cell involves a large number of chemical reactions. If any one of these is carried out outside of the cell, such as in a test tube or flask, and sufficient time is allowed for the reaction to continue until the concentration of the products and the concentration of the reactants become constant, the reaction is said to have reached *thermodynamic equilibrium*. Under similar conditions, the concentrations of the reactants and products will always give the same ratio after equilibrium has been reached, with the concentration of the products being determined by the concentration of the reactants. This is true irrespective of the presence or absence of an enzyme or other catalyst. In fact, enzymes do not alter this equilibrium constant. However, within the living plant cell, the thermodynamic equilibrium is rarely, if ever, reached. To understand why, consider the process of thermodynamic equilibrium and some of the factors which control it. In the hypothetical reaction $2A + 4B \longrightarrow 2C + 4D$, the thermodynamic equilibrium constant is derived from the concentrations of the products and the reactants after the reaction is allowed to continue until there is no change in either the concentrations of the products or of the reactants. At such time, the reaction is said to be at thermodynamic equilibrium, and the equilibrium constant is then equal to the ratio of the concentrations, as follows:

$$\frac{(C)^2 + (D)^4}{(A)^2 + (B)^4}$$

Steady-state Equilibrium

However, such equilibrium will not be reached if more reactant continues to be added, or if one or more of the products is continually removed as soon as formed. Such conditions would result instead in a condition known as *steady-state equilibrium*, where the concentrations of the products are often very low, and the equilibrium constant differs from that of thermodynamic equilibrium, for the same reaction. Such steady-state equilibrium is the situation associated with the chemical reactions in the living plant cell. The reasons for this are that all chemical reactions within the cell are directly or indirectly associated, so that the product of one reaction is often the reactant for another. This offers a continuous source of reactant. This also removes the product as formed. In addition, some products are volatile, such as the essential oil, and some form insoluble compounds, such as calcium oxalate or sporopollenin or cellulose, and some are quickly transported away from the reaction site. Therefore, the concept of the thermodynamic equilibrium constant has little practical value in understanding the living condition within the plant cell. However, such conditions may be different in a cell extract, and this should be considered when interpreting such studies and when trying to transfer such interpretations into the living cell.

REGULATION BY INTERNAL CONTROL SYSTEMS

In addition to the above mentioned factors which regulate metabolism, other types of internal control systems exist. Such internal control systems must do two things. First, they must be able to comprehend the present condition that exists in the cell, and compare this with the desired condition, and second, they must be able to alter the existing condition to make it become more like that required. In addition, there must either be alternate mechanisms of control or the regulatory mechanism must be very resistant to undesirable alterations. In plant cells, alternate mechanisms usually exist.

Negative Feedback

The cell must be able to produce as much as is needed of any one chemical, but little in excess, since excess is wasteful to the cell and may result in intolerable toxicity conditions developing. Therefore, there must be a mechanism that will turn off the machinery that synthesizes a certain amino acid, for instance, when the concentration of that amino acid reaches a defined value. One such mechanism for this type of regulation is known as *negative feedback*. Negative feedback mechanism is a very common mechanism of metabolic regulation in plant cells. Its mechanism, illustrated in Fig. 12.2, is as follows: When an excessive

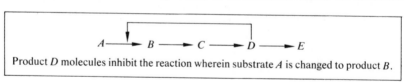

Product D molecules inhibit the reaction wherein substrate A is changed to product B.

Fig. 12.2   Illustration of the mechanism of negative feedback.

amount of a given chemical (D) is synthesized, this chemical — the product of the reaction or series of reactions — goes back and either inhibits one of the enzymes in the reaction sequence or inhibits the production of one of the required enzymes. In any case, the result is the same, namely, that the reaction by which the product is synthesized is slowed down resulting in less product being produced. As this happens, and some of the product is used up on other reactions (E), the product concentration decreases, and now the product is no longer inhibitory to the enzyme or to the enzyme's synthesis, so the reaction rate increases, increasing the amount of product produced and accumulated. Of course, eventually the product concentration will increase so much that negative feedback will once again reduce the rate of product synthesis, and so forth. Indeed, this is a very effective way to quickly control the concentration of a chemical substance within the living cell and one that is very commonly a cause of metabolic regulation.

Negative feedback is a simple control mechanism, whereby the substance actually controlling the reaction is transferred from the site of production to the site of utilization. Indeed, this is much like what happens when a thermostat controls the temperature of the room in a house. Here the heat from the furnace moves to the thermostat, increasing the temperature and thereby causing a break in the circuit which turns off the furnace. With the furnace off, the temperature decreases until it gets to a sufficiently low value that the circuit is again closed, turning the furnace on.

The rate of a chemical reaction within the living cell is also controlled by the concentration of its enzymes at the reaction site. This concentration is a result of enzyme synthesis, enzyme degradation, and enzyme distribution.

REGULATION BY ENZYME CONCENTRATION

Enzyme synthesis is dependent upon the synthesis of the protein constituent of the enzyme. However, before going into the study of protein synthesis, a word of caution is in order. The mechanism to be discussed has been studied only in cell-free extracts, not in the living cell. Such studies indicate that protein synthesis may be the same in the cells of plants, animals, and microorganisms, but are such conclusions justified? Perhaps not.

Amino Acid Activation

Many factors are involved in the control of protein synthesis. To begin with, *amino acid* activation must occur. The needed amino acids must be synthesized and be present at the site of amino acid activation. Also, since their activation involves specific enzymes, these too must be present, as must ATP as an energy source. It has also been reported that gibberellins and red light greatly increase amino acid activation, whereas far red light is inhibitory, implicating the phytochrome system in this activation.

Aminoacyl *t*-RNA

The next step in protein synthesis, is the attachment of the amino acid molecule to its specific *t*-RNA molecule, to form the aminoacyl *t*-RNA molecule. Therefore, the activated amino acids and their specific *t*-RNA molecules must be available at the site of this reaction. Also, both $MgSO_4$ and ATP are required. When such conditions are met, the aminoacyl *t*-RNA molecules are formed, and can then be transported to the site of protein synthesis, namely, the polyribosome. Note that both the amino acid activation and aminoacyl *t*-RNA formation require ATP and therefore occur only when metabolic energy is available.

The subsequent polymerization of the amino acid molecules at the polyribosome is dependent upon the appropriate number and kinds of aminoacyl *t*-RNA molecules. A plant cell protein cannot be synthesized, no matter how many aminoacyl *t*-RNA molecules are present, if all of these molecules are alike. One does not find a protein made of just alanine molecules. It takes the presence of aminoacyl *t*-RNA molecules of all of the types of amino acids that will be in the finished protein to be present at the polyribosome before the synthesis of the protein can be completed. No doubt, this is the explanation for the often-reported observation in man whereby protein synthesis does not occur in the absence of one type of amino acid even if the other amino acids are in excess. Such an observation is not common in plants however, since plants have the ability to interconvert amino acids, using those in excess to synthesize those that are deficient. Nevertheless, if such transformations are not able to occur, protein synthesis will not occur. Just how important this proper balance is to protein synthesis in plants has not been determined.

Energy Requirements

The polymerization of the amino acids at the site of protein synthesis requires not only the *r*-RNA, the *m*-RNA, and the aminoacyl *t*-RNA molecules, but also a source of energy, since this too is an energy-requiring reaction. The source of energy must be GTP, which is synthesized at the expense of ATP. Manganese is the activator for this reaction. The reaction by which the protein is synthesized by amino acid polymerization and its subsequent release from the polyribosome can be inhibited by cycloheximide, which is a good inhibitor of protein synthesis in plant cells and is often used for this purpose when protein synthesis is being studied.

It should also be noted, that all three of the reactions so far discussed, that are necessary for protein synthesis, require a source of energy. This source is usually derived from either respiration or photosynthesis, and indicates the need for these reactions to occur during protein synthesis. Only in the presence of either respiration or photosynthesis, does significant protein synthesis occur.

DNA

Since protein synthesis requires three types of RNA, and since this RNA is synthesized by DNA, it follows that DNA must be present for protein and enzyme synthesis. One apparent explanation for the regulation of metabolism may be changes in the amount of DNA in each cell of a plant. If the amount of DNA changes, then one could explain metabolic regulation as being due to the loss or gain of DNA, which would result in more or fewer enzymes being synthesized. Indeed, there is a difference in the amount of DNA per cell when plants of different species are compared. Also, there are differences in the amount of cellular DNA within an individual multicellular plant. However, these differences are correlated with either the genome number of chromosomes in the cell or with the state of chromosomal replication. As Swift showed in 1950, haploid cells have only one half as much DNA as do diploid cells within the same individual. And diploid cells can be shown to have twice as much DNA as one might expect, if the DNA content is determined just prior to nuclear division at a time when the chromosomes have been replicated. However, in a cell that is not undergoing chromosomal division, the amount of DNA is usually constant during the life of the cell. Therefore, it is not possible to explain metabolic regulation or even the change in the kinds of enzymes within a cell on the basis of changes in the DNA content of that cell.

Since the amount of DNA does not change in any cell during its life span, barring preparation for nuclear division, then perhaps there is a change in the structure of the DNA molecules which results in different enzymes being formed. This does not seem to be the case, since it is possible to extract DNA from a plant at different stages of development and treat it in such a way that it behaves similarly each time.

Associated with the DNA molecules are the genes, so well known for their function in heredity. However, it is not known whether one DNA molecule comprises a gene, or whether several genes are part of one DNA molecule. The latter seems most likely, since the number of genes in a cell is about 200 times the number of chromosomes. Nevertheless, the assumption has been made, and has some good support, that one gene is responsible for the synthesis of one polypeptide chain. Therefore, for each type of polypeptide chain formed, a different gene must exist.

Since m-RNA serves as a template for protein synthesis and thereby

determines the order in which the specific amino acids occur in the protein, and since m-RNA is synthesized by DNA, it stands to reason that
the kind of protein produced depends upon the activity of the DNA
molecule which determines the kind of m-RNA produced.

Messenger RNA is produced at the surface of the DNA molecule.
This is thought to occur by the DNA double-helix partially unwinding
and a base-pairing occurring between the one strand of the DNA and the
nucleotides of the m-RNA molecule that is being formed. Thus, the
position of the bases in the DNA molecule will determine the position
of the bases in the m-RNA molecule and the codon of the m-RNA, and
therefore the sequence of amino acids in the protein molecule being
formed.

The m-RNA is the most short-lived of the RNA molecules. Much of
that formed, but not all, appears to be broken down before it leaves the
nucleus. Actinomycin D is a good inhibitor of RNA synthesis, and it
functions by binding with DNA to prevent the base-pairing between the
DNA and the RNA being formed. Studies with it show that protein
synthesis may continue in the plant cell for several hours after RNA
synthesis is inhibited, indicating some longevity associated with at least
some kinds of RNA molecules. In any respect, for a continual synthesis
of proteins, and enzymes, a continuous synthesis of m-RNA is necessary.
Actually, there is some question as to whether or not m-RNA has ever
been isolated from plant tissues, due to its instability and also since there
are no good criteria for identifying it once it has been isolated.

Hormones

Protein synthesis is also regulated by plant hormones, which do,
in many cases, alter the synthesis and/or function of RNA. These
hormones are growth regulators—chemicals which when present in small
amounts alter the physiology of the plant—that are produced by the
plant, and include the auxins, gibberellins, cytokinins, abscissic acid, and
ethylene, as well as some others.

Auxin has been known for many years, to stimulate that phase of
growth known as cell enlargement. It does this by increasing the plasticity of the cell wall. Such softening of the wall is inhibited by inhibiting
either RNA synthesis or protein synthesis. Therefore, the action of auxin
in cell wall softening is dependent upon the stimulation of RNA synthesis
by the auxin. Indeed, it is known that auxins do cause a stimulation of
RNA synthesis with a concomitant increase in RNA concentration.

Since auxins may function by derepressing a gene, this may explain how they stimulate RNA synthesis. That this increase is not directly related to cell enlargement is evident since increasing the OP of the external solution stops cell enlargement, but not auxin-stimulated RNA increase. This RNA increase is evident with all types of RNA and is not limited to any one. However, the presence of auxin brings about a large increase in polyribosome number.

Auxins have been reported to also increase DNA synthesis. Since they do not stimulate protein degradation, this DNA stimulation plus RNA stimulation result in an accumulation of both the nucleic acids and protein. This may result in a decrease in growth, growth represented both by cell division and cell enlargement. This is due, in part, to the lack of uniformity in the DNA, RNA, and protein stimulation, a lack of uniformity that results in increased RNA:DNA and RNA:protein ratios. Ethylene production is also increased by the presence of auxin.

Contrary to what is said above about auxin effects on intact tissues, when plant tissues or organs are excised, there is a significant decrease in their RNA contents. This loss is aggravated by auxins, with low concentrations of auxin stimulating RNA degradation, and high concentrations of auxin inhibiting RNA synthesis. The former is probably due to a stimulation of ribonuclease activity. Why these results differ from those with intact tissues is not known. Apparently, even the act of excision results in changes in metabolic regulation.

In senescent tissues, auxins also stimulate RNA and protein degradation, while cytokinins and gibberellins increase it. This may be due to their effects on nuclease and protease activities.

Gibberellin causes an increase in both the amount of DNA and the amount and kinds of RNA in plants. This is due both to a stimulation of synthesis and an inhibition of the activities of both nucleases and proteases. It appears to act on the gene level, controlling gene activity by removing the histone repressor from certain genes. Its action on the formation of α-amylase in germinating barley seeds has been well documented. In their aleurone layer, certain hydrolases — especially α-amylase — are formed. This formation is known to be actually initiated, not just stimulated, by gibberellin. The entire α-amylase molecule is synthesized in response to gibberellin, and some evidence indicates that protease is also so induced.

Only two *cytokinins* have been identified in plant tissues, although evidence exists that many more are present. Zeatin, from corn, and dihydrozeatin are the two known to date. The cytokinins appear to be constituents of the *t*-RNA molecule. However, all *t*-RNA molecules do not contain cytokinins, but only about one out of every twenty do. Other types of RNA molecules also lack cytokinins.

The cytokinins also increase the synthesis, and therefore the accumulation, of RNA including that of the nucleolus. In this capacity, they are known to slow down the decrease in RNA content that is normally associated with abscission and senescence of plant structures. Whether this is due entirely to their action on stimulating RNA synthesis or their known action on depressing both nuclease and protease activity is not well known. Other activities may be involved.

Ethylene is a hormone whose synthesis is stimulated by auxin, so it is difficult to differentiate between auxin and ethylene activity. It increases both RNA and protein synthesis in senescent tissues, but not in younger tissues. Perhaps this is one reason why it enhances abscission, since abscission is dependent upon both RNA and protein synthesis. However, this is contrary to the action of auxin, which delays abscission. There is some evidence that the action of ethylene on abscission is due largely to its stimulation of cellulase activity.

Abscissic acid is known to inhibit DNA synthesis and therefore to inhibit both RNA and protein synthesis. Such inhibition results in growth inhibition. RNA and protein syntheses are essential for cell enlargement, and their inhibition also inhibits enlargement. However, a considerable amount of RNA synthesis can be inhibited without inhibiting cell enlargement. Perhaps the effect is largely on *m*-RNA rather than RNA generally.

Abscissic acid also will inhibit the induction of α-amylase formation by gibberellin. This effect is due to the action of abscissic acid in inhibiting a specific RNA needed for the synthesis of the amylase.

In summary, hormones often function by stimulating or depressing RNA synthesis. The induced synthesis of hydrolases by gibberellins, cell enlargement by auxin, and ethylene stimulation of abscission, are associated with increased RNA synthesis not only with total increase in RNA, but often with the types of RNA that accumulate. How such changes are brought about by the hormones is not known but one function of these hormones is associated with the removal of the histone repressor of the DNA, to be discussed below. They may function in other capacities too. They do not appear to serve as prosthetic groups for enzymes.

Gene Repression

One good mechanism for the control of metabolism and therefore for metabolic regulation is that of the repression of gene activity. Studies with isolated chromatin material from plant cells reveal that not all of the DNA of a cell is functioning at any one time. In fact, it appears that only about 5% of the cell DNA is active at a given time. The remainder is *repressed*, or prevented from functioning in RNA synthesis. The mechanism of this repression is not well understood, and its elucidation will be one of the great discoveries of biology, but it appears that there may be certain proteins, the *histones*, which bind to the DNA of the gene, preventing it from functioning in RNA synthesis. Therefore, these repressors control metabolism by determining what polypeptides, and subsequently, what proteins and enzymes will be formed. A very important study for the future is to find out what controls repression. What determines which genes will be repressed — turned off — and which will not? When this mechanism is discovered, this discovery will give us a much better understanding of cell and plant differentiation. More will be said about this subject in future chapters, but RNA seems to be involved. Of course, this leads to the question, what controls the action of the RNA in this capacity. Yes, much remains to be learned.

Induced Enzymes

Enzyme *induction* is the act of removing the repressor from a gene so this gene can produce an enzyme. As stated above, some hormones may function in this capacity. However, it also appears that other molecules may also regulate metabolism in this manner. The plant cell has the potential to produce many more enzymes than are found in it at any one time. These enzymes can be produced by a cell but normally are not, so are referred to as *induced enzymes*, in contrast to the normal cell enzymes, the *constitutive enzymes*. Often the induced enzymes are formed only when the substrate for which they are needed is present. Why should a plant cell contain all of the enzymes of which it is capable of synthesizing, when many of these are not currently needed? The ability to form induced enzymes is a great asset to the cell, since these can be formed only when they are needed, and yet they allow the cell much greater latitude in adjusting to changes in the cell's environment. For instance, if a plant cell is cultured on a medium which furnishes the carbohydrate only in the form of sucrose and then a transfer is made, removing the sucrose and supplying the sugar as glucose, the cell may not continue to grow for some time since it does not contain the enzymes for glucose utilization. However, after a time interval, the glucose induces the synthesis of the needed enzymes, and the cell resumes its growth on the glucose medium. This phenomenon is very common in plant cells.

Since metabolic regulation is dependent upon the enzymes that are produced, and since enzyme production has been shown to be influenced by many factors, there is no wonder that enzyme synthesis is a very important factor in metabolic regulation. That the rate of a chemical reaction is greatly influenced by the enzyme concentration is well known, with the rate increasing almost linearly with enzyme concentration.

Proenzymes

Another factor to consider in enzyme concentration control is the subject of proenzymes. A number of *proenzymes* are known to exist in plant cells. These are precursors of enzymes but do not have any enzyme activity in themselves until they are converted into the respective enzyme. Thus, an enzyme extract may increase in activity after its extraction, due to the conversion of proenzymes into enzymes during the extraction process, complicating the interpretation of *in vitro* enzyme studies.

Inhibitors

Enzyme inhibitors are also important regulators of metabolism. Some of these are non-specific, inhibiting a large variety of enzymes. Such inhibitors would include pH values, *p*-chloromercuribenzoate, which inhibits the SH– site of many enzymes, and factors which denature proteins. There appears to be many chemical molecules within the cell which also are non-specific inhibitors so their presence and concentrations, at the site of reaction, would definitely be factors which regulate metabolism.

In addition to these non-specific inhibitors, a large variety of specific enzyme inhibitors are known — inhibitors which inhibit only one type of enzyme or only one enzyme activity. These may be *competitive inhibitors*, whereby the inhibitor molecule is structurally similar to the substrate, or they may function by *product enzyme inhibition*, which is a very common type of inhibition in plant cells and in which the product inhibits the enzyme that catalyzed its synthesis. An example of the former, is the inhibition of the oxidation of succinic acid by malonic acid. This inhibitor is specific for this reaction, although it is not the only inhibitor of this reaction.

And so, in conclusion, many mechanisms are known that will regulate the metabolism of the plant and its cells. Such regulation is the very essence of life, and is responsible for the chemical composition of the plant as well. In the next chapter, the chemical composition of the plant will be studied, at which time, more will be revealed about the intricacies of metabolic regulation.

Chapter Thirteen

Phytochemistry

The subject of *phytochemistry* can be broadly defined as the study of the chemistry of the plant and its cells. Indeed, it includes, and may appear to be identical with, plant biochemistry. However, phytochemistry is more concerned with what chemicals are present within the plant and what their concentrations are, whereas plant biochemistry is more concerned with how these chemicals are synthesized. However, both subjects do broadly overlap.

Phytochemical Diversity

The subject of phytochemistry is studied by scientists of diverse interests. Physiologists are interested in plant chemicals primarily because of their potential functions within the cell; biochemists for the enzymes that catalyze their synthesis and breakdown; pharmacologists for the potential of these chemicals to serve as drugs and medicines, the alkaloids being particularly well known in this category as are the halucinogenic drugs; nutritionists because these chemicals form the food for man and other animals; industrialists because of the potential of plant chemicals to serve as raw materials in industry, with plant oils falling into this category; taxonomists, since the distribution of specific chemicals among plant species is an indication of the interrelationships of these species; and agriculturists because of their interest in increasing the production of economically-important plant chemicals.

Due to these diverse interests in the subject, a diversity of approaches is possible. However, being a study of plant physiology, this chapter shall be primarily concerned, but not exclusively so, with the functions of the chemicals within the plant, and shall be outlined in this manner.

WHY CHEMICALS ACCUMULATE

The presence and concentration of a chemical within the plant cell is the result of metabolic regulation, as discussed in the last chapter. In addition, if this chemical is found in high concentration, one of two things is indicated. Either the chemical is not broken down within the cell, as is the case with calcium oxalate which forms microscopically-visible crystals within the leaf and root cells of certain plant species, as seen in Fig. 13.1, or the cellulose, lignin, or sporopollenin of plant cells, or else it means that the chemical exists in metabolic pools.

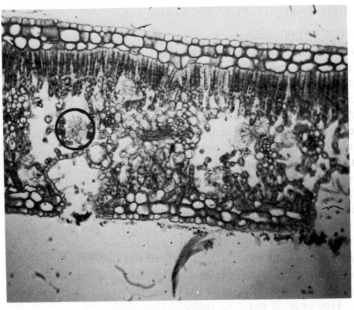

Fig. 13.1 Crystals are often found in the older leaves of plants. Such a crystal is shown below, within the ring, in the photograph of a cross-section of a leaf section of *Nerium oleander*.

METABOLIC POOLS

A *metabolic pool* would be a location within the plant cell where a chemical accumulates. Such accumulation usually occurs because the molecule is separated from the enzymes responsible for its degradation. This is possible because the mechanism of degradation is usually different from that of synthesis, and involves different enzymes. If the chemical molecule remains isolated from its enzymes of degradation, it can accumulate. This is often possible if the molecule accumulates in cell organelles that do not contain its enzymes of degradation. For instance, if a chemical produced in the cytoplasm moves into the vacuole, it can accumulate there since there are few, if any, enzymes in the vacuole. Here the chemical can be stored, and in high concentration, safe from degradation. The vacuole is the largest metabolic pool in the cell, but other cell organelles can also function in this capacity.

SIGNIFICANCE OF PLANT CHEMICALS

About 70,000 different kinds of chemicals have been identified in plants, and no doubt many more exist. How many different kinds are present in any one cell at a given time is also not known, but surely there must be more than three thousand. Some of these may be found in all species, and therefore are considered to be metabolically essential. These would include the proteins, lipids, and nucleic acids. Others are much more limited in their distribution, with some, such as ricinine from *Ricinus communis*, being found only in one plant species. Those of more limited distribution are not generally considered to be metabolically essential but increased knowledge about their function within the cell could very well cause an alteration of this conclusion.

In this respect, the question could be asked, what is the significance of plant chemicals as far as the plant is concerned? No doubt some must be present for the survival of the plant, but is this true of all of them? This is indeed a good and interesting question, but one that will require much more study before it can be answered.

Most of the chemicals in the cell can be identified with certain functions. However, it must be realized that most of these chemicals have more than one function, and in most cases some or all of these functions are unknown to us. Functionally, though, it is possible to say that chemicals within the cell serve primarily as enzymatic, storage (reserve), structural, transitory, or secondary plant products (attractant or protective chemicals), and this is the order in which they shall be discussed in the following pages, with those in the latter two categories being considered as secondary plant products.

Enzymatic

Enzymatic chemicals are constituents of the cell enzymes, and include many of the proteins as well as the prosthetic groups of enzymes. However, since these have been discussed in previous chapters, they will not be further considered here.

Storage

Storage chemicals are of a diverse nature. Some are proteins, and these storage proteins were discussed in Chapter 10. Some are carbohydrates, such as sugars and starch, some are lipids, such as the fats, some are RNA, such as the storage RNA of seeds, etc. Stored chemicals can be defined as those that are present in large amounts, but can be reutilized.

STORAGE CARBOHYDRATES

Storage carbohydrates comprise the greatest amount of stored materials in plants, and the ultimate source of these is the phosphorylated sugars produced during photosynthesis. The sites of carbohydrate storage are found largely in the parenchyma cells of the pith or cortex of the plant, especially in storage organs such as tubers, bulbs, roots, and stems, although some storage may occur in most cells, but being least prevalent in the meristematic and phloem sieve tube cells.

Carbohydrates may be stored as sugars — sugars that are derived directly from phosphorylated sugars by hydrolysis of the phosphate group.

Sugars

The simplest carbohydrates are the simple sugars or the *monosaccharides*. These are the smallest sugar molecules, comprising less than ten carbon atoms per molecule, and are also the structures from which all other carbohydrates are formed. They serve a number of functions in the plant including the source of carbon for most other organic compounds within the plant. Since they are produced as a result of photosynthesis, they contain a great deal of energy in each molecule and serve as sources of energy for the living processes of the plant. Since they are small molecules, and at times present in the vacuole in rather high concentrations, they affect the osmotic pressure and therefore the water relations of the plant. The sweetness of ripe fruit is due to the presence of sugars, largely monosaccharides in the fruit. So, monosaccharides serve many functions within the plant.

Many types of monosaccharides have been found in plant tissues, but only two are present in the free form to any great extent in plants, namely, *glucose* and *fructose*, while the other monosaccharides are found only joined with other monosaccharide molecules to form more complex carbohydrates. Glucose and fructose both have the same chemical formula ($C_6H_{12}O_6$) and contain six carbon atoms per molecule. Glucose and fructose are often found as free molecules, dissolved in both the cytoplasm and vacuole of perhaps all plant cells. They also serve as building blocks for larger carbohydrate molecules. Their greatest concentrations occur in ripe fruits, where, as free molecules, they may comprise as much as 80–90% of the dry weight of the fruit juices. For instance, in cherry fruit juice, over 90% of the sugar is comprised of free glucose and free fructose. Fructose is often present in fruits in higher concentrations than glucose, as for example in grapes, but in plant leaves, glucose is usually present in the greatest amount with sucrose making up about 0.2% and fructose about 0.1% of the fresh weight. This is much less than their concentrations in ripe fruit.

Other monosaccharides are also present in plant cells, but rarely as the free molecule. They serve as building blocks for more complex carbohydrates. This group includes the six carbon galactose and mannose which have the same chemical formula as glucose and fructose (Fig. 13.2).

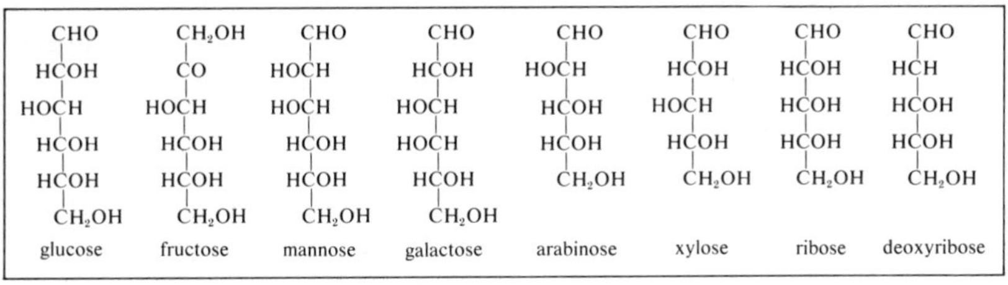

glucose	fructose	mannose	galactose	arabinose	xylose	ribose	deoxyribose
CHO	CH₂OH	CHO	CHO	CHO	CHO	CHO	CHO
HCOH	CO	HOCH	HCOH	HOCH	HCOH	HCOH	HCH
HOCH	HOCH	HOCH	HOCH	HCOH	HOCH	HCOH	HCOH
HCOH	HCOH	HCOH	HOCH	HCOH	HCOH	HCOH	HCOH
HCOH	HCOH	HCOH	HCOH	CH₂OH	CH₂OH	CH₂OH	CH₂OH
CH₂OH	CH₂OH	CH₂OH	CH₂OH				

Fig. 13.2 Structural formulae of the most common monosaccharides.

Two monosaccharide molecules can join together by a water molecule being pulled out between them, to form a *disaccharide*. A disaccharide is a sugar molecule made up of two monosaccharide molecules. The most common representative being *sucrose* – ordinary table sugar. Sucrose has the chemical formula ($C_{12}H_{22}O_{11}$). Note how this formula differs from that of two monosaccharides by the absence of two hydrogens and one oxygen, the formula for water. Sucrose is by far the most common of the disaccharides, and is composed of one glucose and one fructose molecule. It is probably present in every plant cell as the free sucrose molecule dissolved in the water of the vacuolar sap or in the cytoplasm of the cell. It is particularly abundant in the stalk of sugar cane and in the root of sugar beet, both of which contain about 28% of their dry weight as sucrose. Sucrose is also present in ripe fruit, but at a much lower concentration than that of glucose and fructose. In leaves, it comprises 0.05–1.0% of the fresh weight, or about twice the concentration of glucose and 5 times that of fructose. Sucrose functions as a very important storage form of carbohydrates in some species, especially in sugar beet, grasses including sugar cane, and in members of the *Liliaceae* family of plants. It also is the form in which carbohydrates are translocated from one part of the plant to another, for example from the leaves to the stems.

Three monosaccharide molecules can also join together to form a *trisaccharide*. Glucose, fructose, and galactose are so united to form the trisaccharide, *raffinose*. Raffinose is a common sugar in the sap of the phloem cells, particularly in trees. In fact, in elm and ash trees, it is as abundant in the exudate of sieve tubes as is sucrose even though sucrose is the common form of carbohydrate translocation. Raffinose occurs almost as widely in plant tissues as does sucrose, although it is usually present in much lower concentration. Perhaps it functions as a storage form of carbohydrate and may be even an alternate form of carbohydrate translocation.

A *tetrasaccharide* is formed when four monosaccharide molecules join together, with the usual removal of a water molecule between them. The most common tetrasaccharide is *stachyose*, formed from two galactose, one glucose, and one fructose molecule. Stachyose, too, is often found in the phloem exudate of trees but in even much lower concentrations than raffinose. It is also found in many plant leaves in very low concentrations. It is especially common in the phloem exudate of elm and ash, but present in the phloem of other woody species also. It is found in many plant tissues, but is particularly abundant in the seeds of soybean (*Glycine max*) and in the rhizomes of *Stachys sieboldii*. Perhaps the tetrasaccharides function as do the trisaccharides, but they have been studied less intensively than have the mono- and disaccharides.

Many monosaccharide molecules can join together to form a larger molecule. When the number of monosaccharides in the molecule exceeds ten, the molecule is referred to as a *polysaccharide*, meaning a molecule of many monosaccharides. Two polysaccharides are particularly important as storage forms of carbohydrates and will be discussed now. These are starch and inulin. Other polysaccharides are closely associated with cell wall structure, and will be considered in Chapter 17.

Starch

Actually, *starch* should not be referred to as a molecule, but rather as a mixture of molecules. It is, in reality, a mixture of two types of polysaccharides, namely, *amylose* and *amylopectin*. These two, in varying proportions, comprise starch. Both are formed from glucose.

The amylose molecule is formed from about 300 glucose molecules joined together to form a long, unbranched chain. It is soluble in water and therefore is easily dissolved. The amylopectin molecule is formed from 1000 or more glucose molecules, but instead of forming a long, straight chain, the glucose molecules form a branched molecule. This seems to influence water solubility, since amylopectin is not soluble in water. Mixed together, the amylose and amylopectin form starch.

Although the ratio of amylose to amylopectin in starch is often one, meaning that both are present in equal amount, exceptions are rather common, as can be seen in Table 13.1.

Table 13.1   Percent amylose in starch.

Species	% amylose	% amylopectin
Banana	21	79
Wheat	24	76
Sorghum seed	23	77
Waxy sorghum seed	0	100
Field corn	22	78
Waxy corn	0	100
Potato tuber	22	78
Potato leaf	18	82
Potato sprouts	46	54

Starch is the most abundant storage carbohydrate in plants being found in most cells at one time or another. In the cell it is located in the chloroplasts, where it is formed from the sugars during photosynthesis only to disappear when photosynthesis stops, and in plastids called *amyloplasts.* The starch forms grains of various sizes, many so large that they can be seen easily under the microscope. These starch grains appear to have a pattern of concentric circles on their surfaces, and this pattern is often specific for each species, making it possible to identify the species of plant by studying the starch grains it produces. Starch is particularly abundant in roots, tuber, bulbs, stems, and twigs.

Starch is the most universal form of storage carbohydrate, but it is absent in some species, particularly some monocots such as *Iris, Hyacin-*

thus, and *Glaanthus nivalis*. In seeds, starch is particularly abundant and may comprise over 70% of the dry weight of some seeds. Some fruits also contain a great deal of starch, with much more normally being present in green fruit than in ripe fruit because starch is converted to sugar as fruit ripens, adding to the sweet taste of the fruit.

Inulin

Another polysaccharide that serves as a storage form of carbohydrates in some plant species in inulin. *Inulin* is composed largely of fructose molecules, instead of glucose, with 25–28 fructose molecules making up the inulin molecule, although a small amount of glucose is also present. It is soluble in hot water, and can be extracted from plants by suspending the plant tissues in hot water. Within the plant it is normally found dissolved in the vacuolar sap, or as crystals. Inulin is seldom found in the tops of plants but is usually present in the roots, being particularly abundant in certain members of the *Compositae* and *Campanulaceae* families, especially in jerusalem artichoke (*Helianthus tuberosus*), *Dahlia variabilis*, chickory, dandelion (*Taraxacum* sp.), salsify, and goldenrod. It is also found in a few species of the *Liliaceae* and *Amaryllidaceae* families. Both inulin and starch may be found in the same plant, with inulin being most abundant in the roots, and starch in the shoots. Other polysaccharides, composed of fructose molecules (*fructosans*) have also been found in certain plant species.

Carbohydrate Synthesis

The synthesis of carbohydrates is mediated through the phosphorylated sugars. During photosynthesis, 3-phosphoglyceric acid is formed and is subsequently reduced to 3-phosphoglyceraldehyde by the energy carriers formed during the light reaction of photosynthesis, as seen in Fig. 6.7 and Fig. 13.3. Whenever 3-phosphoglyceraldehyde is present, some of its molecules will be converted to the 3-carbon dihydroxyacetone phosphate, after which one molecule of dihydroxyacetone phosphate and one molecule of 3-phosphoglyceraldehyde will combine to form one molecule of the 6-carbon phosphorylated sugar, fructose-1,6-diphosphate. This molecule then loses one phosphate group to form fructose-6-phosphate which is converted to glucose-6-phosphate, and the glucose-6-phosphate transfers its phosphate group to the number 1 position to form glucose-1-phosphate. These phosphorylated sugars are used for the synthesis of sugar, starch, and inulin.

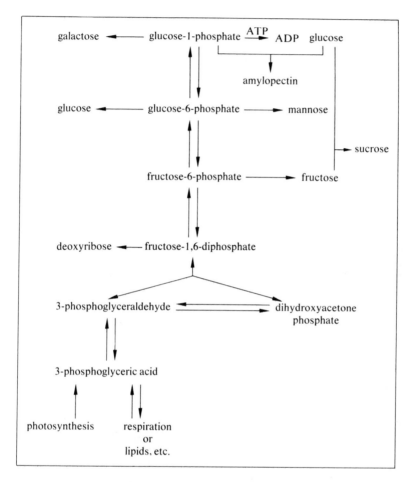

Fig. 13.3 The synthesis of some carbohydrates.

FAT STORAGE

Although starch, sugars, and inulin represent forms of carbon and energy storage in some tissues and in some very common seeds, such as in seeds of wheat and corn, there is another storage form that is, in many respects, better, and that is fats. Fats are lipids. *Lipids* are those substances that are soluble in fat solvents, solvents such as ether, chloroform, and benzene, but are not soluble in water, and therefore are not dissolved in the vacuolar sap.

Distribution

Fats occur in all plant cells where they may be found either as small droplets suspended in the ground substance of the cell, or its organelles, or they may be found in the cell sphaerosomes. These sphaerosomes are small spherical structures that often are very numerous in the cell cytoplasm. They appear to be limited by a single membrane, and to consist almost entirely of fats. As one looks at a living cell through the microscope, the sphaerosomes can be seen in constant motion, moving with protoplasmic streaming or by Brownian movement. They are smaller than the mitochondria and plastids, but are large enough to be seen with the high power objective of the light microscope.

Function

Fats represent a better form of energy storage than do the carbohydrates, because much more energy is stored in a given volume of fats than in a similar volume of starch. One gram of fats contains over 9 kcal of energy whereas 1 gram of carbohydrates will contain only 4 kcal. Therefore, for the same amount of energy stored, less than half the volume is needed for fats as for starch. Perhaps this is one reason why small seeds more often store fats than carbohydrates.

Seed and Fruit Fats

Although all plant cells contain fats, the highest concentrations are found in the cells of seeds or fruits, as seen in Table 13.2, whereas other plant tissues contain less than 5% fats on a dry weight basis. In seeds, an inverse relationship exists between the fat and starch content. Those

Table 13.2   The fat content of some seeds, given as percent of the dry weight of the seed.

Species	Fat content
Pecan and brazil nuts	70
Coconut	65
Castor bean	60
Sunflower	50
Soybean	20
Corn	5
Wheat, barley, rice	2

seeds high in fats are low in starch content, whereas those high in starch are low in fat content. Seeds must store carbon and energy for the initial stages of seedling development and these are stored either as starch or fats. However, this does not imply that seeds contain either starch or fats but not both. They do contain both. It is impossible to find seeds that do not contain some fats.

The fats of seeds are very important commercially, where they are referred to as vegetable oils. Soybeans and peanuts supply much oil for commerce, followed by the seeds of cotton, sunflower, rape, flax, coconut, sesame, palm, olive, and castor bean. A great deal of corn seed oil is also available commercially even though corn seeds do not accumulate fats to the extent that most seeds do. The vegetable oils of commerce are used mostly as food by man, although some are used for the manufacture of soaps, surfactants, and drying oils, and for other purposes.

Bark Fats

Fats are also stored in the bark of trees to be utilized during the growing season. Thus, the fat content of tree bark fluctuates with the season, being low in the Spring and early Summer.

Fat Structure

Fats are *triglycerides* of fatty acids. In other words, each fat molecule is made up of one glycerol molecule to which three fatty acid molecules are attached to the OH groups of the glycerol molecule, as shown in Fig. 13.4.

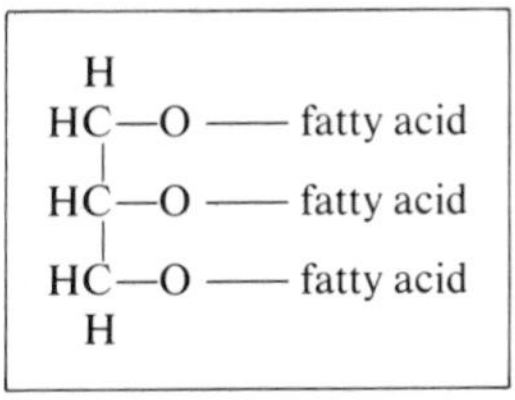

```
    H
   HC—O ——— fatty acid
   |
   HC—O ——— fatty acid
   |
   HC—O ——— fatty acid
    H
```

Fig. 13.4   The structure of a fat molecule.

Fatty acids are long chain molecules of carbon and hydrogen with a carboxyl group on one end. Each molecule contains even numbers of carbon atoms but the number present varies with the type of fatty acid. Only in the COOH group is oxygen present, so fatty acids are definitely oxygen deficient. The fatty acids are further classified by the presence or absence of double bonds in the carbon chain. If no double bonds are present, the fatty acid is said to be saturated. If one or more double bonds is present, the fatty acid is said to be unsaturated. The unsaturated fatty acids are more abundant in plants than are the saturated fatty acids The most abundant fatty acids in plants are listed in Table 13.3. Note that stearic, oleic, linoleic, and linolenic acids each have the same number of

Table 13.3   Common fatty acids in plants.

Fatty acid	Chemical formula	No. of double bonds
Saturated:		
palmitic	$C_{16}H_{32}O_2$	0
stearic	$C_{18}H_{36}O_2$	0
Unsaturated:		
oleic	$C_{18}H_{34}O_2$	1
linoleic	$C_{18}H_{32}O_2$	2
linolenic	$C_{18}H_{30}O_2$	3

carbon atoms per molecule. They differ only in the number of double bonds. Note also, for each double bond, two hydrogens are lost from the molecule. The reason for this is that a double bond is formed by removing two hydrogens from the molecule and a double bond is lost by adding two hydrogens per molecule (Fig. 13.5).

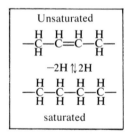

Fig. 13.5   Illustration of how a molecule can become saturated or unsaturated.

Fatty acids are synthesized from the acetyl groups of acetyl CoA, formed during the breakdown of pyruvic acid in the mitochondria, as seen in Fig. 13.6. Since acetyl is a 2-carbon group, this explains why plant

acetyl CoA + CO_2 + ATP $\longrightarrow$ malonyl CoA + ADP + Pi

8 malonyl CoA + 14 $NADPH_2$ $\longrightarrow$ palmityl CoA + 8 CO_2 + 7 CoA + 14 NADP + 7 H_2O

Fig. 13.6   Fatty acid synthesis.

lipids contain only even numbers of carbon atoms per molecule. The fatty acid molecule is synthesized first by acetyl CoA joining with carbon dioxide to form malonyl CoA. Malonyl CoA's will join, giving off CO_2, and this will continue until a molecule of the desired length is formed, with CoA on the end. Then the CoA will be hydrolyzed off, leaving the fatty acid, or the fatty acyl CoA will participate directly in the synthesis of fats. Note that the synthesis of fatty acids requires a great deal of energy and this energy is stored in the molecules making fats energy rich. Eight molecules of ATP are required to synthesize one of palmitic acid.

The glycerol molecule of a fat, is formed from dihydroxyacetone phosphate formed during glycolysis, as shown in Fig. 13.7. This too is an energy-requiring process.

dihydroxyacetone phosphate $+ NADPH_2 \longrightarrow$ glycerophosphate $+ NADP$

glycerophosphate $+ H_2O \longrightarrow$ glycerol

Fig. 13.7   Glycerol synthesis.

The final synthesis of the fat involves three fatty acetyl CoA molecules combining with glycerol, with three molecules of CoA being released to again function in the synthesis of more fatty acids.

In plant fats, most of the fatty acids are unsaturated, consisting mostly of oleic and linoleic acids. This contrasts with animal fats, where most of the fatty acids are saturated.

The fat molecule may contain all three fatty acid molecules of the same type, but this is unusual and each such molecule usually contains two or three different fatty acids. Also, vegetable oils are not composed of only one type of fat molecule, but are a mixture of different kinds of fats. These fats differ both in the types of fatty acids that comprise the individual molecule, and in the sequence of the fatty acids within the molecule.

Fats are normally synthesized from carbohydrates, and when broken down can either be used directly in respiration, or, by a reversal of glycolysis, can be used to form carbohydrates. The first step in such breakdown involves the enzyme lipase, which hydrolyzes the fat molecule, releasing the glycerol and the fatty acids. The fatty acid is then broken down to form acetyl CoA molecules, which enter the glyoxylate cycle, as shown in Fig. 13.8.

This completes the discussion of those chemicals, in the plant cell, that are most important storage forms. However, others, such as the

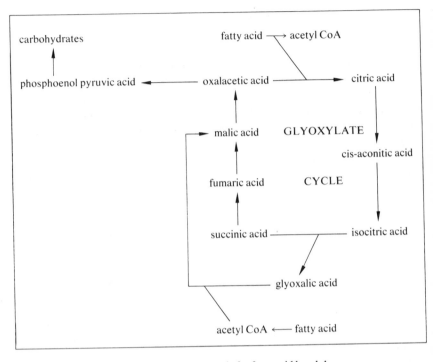

Fig. 13.8   The glyoxylate cycle for fatty acid breakdown.

glycosides, can be reutilized and therefore may be considered as storage materials, but since they also appear to have other functions, they will be considered later. The chemicals to be discussed in the remainder of this chapter are not primarily enzymatic, structural, or storage in function, and are therefore known as *secondary plant products.*

SECONDARY PLANT PRODUCTS – WAXES

Among the secondary plant products are a number of chemicals that are protective in function, although they may have other functions too. Not the least of these is the waxes.

Although rare reports of the occurrence of plant waxes in the cytoplasm are available, by far the majority of waxes are found on the surface of plant organs such as stems, fruits, flower petals, and leaves. These waxes coat the surface of these organs that is exposed to the atmosphere, and are located on the outer surface of the cuticle. This wax is what is polished when the proverbial apple is polished for the teacher. It gives the shiny appearance to the apple fruit, after the fruit surface has been rubbed.

Functions

The wax probably functions in several capacities. First, it serves to preserve the water content of the plant, since water cannot easily pass through it even in the vapor state. Second, wax may reduce mechanical injury to plant tissues such as the leaves as they are moved about by the wind. Third, waxes may reduce fungal and insect attack. Fourth, due to their ability to scatter light and to absorb light, waxes may help to protect the plant from excessive radiation, particularly from injurious ultraviolet radiation. Fifth, the wax alters the wetability of the leaf and thereby the efficiency of agricultural sprays, which may be advantageous or disadvantageous. These functions may be due either to the physical arrangement of the wax or to its chemical composition.

Composition

Chemically, the wax is composed of a mixture of many chemical compounds, compounds that are soluble in fat solvents, such as ether, but not soluble in water. Waxes are composed of long chain alkanes, alcohols, esters, ketones, aldehydes, acetals, and acids, that may consist of over twenty carbon atoms per chain. Many of these wax molecules are hydrocarbons, and all are composed only, or largely of, carbon and hydrogen. Attempts have been made to study the chemical composition of the individual waxes to learn if their chemical composition is related to taxonomic classification, i.e., do the waxes of all species of a given genus have the same composition? However, the composition varies with so many factors that most such studies have not given the results hoped for.

The amount of wax produced seems to be related to the species; to the light intensity, with less wax produced in the dark; to climate; and to the age of the plant, with the wax content increasing with age.

The waxes must be synthesized in the protoplasm of the cells but just where in the protoplasm is unknown. They penetrate through holes in the cuticle, probably dissolved in a volatile, fat solvent. The volatile solvent then evaporates leaving the wax behind on the surface of the cuticle.

Waxes once formed cannot be reutilized by the plant and therefore do not represent reserve food materials. They remain on the plant structure until the part abscises and falls to the soil, following which these waxes may be slowly broken down by soil organisms. However, they are resistant to decomposition, and plant waxes can even be found in fossils, such as coal and shale, where they have remained without decomposing for eons of time.

Synthesis

The pathways of wax synthesis are not known at the present time. At least some of the molecules seem to be synthesized from acetyl CoA, much as fatty acids are. Much more work needs to be done to understand the metabolism and the synthesis of waxes by plants, but they are important components of most higher plant species.

ESSENTIAL OILS

One very interesting mixture of chemicals produced by some plants are the *essential oils*. An essential oil is a mixture of terpenoids, in which are dissolved hydrocarbons, such as the paraffins, organic sulfur compounds, amines, and indole and anthranilic acid esters. These are found in some 2000 of the 400,000 species or about $\frac{1}{2}$% of the plant species, and are especially prevalent in members of the *Pinaceae, Umbellifireae, Myrtaceae, Lauraceae, Labiateae*, and *Compositae* families.

The essential oils are volatile oils and are soluble in fat solvents. They contribute to the aromas that are associated with many plant species.

Essential oils are produced only in specialized gland cells, including the glandular hairs of stems or leaves, and therefore are not produced in all cells even of those species which contain essential oils. In these cells, they can be seen as globules suspended in the cytoplasm, and from there the oils are secreted into ducts, which may be found in the bark, wood, or leaf tissues of trees, or they may be secreted to the external surface of the cells, as is true in flowers, leaves, and in some fruits.

As far as studies have revealed, the essential oils are not essential for plants. These are used commercially as flavors, perfumes, and solvents, and serve some aesthetic values associated with certain species. A conifer forest can be detected by the "pine" fragrance of the volatilized essential oils.

The essential oil terpenoids are the mono-, sesqui-, and di-terpenes, the latter which also includes vitamin A and the phytol of the chlorophyll molecule. They are synthesized from mevalonic acid, and include such substances as menthol, pinene, and farnesol, as shown in Fig. 13.9.

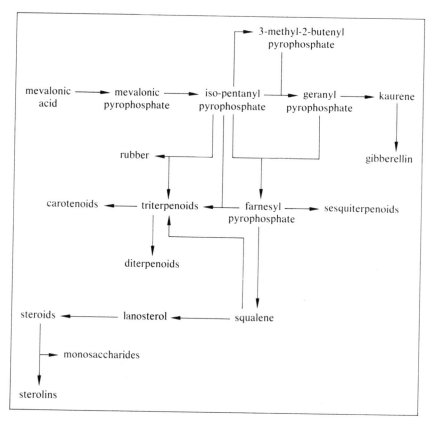

Fig. 13.9   A pathway for terpenoid synthesis.

GLYCOSIDES

Within most plant species, is a group of organic compounds called *glycosides*. These are made up of one or more sugar molecules plus another organic compound. The sugar is usually a monosaccharide but the other molecule (the aglycone) can be one of many non-sugar molecules. The sugar molecule is attached in such a way that it can be easily removed by hydrolysis, yielding one or more sugar molecules and one aglycone per glycoside molecule. These glycosides are further subdivided into glucosides, fructosides, rhamnosides, etc. depending upon what the sugar component of the molecule is.

Functions

The functions of glycosides are not well understood. However, it appears that they may serve several functions. They may, in times of dire need, release the sugar molecule to be used by the plant for energy or carbon sources. They may serve as a means of detoxication because many of the aglycones are toxic to the plant in the free form, but when a sugar is attached to the molecule it is no longer toxic. Perhaps the aglycones are produced as a result of necessary metabolic activities of the plant and some means must be available to prevent their injury to the plant.

The glycosides are present largely in the water of the vacuole, away from the essential metabolic activities of the cytoplasm. The sugar molecules make the aglycone water-soluble, so that it can move around in the plant more readily. If the sugar was not attached, the aglycone could not dissolve in the vacuolar sap.

Some glycosides protect the plant from infection. Plant disease organisms are often discouraged from invading, or otherwise damaging, a plant because of the presence of certain glycosides in the plant. Although these glycosides are not injurious to the plant, they are injurious or distasteful to certain insects, etc. Perhaps this is one reason why some species are resistant to certain diseases.

Types

There are many groups of glycosides, if grouped according to the nature of the aglycone. The most common are the flavonoids, cyanogenic glycosides, mustard oil or thioglycosides, the saponins, tannins, and even certain proteins, lipids, and the nucleosides. The latter three groups were discussed in other chapters so will not be considered here although technically they are glycosides, but they function in ways different from other glycosides.

CYANOGENIC GLYCOSIDES

The cyanogenic glycosides are important not only because of their functions within the plant as glycosides, but also because upon hydrolysis they yield a sugar and *hydrogen cyanide*. Hydrogen cyanide is poisonous to animals, including humans. In fact, amygdalin, a cyanogenic glycoside from the seed of bitter almond, peach, and other *Prunus* sp., has been used in the past as a poisonous drink for the execution of criminals. Upon hydrolysis it yields hydrogen cyanide, benzaldehyde, and glucose.

Distribution

Actually, the cyanogenic glycosides are rather widely distributed in the plant kingdom, being found in at least 13% of the families of *Tracheophyta*, and in some of the common species, such as sorghum, yams, lima beans, peaches, Johnson grass, sudan grass, wild plum, and chokecherry. They may be found in the seeds or in the leaves, as in the wilted leaves of the cherry, peach, or plum, and the concentration of HCN released when the cells are ruptured may be very high. For example, sorghum may release 1000 to 2000 ppm HCN.

Concentration

The amount of the cyanogenic glycoside actually present will vary with the conditions under which the plant is growing as well as with the species of plant and with the specific organ on the plant. Any condition of the environment which causes a reduced growth of the plant is conducive to high cyanogenic glycoside accumulation. In the Fall of the year, when the temperature is cool, shoots of sorghum, Johnson grass, etc. may contain so much cyanogenic glycoside that cattle feeding on the leaves may die from HCN poisoning. Drought or excessively high temperatures can result in similarly high concentrations. On the other hand, optimum growth of the plant will result in low cyanogenic glycoside accumulation even in those species which have notably high concentrations, and such species, under these conditions, may form good forage food for livestock.

Location

The cyanogenic glycosides are water-soluble and probably accumulate largely in the vacuole of the cell. The aglycone is synthesized from amino acids and is released from the sugar by hydrolysis. Examples of common cyanogenic glycosides, their aglycones, sugar constituents and the plant in which they are found are given in Table 13.4.

Table 13.4   Common cyanogenic glycosides – their structure and distribution.

Glycoside	Aglycone	Sugar	Species
Amygdalin	benzaldehyde	glucose	*Amygdalus nana*
Dhurrin	p-hydroxybenzaldehyde	glucose	sorghum
Prunasin	benzaldehyde	glucose	*Prunus* sp.
Linamarin	acetone	glucose	*Linum* sp.
Lotaustralin	methylethylketone	glucose	*Lotus* sp.

MUSTARD OIL GLYCOSIDES

The mustard oil glycosides, or *thioglycosides* as they are sometimes called, release upon hydrolysis a sugar plus an aglycone that contains sulfur. These glycosides are especially common in members of the *Cruciferae* family but are occasionally found in species of other families. Some of the common thioglycosides are given in Table 13.5.

Table 13.5   Some thioglycosides – their structure and distribution.

Glycoside	Aglycone	Sugar	Species
Glucoraphenin	sulforaphene	glucose	radish
Sinigrin	allylisothiocyanate potassium hydrogen sulfate	glucose	*Brasica nigra*

The thioglycosides function to protect the plant against parasites and some are known to be antibiotics.

SAPONINS

Another interesting group of glycosides are the *saponins*. These glycosides have been extracted from plants and used by natives for many years as fish poisons. They enter the stomachs and intestines of the fish and kill it without accumulating in the flesh. Therefore, although saponins are also poisonous to man, the fish so killed can be eaten without harm. They, like all of the other glycosides, are soluble in water so they can be added to pools of water inhabited by fish. Actually, these glycosides are rather widely distributed in many plant species representing our most common plant families, such as the *Compositae, Gramineae, Leguminoseae, Rosaceae, Liliaceae, Rutaceae*, etc. The aglycone is called a *sapogenin*, examples of which are trillin or sarsasapogenin.

PHENOLIC GLYCOSIDES

In plants, there are numerous glycosides whose aglycone is a ring structure chemically derived from phenol. These aglycones may be free but are usually attached to one or more sugars to form *phenolic glycosides*. Actually, little is known about these, and it has only been within the past few years that the techniques have been available for their serious study. Much remains yet to be done even to determine their distribution in plants.

Some of the phenolics are very widely distributed in plants among numerous plant species, but many of them are limited in their distribution to one or a few species of plants. These of limited distribution are often useful for the study of the taxonomic classification and relationships among plants.

Although much yet remains to be learned about the distribution of the phenolic glycosides in plants, less is known about their functions within the plant. Although the phenolic compounds are found free in plants, when so found they are, except for the anthocyanins, limited in distribution to certain plant tissues, such as the seed, flower petals, or berries, or to dying or dead tissues, such as the heartwood of trees. Most often they are found as glycosides. The phenolic aglycones are not soluble in water, and some are detrimental to the well-being of the plant. However, when a sugar molecule becomes attached to form the glucoside, this greatly increases the water-solubility, making them soluble so they can dissolve in the cell sap. The sugar also eliminates the toxicity of these aglycones. Perhaps the reason these are found combined with sugars is that this is a mechanism by which the plant can tolerate these phenolic compounds which it must produce as by-products of essential metabolic functions. Thus they can be stored in the cell without interfering with vital cellular mechanisms of metabolism.

Another possible function of the phenolics may be in disease control, either because they may be toxic to the pathogen or they may serve as substrates in the synthesis of other organic compounds that are toxic to the invader. Just how extensive their function in disease resistance is, is still a matter of some conjecture and more work needs to be done on this subject. In summary, it can be stated that the phenolic glycosides may function either as a means of detoxication of the phenolics produced as by-products of essential metabolic reactions, in disease control, or may serve as colors of plant petals to attract insects and other animals for pollination of the flowers.

Perhaps the best studied of the phenolic aglycones has been the anthocyanidins, but phenolic aglycones also include the simple phenols, phenolic acids, acetophenones, phenylacetic acids, cinnamic acids, coumarins, isocoumarins, chromones, isoflavones, isoflavonoids, flavonols, chalcones, aurones, dihydrochalcones, biflavonyls, benzophenones, xanthones, stilbenes, benzoquinones, naphthoquinines, and anthraquinines. These represent groups of variable numbers of representative chemicals in each group. All have not been studied equally well so our knowledge of some of these is very limited. However, the following topics will briefly introduce the reader to some of these chemicals.

PHENOLIC ACIDS

The phenolic acids appear to be widely distributed in angiosperms, and have been isolated and detected in most plant leaves studied. Examples of the phenolic acid aglycones include *p*-hydroxybenzoic, gentisic acid, vanillic, syringic, protocatechuic acid, ellagic acid, and the aldehydes hydroxybenzaldehyde, vanillin, and salicylaldehyde. Most of these can be found in most plant species, and *p*-hydroxybenzoic acid, vanillic acid, and syringic acid are present in lignin, one of the components of plant cell walls. Gallic and ellagic acids are found in the tannins that give a brown color to plants and are used commercially for tanning leather. The structure of two of these phenolic acids is given in Fig. 13.10.

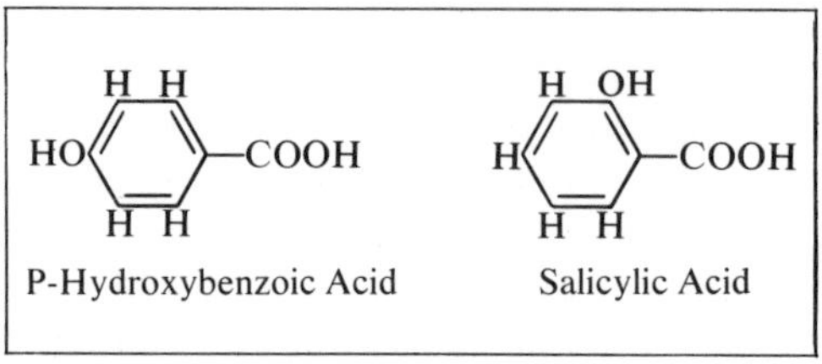

Fig. 13.10   Two phenolic acid derivatives.

CINNAMIC ACIDS

The cinnamic acids are found in the leaves in probably every higher plant. Particularly abundant are *p*-coumaric acid, caffeic acid, ferulic acid, and sinapic acid. Chemical structures of two of these can be seen in Fig. 13.11.

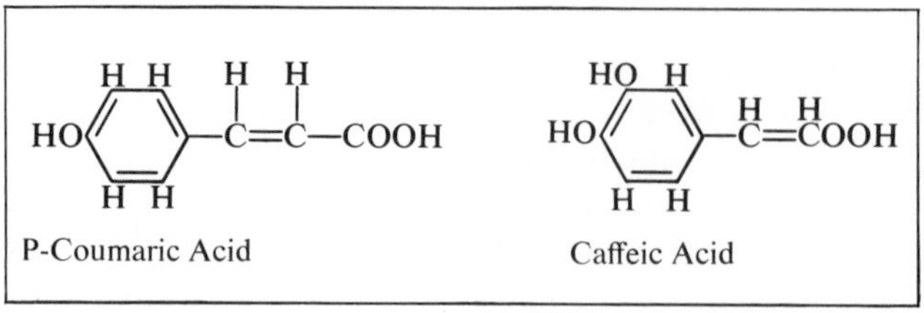

Fig. 13.11   Two cinnamic acid derivatives.

COUMARIN

Coumarin is a common volatile constituent of many plants. Being volatile it vaporizes into the atmosphere and often can be detected by the odor. Coumarin smells like new-mown hay, for those of you who are farmers. Coumarin has been shown to be an inhibitor of plant development, about which more will be said in a later chapter. The chemical structures of two coumarins can be seen below in Fig. 13.12.

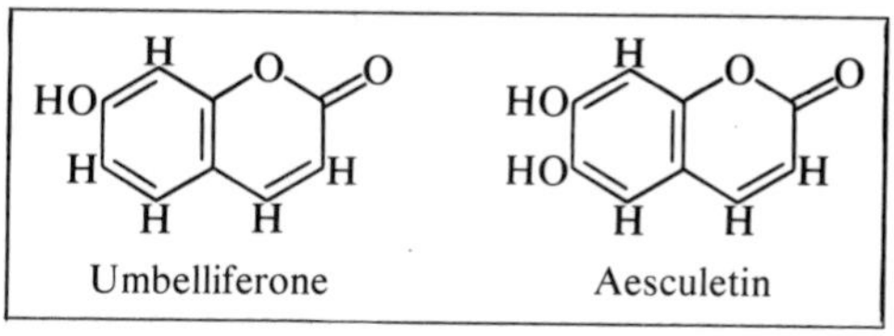

Umbelliferone   Aesculetin

Fig. 13.12   The structure of two coumarins.

FLAVONES

Only three flavones occur widely distributed although others are known. These three are apigenin, luteolin, and tricin. The structure of two of these are given in Fig. 13.13. They occur in leaves but the first two are probably more abundant in flowers.

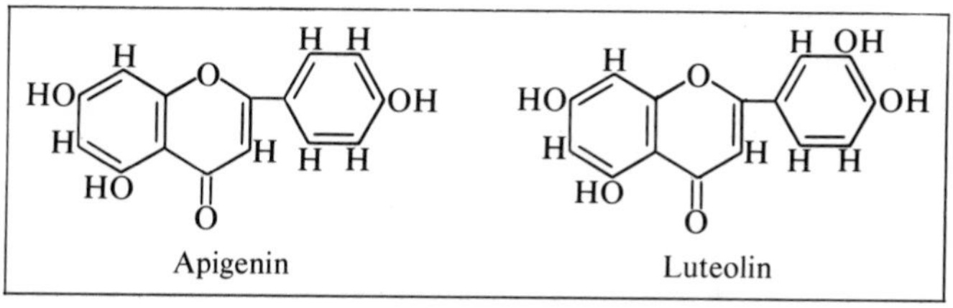

Apigenin   Luteolin

Fig. 13.13   The structure of two flavones.

FLAVONONES

Two flavonones are quite common. These are naringenin and eriodictyol, whose structures can be seen in Fig. 13.14. It can be seen that they correspond with apigenin and luteolin.

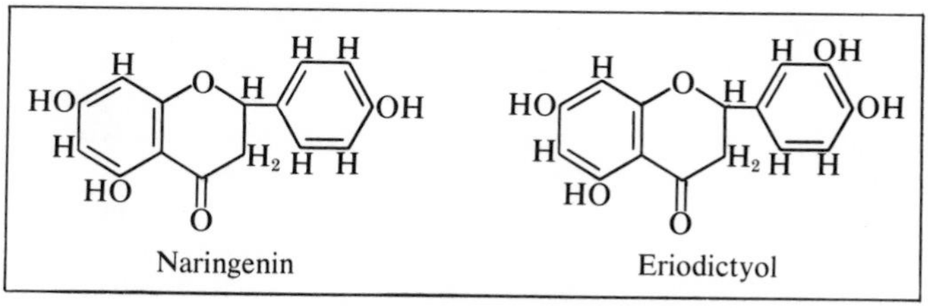

Naringenin   Eriodictyol

Fig. 13.14   The structure of two flavonones.

FLAVONOLS

The flavonols are widely distributed in most plant leaves, and flower petals, with kaempferol and quercetin being the most common, followed by myricetin. The structures of kaempferol and quercetin are given in Fig. 13.15.

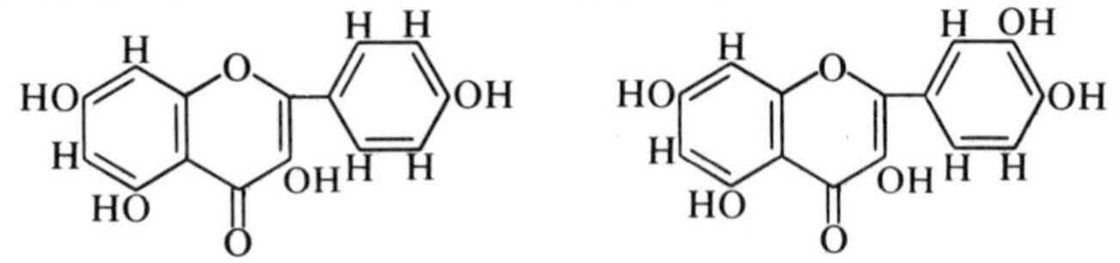

Fig. 13.15   The structure of two flavonols.

ANTHOCYANINS

Perhaps the most common flavonoid is the group known as the *anthocyanins*. These are the common pink, red, blue, and purple pigments of fruits, leaves, and twigs of many species. Anthocyanins are water-soluble pigments that are located within the vacuolar sap of the cell. Since they are water-soluble, they can be extracted from plant tissues by grinding the tissues underwater.

Structure

Structurally, the anthocyanins are composed of two components, namely, an anthocyanidin (aglycone) and one or more monosaccharide molecules. The sugar may be glucose, galactose, gentiobiose, rhamnose, or xylose, and the anthocyanidin one of those listed in Table 13.6.

Table 13.6   Common anthocyanin constituents.

Anthocyanidin (aglycone)	Sugar	Species
Delphinidin, cyanidin	rhamnose	*Lathyrus odoratus* petals.
Cyanidin, pelargonidin	rutinose	*Antirrhinum majus* petals.
Pelargonidin, cyanidin	glucose	*Dahlia variabilis* petals.
Cyanidin, pelargonidin	glucose	strawberry fruits
Cyanidin	galactose	apple fruit
Cyanidin	glucose, rutose	cherry fruit (sweet or sour)
Cyanidin	glucose	peach fruit
Cyanidin	glucose	blackberry fruit
Cyanidin	glucose, sambubiose	elderberry fruit
Cyanidin	glucose	mulberry fruit

Characteristics

There are numerous anthocyanidins and therefore numerous anthocyanins, and the blue, red, or purple colors of plants are caused by mixtures of anthocyanins rather than by any one being present alone. However, cyanidin is the most common anthocyanidin found in leaves and fruits, whereas more pelargonidin and delphinidin are found in flower petals.

The anthocyanins give essentially all of the water-soluble red, scarlet, blue, violet, and purple colors to higher plants, such as the purple or red colors of cherry and grape fruits, the red color to red cabbage, the red flower petals of poinsettia, the red maple leaves, and the red leaves of copper beech. Actually, one anthocyanin pigment can give different colors to plant structures because they are acid-base indicators meaning that they change their color as the acidity of the solution in which they are dissolved changes. In an acid solution with a low pH, they are red or pink, whereas in a basic solution with a high pH, they become blue or purple. For this reason, one can often darken the colors of flower petals by exposing the flower to ammonium hydroxide fumes.

Concentration

The amount of anthocyanin present in the plant is influenced by four main factors of the environment: (1) high or low temperatures, (2) excessive sunlight, (3) nitrogen deficiency, and (4) a deficiency of phosphorus. It appears at times that a high sugar content in the plant tissues causes a stimulus to anthocyanin synthesis, but this has never been definitely proven.

Anthocyanin content can be reduced by adding nitrogen, perhaps due to the conversion of sugars to nitrogen compounds, reducing the sugar concentration.

Certain antibiotics, such as chloramphenicol, terramycin, streptomycin, or aureomycin increase the anthocyanin concentration in certain plants. The reason for this is apparently unknown.

Light stimulates, and often is necessary for, anthocyanin formation. Studies of the spectrum utilized show that both red and blue light are useful, with the red light within the range of 650–750 nm. The blue light reaction is referred to as photoreaction I, and the red light reaction as photoreaction II. In some species, photoreaction II involves phytochrome and is photoreversible, but this is not true for all plant species. The mechanism by which light controls anthocyanin synthesis is not known. Light is not needed for the synthesis of the simple phenols, that are synthesized via the shikimic acid pathway, but chlorogenic acid formation from quinic and caffeic acids is stimulated by light, as is the synthesis of phenylalanine, and tyrosine.

ANTHOCHLORS

The anthochlor pigments (chalcones and aurones) are phenolic compounds and are common yellow flower pigments, and include isoliquiritigenin, butein, sulphuretin, aureusidin, leptosidin, and maritimetin. They are found primarily in the *Compositae, Scrophylariaceae, Leguminoseae, Oxalis* sp., and *Didymocarpus* sp. A structure of a *chalcone* and of an *aurone* is given in Fig. 13.16.

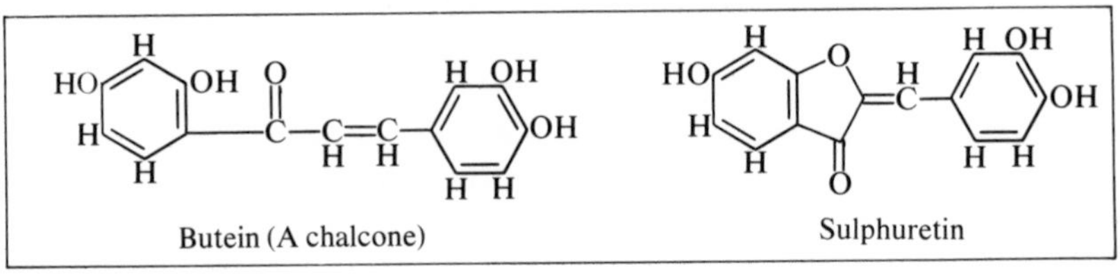

Butein (A chalcone)   Sulphuretin

Fig. 13.16   The structure of two anthochlor pigments.

BIFLAVONYLS

The biflavonyls are limited in distribution to the gymnosperms, except they seem to be absent in the *Pinaceae*. An example of the structure of one of these is given in Fig. 13.17, but other biflavonyls include ginkgetin, sotetsuflavone, and isoginkgetin.

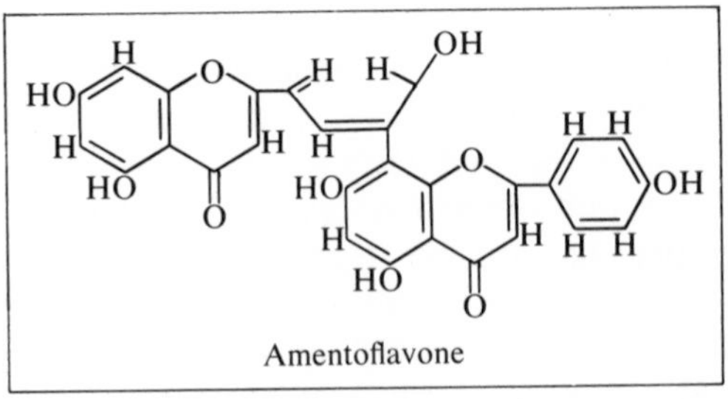

Amentoflavone

Fig. 13.17   The structure of a biflavonyl.

BENZOQUINONES

Benzoquinones are rather widely distributed in plants and animals, but are more common in fungi than in the higher plants. Ubiquinone is perhaps the most abundant of these.

ANTHRAQUINONES

The anthraquinones are rather abundant in plants but occur more frequently in the *Rubiaceae*. Alizarin, anthragollol, purpurin, chrysophanol, and emodin are examples.

The accumulation of flavonoid compounds is influenced by light treatment. If plants are grown in the dark, some flavonoids are produced but their concentrations are very low. Such concentrations are tremendously increased a day or so after exposing these plants to red light. Since this is far red reversible, it is evident that phytochrome plays a role in this phenomenon. In addition, higher concentrations accumulate if the plant is exposed to rather high intensity blue light a few hours before the red light exposure. The mechanism by which light alters the accumulation of the flavonoids, including the anthocyanins, flavonols, and glycoflavones is not understood, and needs further investigation.

Chalcones are precursors to all flavonoids studied.

BETALAINS

In addition to the flavonoid pigments, there is another group of water-soluble pigments found in the vacuoles of about ten plant families, known as the betalains. Perhaps the best examples of betalains are the red pigment of the table beet and the red pigment of the berries of the pokeberry plant. Other examples would be those giving the red or yellow colors to fruits and flowers of cacti.

The betalains are found in the flowers, fruits, and leaves, as well as in beet roots. They differ in several ways from the anthocyanins, such as their molecules contain nitrogen and migrate to the cathode instead of the anode.

The betalains are further grouped into two groups, namely, the betacyanins, which are the red-violet pigments, and the betaxanthins, which are yellow in color.

BETACYANINS

The betacyanins were formerly considered to be nitrogenous anthocyanins. They too are plant pigments, the best known being the red pigment of the table beet, namely, betanidin. Its structure is given in Fig. 13.18, and it is synthesized in the plant from two molecules of dopa. Its distribution is rather restricted.

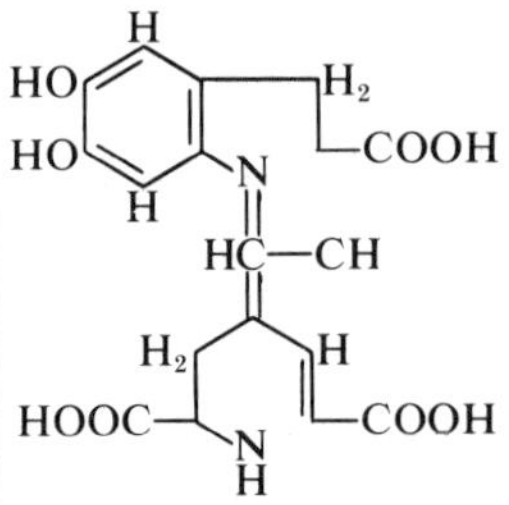

Betacyanin

Fig. 13.18 The structure of betacyanin.

Their synthesis occurs probably by the conversion of 5,6-dihydroxyphenylalanine to betalamic acid, with the further addition of either cyclodopa to the betalamic acid to form the betacyanins, or the further addition of other amino acids or amines to the betalamic acid to form betaxanthins. Then these molecules are altered to form other species of each group. For example, the betacyanins appear to differ either by the stereochemistry at C-15, by the nature of the attached sugar, or by the esterification of the sugar or carboxyl groups.

The structural formula for betalamic acid is given in Fig. 13.19. These betalain molecules may or may not include an esterified sugar molecule.

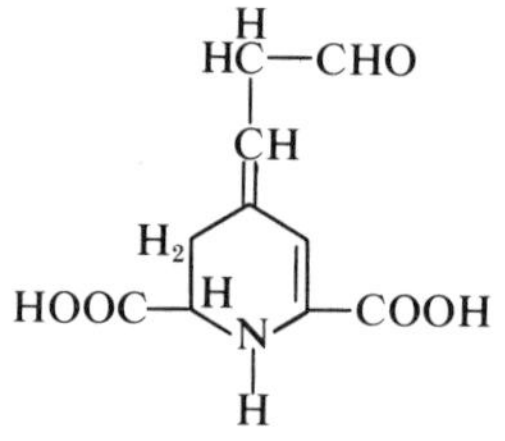

Betalamic Acid

Fig. 13.19 The structure of betalamic acid.

SYNTHESIS OF PHENOLIC GLYCOSIDES

The phenolic aglycones are synthesized, in higher plants, from phosphoenol pyruvic acid, an intermediate of glycolysis, and from either erythrose-4-phosphate, a phosphorylated sugar, or from acetyl CoA, as seen in Fig. 13.20.

Five monosaccharides which most often form the sugar of the phenolic glycosides are glucose, galactose, xylose, rhamnose, and arabinose. However, disaccharides may also be found occasionally, especially is this true of rutinose, sophorose, and sambubiose.

ALKALOIDS

A consideration of the secondary plant products would not be complete without mention of that very important group of drugs known as the *alkaloids*. Although the alkaloids are common in plants, being found in about 20% of the plant species, they are not found in animal species. In the plants, they are not known to have important functions, although they may help protect the plant from parasites and predators, serve as growth inhibitors, or maintain cell ionic balance due to their basic property. In animals they function as important drugs, but they are not drugs to plants that produce them (you do not see a "high" hemp plant although they do grow rather tall).

Types

Some of the common alkaloids, their drug functions, and their source, are listed in Table 13.7. The hallucinogenic drugs, at least most of them, including LSD and marijuana, are also primarily alkaloids.

Properties

The alkaloids are nitrogenous bases just like the purines and pyrimidines. Nitrogen is part of their molecule and they give a basic reaction when dissolved in water. Their molecules include a nitrogenous ring structure and are sometimes rather complex. Although they give a basic reaction, when dissolved in water, their water-solubility is low.

The alkaloid molecule is surprisingly stable, with some alkaloid molecules being reported to be intact in plant residues at least 1300 years old. Therefore, it is not surprising to find that they are usually not reutilized by the plant that produces them, but accumulate within the cells.

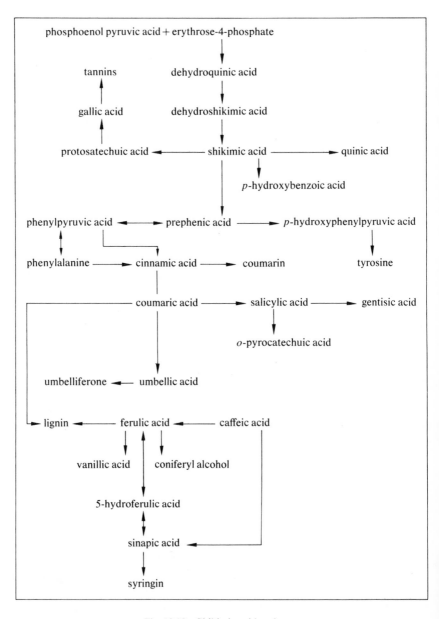

Fig. 13.20   Shikimic acid pathway.

Table 13.7   Some important alkaloids and a source of each.

Name	Importance	Source
Atropine	antispasmodic	*Datura* and other *Solanaceae*
Coniine	paralysis and death	Poison hemlock
Hyoscyamine	antispasmodic	*Datura* and other *Solanaceae*
Marijuana	poison	Hemp (*Cannabis sativa*)
Nicotine	toxic stimulant of the central nervous system	*Nicotiana* sp.
Nornicotine	no effect on man	*Nicotiana* sp.
Cocaine	local anesthetic	*Coca*
Emetine	emetic	*Ipecacuanha* sp.
Hydrastine	antihemorrhagic	*Hydrastis* sp.
LSD	hallucinogen	*Ipomoea violaceae*
Morphine	narcotic	Opium poppy
Pelletierine	vermifuge	Pomegranate
Pilocarpine	diaphoretic	*Philocarpus pennatifolius*
Quinine	cardiac depressant	*Cinchona* sp.
Reserpine	tranquilizer	*Rauwolfia serpentina*
Strychnine	nervous stimulant	*Strychnos nux-vomica*
Tubocurarine	muscle paralysant	*Chondodendron* sp.

Perhaps there is no more active field of phytochemistry today than the search for and study of plant alkaloids. Each month, reports appear about newly-discovered alkaloids. Therefore, just how many different kinds of alkaloids exist is not known, but to date, their numbers run into the thousands.

Interesting correlations have been found between the chemical activity and structures of the alkaloids, and the natural environment of the plant producing them. For instance, tropical alkaloids have a higher boiling point than those obtained from temperate regions. The number of carbon atoms in each alkalid molecule increases from the tropics to the temperate regions, the number of oxygen atoms per molecule is lowest in the tropics and highest in temperate regions, and the alkaloids of the tropics are generally less toxic to man than are those of the temperate regions. Are these differences due to temperature regulation of metabolism?

Synthesis

The synthesis of alkaloids no doubt involves a number of pathways, the one used in each case depending upon the nature of the structure of the molecule. Generally, they are synthesized from amino acids and fatty acids, or at least from fatty acid intermediates. Also, there seems to be some relationship between the synthesis of alkaloids and of the terpenes, since plants that accumulate alkaloids do not usually also accumulate terpenes.

General Comments About Secondary Plant Products

Perhaps, at this point, it would be well to point out some generalizations that can be made about the secondary plant products discussed in this chapter. When interpreting the results of chemical analyses of a plant, it is essential that factors that may alter the chemical composition be considered, and these may be numerous. The base for expressing the results is one factor, the environmental conditions which prevailed prior to the analyses is another, the precision of the analytical methods used, stage of development of the plant, age of the plant, and the fact that diurnal variations in the concentrations of such chemicals occur, with a minimum concentration at night and maximum during the day, changes also occur during the growing season even in perennial plants, and one must also consider the rate and direction of translocation of the chemical within the plant. Lipophilic substances, such as essential oils and resins, are not translocated whereas other chemicals may be.

The greatest production of these secondary products has been found to be very markedly correlated with growth, with their greatest synthesis occurring during periods of intense growth activity.

Indeed, this chapter has not attempted to cover all of the chemicals one may find in a given plant, but has been an introduction to those most often encountered, or to those of most interest. Remember that the concentration of the chemical within a plant can be determined by many factors which can be grouped into those altering metabolic regulation and those associated with metabolic pools. Such metabolic pools are associated with the many membranes of the plant, a subject which will be considered in the next chapter.

Chapter	Plant Membranes and their
Fourteen	Permeability

Chapter 1 showed that the cell membranes are very important structures of the cell, and indeed that the cell would not exist in their absence. All of the cell organelles are basically composed of membranes. Therefore, it is not surprising to find that no subject in plant physiology has been studied more extensively than has that of the structure and function of plant membranes. And yet, in spite of this long and arduous study, very little is actually known about either the structure or the function of these membranes.

The plant cell membrane is a very elusive structure. It can be seen in electron micrographs and its presence can be perceived by microsurgical studies, but to isolate and to purify it is virtually a physical impossibility. And yet, that is what must be done before its structure, both chemical and physical, can be completely understood.

CHEMICAL STRUCTURE OF THE MEMBRANE

The chemical composition of plant cell membranes is difficult to determine since it is extremely difficult or perhaps impossible to separate and to purify the membranes to the extent required for a concise chemical analysis. What data are available come mostly from cytochemical studies or from studies of animal cell membranes, such as those of erythrocytes or sea urchin eggs, membranes that can be obtained in the apparently pure form.

That plant cell membranes are composed predominantly of proteins and lipids has been well documented. The proteins comprise about

50–70% of the dry weight of the membrane and the lipids comprise about 30–50%. In addition, some carbohydrates appear to be present as glycolipids.

Proteins

Not much is known about the proteins of the plant cell membranes. The difficulty in purifying them does not allow the characterization of these proteins or even certainty as to whether they are only structural or both structural and enzymatic proteins. Certainly, at least some of these proteins of cell membranes are structural, but probably some are both structural and enzymatic. For instance, the blue-colored protein phytochrome is believed to be bound to cell membranes and to function by controlling membrane permeability.

Phosphatidic Acids

Most of the membrane lipids are phospholipids, although some glycolipids, sulfolipids, and sterols are present in small amounts. The phospholipids can be considered to be either phosphatidic acid or one of its derivatives. *Phosphatidic acid* is very similar to a fat, although one of the three fatty acids has been removed and replaced with a phosphoric acid molecule, as seen in Fig. 14.1. Many types of phosphatidic acids can exist depending upon whether the phosphate group is in the number 1 or number 2 position, and depending upon the types and location of the two fatty acid molecules.

Phosphatidic acid is never present in high concentrations, but is the least abundant of the phospholipids in plant cells. It is also not found as the free molecule, but forms a salt with one of the inorganic cations,

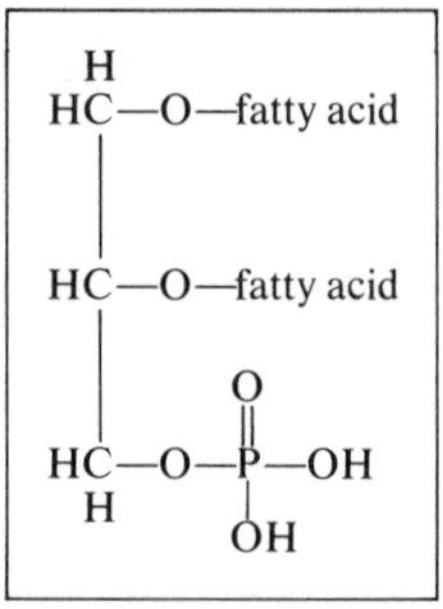

Fig. 14.1   The structure of a phosphatidic acid.

usually either calcium, magnesium, or potassium. Perhaps this is why calcium is so important in the control of membrane permeability.

More commonly, the phospholipids are more complex than phosphatidic acid, with their molecules altered by adding an amine to the phosphatidic acid molecule. For instance, lecithin and cephalin are very common phospholipids in plant cells.

Lecithin

Lecithin (phosphatidyl choline) is similar to phosphatidic acid except the amine choline is attached to the phosphoric acid group, as shown in Fig. 14.2. It is still possible to have a number of different types of lecithins depending again upon whether the phosphate is in the number 1 or the number 2 position, and upon the location and types of fatty acids that are part of the molecule.

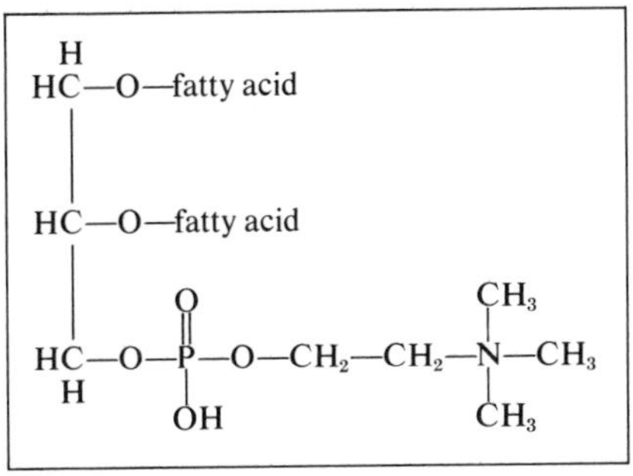

Fig. 14.2   The structure of a lecithin.

Cephalin

Cephalin (phosphatidyl ethanolamine) is similar structurally to phosphatidic acid except ethanolamine is attached to the phosphoric acid group.

Other common plant phospholipids include phosphatidyl serine, phosphatidyl glycerol, and phosphatidyl inositol.

Synthesis

Although one might expect to find any of the plant fatty acids as constituents of the phospholipid molecule, the predominant ones are palmitic, oleic, linoleic, and linolenic acids.

The synthesis of the phosphatidic acids may occur through either of two pathways. Either

$$ATP + glycerol \longrightarrow glycerophosphate \xrightarrow{\text{2 acyl-S-CoA}} phosphatidic acid$$

or

$$\alpha,\beta\text{-diglyceride} + ATP \xrightarrow{\text{Mg}} phosphatidic acid$$

A more complex phospholipid may be formed by then adding an amine to the phosphatidic acid molecule. For instance, ethanolamine can be formed by the decarboxylation of L-serine, as shown in Fig. 14.3. This

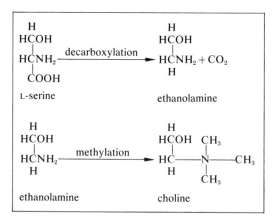

Fig. 14.3   Pathways of choline and ethanolamine synthesis.

can then be used to form cephalin, as shown in Fig. 14.4, or the ethanolamine can be methylated to form choline, which can be used to form lecithin.

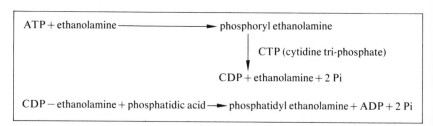

Fig. 14.4   Synthesis of cephalin from phosphatidic acid.

Glycolipids

Glycolipids also form an important fraction of the plant cell membrane lipids. These are glycerides, in which galactose is the predominant sugar, with the principal fatty acids being palmitic, oleic, and sometimes linoleic acids, as in the phospholipids. One or two sugar molecules may be attached to the molecule, but if two are present they are usually joined together. In essence, the glycolipid molecule is a fat molecule with one fatty acid removed and one or more sugars substituted in its place, as seen in Fig. 14.5. Therefore, the glycolipid can vary with the number and kinds of

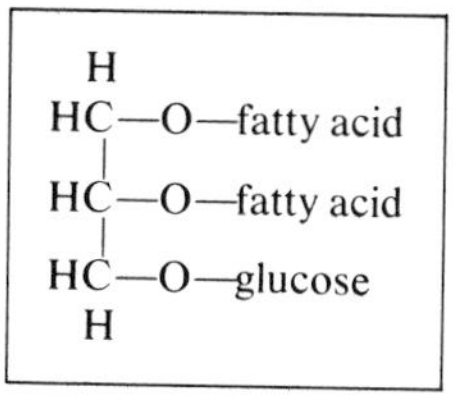

```
         H
        HC—O—fatty acid
        HC—O—fatty acid
        HC—O—glucose
         H
```

Fig. 14.5   The structure of a glycolipid.

sugar molecules attached, their location, whether in the number 1 or number 2 position, and in the location and types of fatty acids on each molecule.

Sulfolipids

Sulfolipids are also common in plants, but are most often present in green plant cells where they are found in the chloroplasts as part of the chloroplast membrane structures. Although they have not been studied extensively, the sulfolipid molecule appears to be similar to the glycolipids, with an SO_3 group attached to the free end of the sugar, as seen in Fig. 14.6.

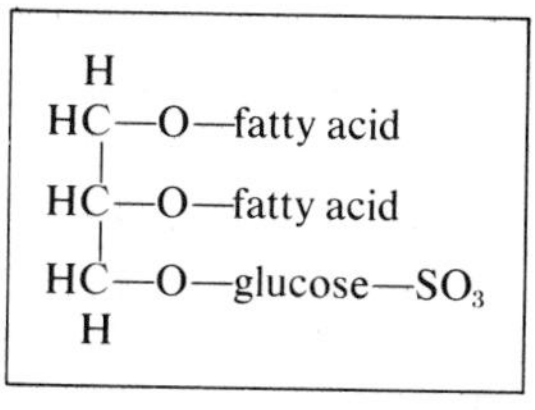

```
         H
        HC—O—fatty acid
        HC—O—fatty acid
        HC—O—glucose—SO₃
         H
```

Fig. 14.6   The structure of a sulfolipid.

Sterols

Sterols are common in plant cell membranes, with β-sitosterol, spinasterol, and stigmasterol being the most prevalent. These are alcohols of high molecular weight and are found either free or as esters or glycosides. The chemical structure of β-sitosterol is given in Fig. 14.7. Sterol synthesis begins with mevalonic acid.

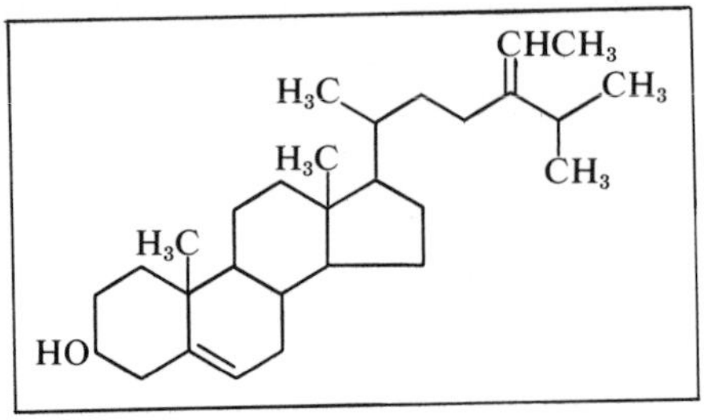

Fig. 14.7   The structure of β-sitosterol.

PHYSICAL STRUCTURE OF THE MEMBRANE

A number of new techniques and fields of study have been combined recently to throw some light on how these chemicals unite to form the physical structure of the cell membranes. These include use of polarized light, electron and ultraviolet microscopy, X-ray diffraction analysis, and the study of surface chemistry and monomolecular films. In spite of the extensive employment of these techniques and these fields of study, it is not possible to give the definitive physical structure of any of the plant cell membranes. In fact, it is not even possible to say whether or not there is a universal physical structure of such or whether each membrane of a cell differs significantly enough to say that all of the membranes of a cell are different. Perhaps the latter view is more nearly correct.

Lipid-water Interactions

As can be seen from the discussion of the chemistry of cell membranes, they are composed of various kinds of lipids and proteins. When lipids are suspended in water, the lipid molecules at the water-lipid interface rapidly become oriented so that their polar groups face the water and their lyophilic groups (fatty acid groups) face the oil, as shown in Fig. 14.8. When bimolecular layers of lipids are formed, on a water surface, the lipid molecules line up in such a way that their polar ends are opposite to the

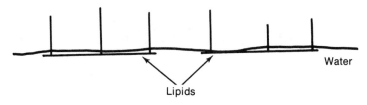

Fig. 14.8   Monolayer of lipid on a water surface.

water and their lyophilic ends face each other, as seen in Fig. 14.9. There-fore, in the cell membrane, one might expect that the lipid molecules are oriented in such a way that their polar groups are toward the membrane surfaces and their lyophilic groups face each other at the center of the membrane. However, such a double-layer would be only about 6 nm in thickness whereas the cell membrane is about 7–10 nm in thickness. However, this additional thickness could be accounted for if two protein layers were present in the membrane.

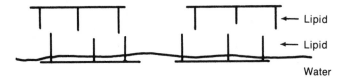

Fig. 14.9   The orientation of a bimolecular lipid layer on a water surface.

Membrane Structure

The next question that one needs to ask is where do the protein mole-cules occur within the membrane? Two possibilities exist. Either they form layers adjacent to and outside of the polar ends of the lipid molecules or they lie adjacent to and among the polar ends of the lipids. When cell membranes are observed in electronmicrographs, they often appear as two dark lines about 7–10 nm apart, and these lines are separated by a clear space. Based on such observations, and the assumption that such observations represent the membrane in the living cell, which may not be true, attempts have been made to explain how the proteins and lipids are positioned in the membrane. The most widely accepted theory states that the outer and inner surfaces of the membranes are composed of a layer of protein molecules, and these two protein layers are separated by two layers of lipid molecules, as shown in Fig. 14.10. Such a structure would be compatible with the observations with the electron microscope, and

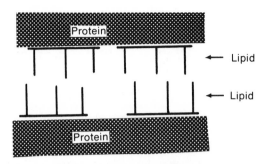

Fig. 14.10   The proposed structure of a plant cell membrane monolayer.

could explain the characteristics of the membrane relative to its permeability. However, this does not explain how membrane permeability can be altered by changes in the rate of respiration, nor does it explain the nature of the carrier that is believed to be associated with the uptake of ions, as discussed later. However, it is a hypothesis from which to work and one on which further discussion can be based.

With this background of information with which to work, the next problem of interest to the physiologist is how the membranes are synthesized. From whence do they arise?

Membrane Synthesis

Membranes may be formed at interfaces, due entirely to differences in the physical and chemical state of the substances forming the interface, and no living activity is necessary for this mode of formation. A good and interesting example of this type of membrane formation occurs when a small crystal of potassium ferrocyanide is dropped into a 5% solution of $CuSO_4$. A differentially-permeable membrane forms spontaneously, swells as water enters the membranous bag, which eventually ruptures and an additional membrane forms around the ruptured area, continuous with the original membrane. This is surprisingly like the situation in the plant cell, and if the student has not observed this experiment, he should do so.

Since such a chemical membrane is so similar in behavior to the cell membranes, what evidence indicates that this is not the normal method of cell membrane formation? First, the differential permeability of cell membranes disappears with the death of the cell. Second, an increase in the number and area of cell membranes occurs only when the cell is alive and not when it is dead. Thus, it appears that membrane synthesis in the plant cell is dependent upon the living activity of the cell.

ORIGIN OF PLANT MEMBRANES

If membrane formation is a living phenomenon, then how does it occur? First, it is necessary to realize that such synthesis requires the prior or subsequent synthesis of proteins and lipids. Proteins are synthesized at the polyribosomes found either within the mitochondria, chloroplasts, or in the ground substance of the cytoplasm, whereas lipid synthesis seems to occur in the mitochondria and chloroplasts. But, where do these proteins and lipids come together for membrane synthesis?

One explanation is that the endoplasmic reticulum and the vesicles produced by the Golgi apparatus are the sources of cell membranes, especially of the plasma membrane. According to this theory, it can be demonstrated that at the time of cytokinesis, the ER and vesicles accumulate between the two nuclei and there they fuse, forming the cell plate and the plasma membranes. Then, during cell enlargement, the vesicles appear to unite with the existing plasma membrane, extruding their contents to the outside of the membrane, and themselves uniting with the plasma membrane to become part of it. This would then locate the site of membrane formation at the site of vesicle formation, namely, at the endoplasmic reticulum or at the Golgi apparatus, as shown in Fig. 14.11.

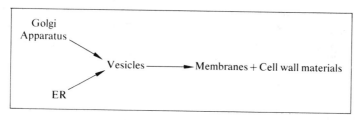

Fig. 14.11   Proposed mechanism of cell membrane formation.

MEMBRANE FUNCTIONS

Perhaps the function of cell membranes is to retain specific environments that are conducive to the functions of the systems contained within the volumes these membranes surround. No doubt, the environment within the plant cell is different from that which surrounds it, but the environment within each cell organelle is also different from that of the surrounding ground substance.

Permeability

Nevertheless, chemicals must be able to enter and leave through these membranes too, in order that the required nutrients be available, that the waste materials be discharged, and that chemicals may be transported from one location to another within the cell or between cells. Therefore, the function of cell membranes can be said to be to maintain a favorable environment in the volume they surround, and to control the entrance and exit of chemicals into and out of this volume. Since maintenance of the favorable environment is done by controlling the passage of chemicals into and out of the volume, the important function of such membranes is to control the movement of chemicals through it, a phenomenon known as *membrane permeability*. Permeability is a property of the membrane, not of the chemical passing through it. It is proper to speak of the permeability of a membrane to a solute, but not of the permeability of a solute.

Flux

The term *flux* has been used to express the rate at which a chemical passes through a membrane and it represents the number of molecules of the solute that pass through a unit area of a membrane per unit duration of time. This value will vary with the type of solute, the concentration of solute on both sides of the membrane, and with the specific membrane used, and the conditions under which the determination was made. Using the cgs system, which should always be used when expressing flux, the dimensions of flux would be the number of molecules penetrating one cubic centimeter of the membrane per second. The terms inward flux and outward flux refer to the direction of net movement across the membrane.

Permeability Characteristics

The permeability of plant membranes has been studied very intensively over a period of many years. If one summarizes the results of these studies, the following conclusions can be drawn from them:

1. Chemicals vary in the ease with which they pass through membranes. Some chemicals appear to move through the membranes very rapidly. These include water, carbon dioxide, oxygen, ammonia, ethanol, acetic acid, and lactic acid.

2. Lipid-soluble chemicals enter more rapidly than water-soluble chemicals.

3. The larger the molecule, the more slowly it moves through the membrane.

4. Undissociated molecules pass through more rapidly than dissociated molecules and ions.

5. Monovalent ions move faster than divalent, which move more rapidly than trivalent ions, etc.

6. Sodium excess increases permeability while calcium excess decreases it.

7. Respiration inhibitors decrease the entrance of many substances, as does poor aeration.

8. Death eliminates any inhibition to permeability.

Measurements

For permeability studies, permeability is determined by measuring the concentration of a given chemical on both sides of a membrane. However, it is not possible to make such measurements adjacent to the two membrane surfaces, so measurements are made at various distances therefrom. This could alter the results because one then cannot be sure that the concentration measured is the same as that adjacent to the membrane surface.

Permeability and Solute Characteristics

The permeability of cell membranes differs with the type of chemical being taken up. Especially does this difference depend upon the magnitude of the electrical charge, either negative or positive, associated with them. However, not all molecules have such a charge, and with some the charge is of such magnitude as to be negligible. Nevertheless, there is a difference in the rate of uptake of a molecule depending upon the effects of the associated charge or its absence.

If the molecule has no electrical charge associated with it, or such charge is negligible in magnitude, then the molecule is referred to as a *non-electrolyte*. Such non-electrolytes include water, oxygen and carbon dioxide, ethylene, sugars, and some amino acids. Certain generalizations can be made about the permeability of non-electrolytes. First, they pass through the cell membranes much more rapidly than do electrolytes. Second, their rate of entrance is indirectly correlated with the size of the molecule. In other words, the smaller the molecule, the more rapidly it passes through the membrane. Small molecules, such as water, oxygen, carbon dioxide, ethylene, and ammonia enter the cell extremely rapidly, whereas larger molecules, such as sucrose, move through much more slowly. This observation tempts one to conclude that membranes have openings or pores in them through which the chemicals pass and these pores are of a limited size so that the smaller the molecule, the more pores large enough to allow its entrance, whereas the larger molecules can find but few pores through which they can pass. This is the basis of the membrane sieve theory which views a membrane as a sieve through which molecules pass through the pores. However, molecular size is not the only factor that determines membrane permeability.

The rate at which a chemical will pass through a plant membrane is determined not only by the size of the molecule, but also by its ability to dissolve in lipids or in lipid solvents. The more lipid-soluble (*lipophilic*) the molecule, the more rapidly it will pass through the membrane. Chemicals such as oxygen, ether, acetone, and the essential oils will enter more rapidly than water-soluble (*hydrophilic*) molecules such as sucrose, glucose, and certain alcohols. Such observations have been demonstrated many times, and have led to a theory that explains membrane permeability as the dissolution of the solute in the lipids of the membrane, and their subsequent passage to the other side of the membrane. This too gives support to the belief that membranes do indeed contain lipids and that these lipids do function in membrane permeability.

It should be noted that all gas molecules pass through membranes very rapidly since they are small molecules and often are also lipophilic. Also, it should be noted that although membranes are less permeable to large, hydrophilic molecules, such as sucrose, such molecules do pass through these membranes as indicated by the fact that sucrose is the carbohydrate translocated in plants.

Superimposed upon the facts that the permeability of cell membranes to non-electrolytes is greatly influenced by both the size of the molecule and its lipid solubility, is the observation that the rate of entrance of solutes into the cell depends upon the rate of respiration of the cell. This relationship does not appear to be a straight line function, but if respiration rate is greatly decreased, the membrane permeability greatly decreases. This seems to be related to ATP formation during respiration, but just how ATP is coupled to membrane permeability is not understood. More will be said about this interesting phenomenon in the subsequent discussion of the relationship between membrane permeability and electrolytes, but an obvious observation related to this subject is the wilting of leaves when mesophytic species are growing in the soil that becomes inundated with water. Flooding a field will result, in a fairly short time, in the plants wilting, even though water is present in the soil in excess. Such plants do have a decreased water content, indicating that flooding decreases the permeability of the cell membranes to water uptake.

Permeability and Metabolic Energy

If studies of the permeability of non-electrolytes emphasized only the correlations between permeability and the size or lipid solubility of the molecule and ignored the relationship to respiration, one would be tempted to conclude that the passage of such non-electrolytes through the membrane was entirely a passive phenomenon, *passive* referring to uptake controlled entirely by physical factors and unrelated to metabolic processes. Indeed, for a long time, such an assumption was made, and there is evidence to support it. Perhaps, some such uptake is passive, but it now seems clear that *active* uptake is also involved, although the mechanism of such active uptake has not been elucidated.

Electrolytes vs. Non-electrolytes

Electrolytes, such as ions and other electrically-charged molecules and atoms, pass through cell membranes much less rapidly than do non-electrolytes. There are a number of reasons for this. First, it is not possible to maintain a cell that is not near neutral electrically, due to the vast amount of energy needed to separate charges. However, a difference in electrical potential exists between the two sides of a given membrane. The magnitude of this potential varies from time to time, but is often about 100 millivolts, with the interior of the membrane being more negative and the exterior more positive. Monovalent cations lower this electropotential difference whereas divalent cations increase it. Just what effect this difference has on solute transport across the membrane is not known, but certainly the transport of ions could be much affected by it. Also, the source of this potential difference is not well known. But generally the cell must contain the same number of negatively-charged chemicals as positively-charged chemicals. Therefore, the uptake of a charged particle must either be accompanied by the simultaneous uptake of a particle of opposite charge, or the one being taken up must be exchanged by adding a particle of similar charge to the environment. If a cation, such as K^+ is taken up, then the cell must simultaneously either give off another cation, such as H^+, or must synthesize an anion, such as an organic acid. Indeed, it has often been observed that if plants are grown in a water culture, the pH of the culture continuously changes due to this evolution of ions from the plant in exchange for ions taken up. The ions often reported are predominantly H^+ and HCO_3^-.

Also, it has been often reported that if a plant is taking up cations more rapidly than anions, there is an increase in the organic acid content of the roots, whereas if excess anions are being taken up, such as NO_3, the organic acid content of the root cells decreases.

Whereas the uptake of non-electrolytes is greatly influenced by both the size and lipid-solubility of the molecule, these have little effect on ion uptake. On the other hand, such ion uptake is much more influenced by the metabolic rate of the root cells or of other cells which are taking up the ions. Therefore, ion uptake is predominantly active uptake and occurs only with the expenditure of metabolic energy.

The rate of ion uptake can be determined by placing plant cells in a concentrated salt solution, which would, as you know, result in rapid cell plasmolysis, and then determine the rate at which deplasmolysis occurs. Such studies show that monovalent cations, those with one electrical charge per ion, enter more rapidly than divalent, and divalent enter more rapidly than trivalent, etc. However, differences also exist among ions of similar valences, for example potassium enters more rapidly than sodium ions.

It should also be noted here, that caution should be used in interpreting the results of such studies, since it has been shown that monovalent ions greatly increase the permeability of membranes to all solutes, whereas calcium and other divalent ions, have opposite effects.

Although the cell membranes of most plant cells are only slightly permeable to ions, there appear to be some exceptions, as is usually the case in biology. These should be mentioned in view of their potential for further studies of ion uptake and membrane permeability. Certain brown algae and numerous marine diatoms, have cells whose plasma membranes appear to be extremely permeable to ions. However, most plant cells do not exhibit this great permeability.

Acids and Bases

Studies of the permeability of cell membranes to acids and bases present other problems, and give results that must be especially interpreted. Such studies must involve special considerations. First, acids and bases are often toxic to the cell. Second, depending upon the acid or base, and both the internal and external pH, the acid or base may pass through the membrane either as an ion, as an undissociated molecule, or as both at the same time. Since the dissociated acids and bases are ions, and since membranes are nearly impermeable to ions, it might be expected that the uptake of dissociated acids and bases would be much slower than the uptake of their undissociated molecules. That such is so can be seen in Fig. 11.2, where the toxicity of transcinnamic acid increases with decreased pH, since at decreased pH the hydrogen ion concentration increases, increasing the percentage of the acid molecules that are undissociated and therefore permitting more rapid uptake of these molecules.

Further evidence is the fact that strong acids, and bases, such as HNO_3, H_2SO_4, HCl, NaOH, KOH, and $Ca(OH)_2$, penetrate into undamaged membranes very slowly, and their molecules are almost entirely disassociated.

Some acids and bases penetrate cell membranes very rapidly. These include CO_2, H_2CO_3, acetic acid, pyruvic acid, ammonia, and ammonium hydroxide, and many of the alkaloids. In these cases, one can correlate their solubility in lipids with permeability. And generally it can be said that as the lipid-solubility of the molecule increases, the rate of penetration through the membrane increases and as the size of the molecule increases, the rate of penetration decreases. Acids such as citric, oxalic, malic, and tartaric penetrate very slowly and have low lipid-solubilities.

An interesting method to use in the determination of the permeability of membranes to acids or bases is one devised by Jacobs who found that the flower petals of *Symphytum peregrinum*, a member of the *Boraginaceae* family, have pigments that are very sensitive to acid-base changes and their colors change quite rapidly when the petals are exposed even to weak acids or bases. This should be a useful species for laboratory demonstrations of membrane permeability.

ENVIRONMENTAL CONTROL OF PERMEABILITY

Temperature

Although the permeability of the membrane to chemicals is well correlated with the size and lipid-solubility of the chemical, there are other factors too that will determine the rate of penetration of a chemical through the membranes. One of these is temperature, a factor that varies greatly from time to time and one that would therefore cause the permeability of a membrane to vary from time to time. Generally, the Q_{10} for membrane permeability varies from 2 to 5, but is constant over a wide temperature range. However, this value does vary with the size of the molecule, being lower with smaller molecules and higher with larger ones.

Light

Light has been shown to alter membrane permeability. This may be due to two causes. First, for photosynthetic cells, and especially for their chloroplasts, permeability may be increased in the light due to the presence of more ATP synthesized during photosynthesis – ATP that is needed for active uptake. Second, it is postulated that phytochrome is attached to the membranes and does control membrane permeability at least in some cells. This permeability then varies with the form of the phytochrome attached to the membrane.

Oxygen

Oxygen deficiency, respiration inhibitors, and carbohydrate supply will often alter membrane permeability. This might be expected since such a deficiency of respiration caused by any of these would reduce the rate of respiration and therefore the amount of ATP for active uptake. At the same time, oxygen deficiency seems to increase the physical properties of the membrane to allow greater permeability. How this is accomplished is unknown.

Stimulation

Certain types of stimulation alter membrane permeability. The best example of this is shown by the sensitive plant, *Mimosa pudica*. When one of its leaves is mechanically stimulated, such as by touching it, the leaf collapses very rapidly. This collapse is due to the transmittal of a stimulus down the leaf from the area touched to the large cells at the leaf base. The nature of this stimulus is unknown, but it does cause a tremendous and very rapid increase in the permeability of these large cells to water so that the water moves out, causing the cells to lose turgor and the leaves to droop. After a few minutes, the permeability of the membrane changes, water is again accumulated by these cells, increasing their turgor pressure, and again restoring the leaf to its normal position.

Other

As the seasons change, changes in membrane permeability also change. There are many reasons for this, such as changes in respiration rate, changes in cell viscosity, changes associated with frost and drought hardening, age of the cell, etc.

It would be well to consider the change in membrane permeability of the cell at death. At this time, all differential permeability is lost and the membranes become very permeable to all substances. More will be said about this in a subsequent chapter.

MECHANISMS OF PERMEABILITY

To derive, from the facts known, one mechanism which explains how chemicals move through plant membranes is impossible. There are two reasons for this. First, membrane permeability changes from time to time and from one membrane to another and second, even at a given time, the mechanism of permeability is not the same even within a given membrane. There is really no good reason to expect only one mechanism by which chemicals pass through cell membranes. Rather, this occurs through several mechanisms often working simultaneously. What might these mechanisms be?

Diffusion

No doubt diffusion plays an important role in solute transport across membranes. In the case of a few small molecules, such as water, carbon dioxide, oxygen, and ammonia, perhaps much of the transport is by diffusion. With others, diffusion plays a role in moving the chemicals to and away from the membrane surface, but its importance in transport across the membranes is probably insignificant. If diffusion was the only mechanism involved, the concentration of a chemical at equilibrium would have to be essentially the same on both sides of the membrane, unless the chemical formed a precipitate or other stable large molecule or structure on one side of the membrane and not on the other side. That such is not the case with ions has been shown by extracting the vacuolar sap of the cell and determining its electrical conductivity. (Why should this indicate elements present as ions?)

Indeed, if a plant tissue is placed in a solution containing an ion that is not present in the tissue, such as bromide, there is a rapid increase of bromide within the tissue. This occurs within a few seconds and levels off with equilibrium being reached within a minute or two. However, such uptake is believed to be due to the diffusion of the ion into the intercellular or free space. This space includes the volume between the cells and in the cell walls. Such solutes may actually show a higher concentration on the cell wall than in the surrounding medium. This is due either to their being bound by the wall structures, or to their precipitation within the wall. Such rapid uptake into the protoplast does not occur.

Within the cell, an increase in the concentration of a solute as compared with its concentration outside the cell may be due to either Donnan equilibrium, precipitation of the solute in the cell, its combination with other molecules to form an entirely different molecule, or an expenditure of energy to maintain concentration differences on the two sides of the membrane. The latter involves the mechanism of active transport.

Active Transport

The mechanism of active transport is not known but perhaps there is a carrier involved as discussed in Chapter 5. In fact, such active transport, which involves an expenditure of metabolic energy, has been implicated in the permeability of membranes to such substances as inorganic ions, sugars, and certain other organic compounds.

Pinocytosis is another method by which substances may enter the cell. This unique phenomenon may explain the entrance of large molecules, such as proteins, enzymes, and even particles into the cell. It has been reported to occur in some plant cells but the extent of its occurrence and its importance to the cell are still to be determined.

Pinocytosis could best be illustrated by visualizing some force inside of the cell taking hold of a small area of the plasma membrane and pulling it inward, forming a long, narrow channel into the cytoplasm. Then portions of this channel break off, forming small vacuoles or vesicles. These vesicles, when first formed, would contain a solution whose concentration is identical with that of the solution outside the cell. Which means that materials which could not pass through the plasma membrane could enter the cell in this manner. More research is needed to learn the importance of this interesting phenomenon to the cell.

Another mechanism by which materials move out of the cell is by active secretion. *Secretion* is common in plant cells generally. It is defined as a process by which chemicals in the cytoplasm are transferred to the exterior of the cell. Perhaps there are several mechanisms by which secretion occurs, but the best known is that involving the Golgi apparatus. The Golgi apparatus is made up of a number of *dictyosomes*, seen in Fig. 14.12, and each dictyosome is made up of a number of rather distinctly morphological membranous tubes called *cisternae*. These cisternae are

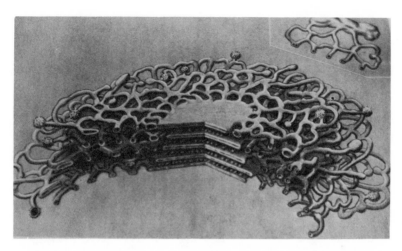

Fig. 14.12   Illustration of a dictyosome, of which a varying number combine to form one Golgi apparatus. This dictyosome is composed of five cisternae, with a cross-section of one seen in the insert. (From Mollenhauer and Morre, 1966. *Annual Rev. Plant Physiol.* **17:** 27–46.)

always smooth and, like other cell membranes, are composed of lipids and proteins. However, the composition of these lipids and proteins is not known.

The dictyosomes are made up of two to about twenty cisternae and are several microns in length, allowing them to be seen with the light microscope. They are often moving about within the cell, but this movement is not always associated with protoplasmic streaming. Rather, their movement is often rotational motion, much different than that associated with protoplasmic streaming. However, this does not mean that protoplasmic streaming is without effect on the Golgi apparatus and indeed, they can often be seen to move passively with this streaming.

Golgi apparatus are found in nearly all eukaryotic plant cells. In quiescent and replicating cells, they are few in number and are not producing vesicles. However, in secretory cells, they are much more numerous and are producing large vesicles which presumably participate in secretion. They do this by the ends of the cisternae pinching off to form spherical, membranous structures. It is believed that these are filled with secretory products and they move to the plasma membrane where the vesicle membrane unites with the plasma membrane, discharging its contents to the outside. How the products get into the vesicles is unknown, but the Golgi apparatus does not synthesize most of these.

Just how energy is implicated in this mechanism is unknown, but it is required. Perhaps it is needed for vesicle formation, since vesicle formation is inhibited by respiration inhibitors.

However, not all secretory mechanisms in the plant cell involve the Golgi apparatus. In fact, some of the most notable examples, such as nectar secretion and the secretion of hydrolytic enzymes, involve endoplasmic reticulum and special cell membrane forms instead of the Golgi apparatus. The mechanism by which these function is not well known.

The materials that enter a plant cell or leave it, must be transported to or from the cell. Such transport is the subject of the next chapter.

Chapter Fifteen

Intercellular Communication

Intercellular communication refers to the movement of chemicals or signals from one cell to another within the living plant. These may be either electrical signals, such as traverse the nerve cells in animals and as appear to function in *Mimosa* sp. of plants, or they may be chemicals of various types.

RAPID COMMUNICATION

One interesting type of intercommunication in plants is exhibited by rather rapid movements of structures in some species, movements that are often caused by stimulation of areas of the plant some distance from the location at which movement occurs. Examples of such types of movements are found in the insect-trapping mechanisms of some insectivorous plants; in the bending of the stamens, when touched, in certain plants such as barberry; and in the collapse of the leaf form of the sensitive plant such as *Mimosa* sp. How such mechanisms work can be demonstrated with *Mimosa*. At the base of the leaf or leaflet, are found *pulvini*, a group of very large cells. These cells have numerous contractile vacuoles, and when any part of the leaf is touched, these contractile vacuoles rapidly expel water, and its solutes, from the cell. This results in a decrease in turgor pressure of the cell and in a rapid decrease in the cell volume. Such a decrease in cell volume reduces the turgidity of the pulvinis, causing the leaf petiole to bend downward. Such a response, is quite rapid, and can occur within a few seconds after the leaf has been touched.

What causes the contractile vacuoles of the leaf to expel their contents is not entirely known, but such action is known to be preceded by a change in the electrical potential of the cells. Such electrical potential changes can be observed throughout the length of the petiole, indicating that the form of intercellular communication in these cases is an electrical signal, generated at the site of touch, which travels down the phloem and protoxylem parenchyma cells to the pulvini, where the cells are caused to contract. The velocity of this propagation is quite fast, and may be as rapid as 17 cm/sec. How the contractile vacuoles are made to contract is unknown, but the ATP-ATPase system appears to be implicated.

TRANSLOCATION

The remainder of this chapter will be concerned with the movement of chemicals within the plant, a movement referred to as *translocation*. Some aspects of this topic have been discussed in previous chapters. These chemicals that are translocated may function as nutrients, hormones or waste materials in the plant.

Evidence of Translocation

That chemicals are translocated between cells and between tissues within the living plant is easy to demonstrate. Oxygen enters the plant from the atmosphere and moves to most of the cells of the plant. Carbon dioxide, given off by all plant cells, moves into the atmosphere surrounding the plant, as does the gas ethylene, and many volatile chemicals, such as the essential oils.

If a leaf is left on the plant overnight, its carbohydrate content will decrease greatly, but if the same leaf is excised, there is little such loss. Also, with senescence, the mineral content of a leaf greatly decreases, indicating the movement of these substances from the leaf.

During fruit formation, the carbohydrate content of the developing fruit increases greatly, but most of these carbohydrates are not produced by the cells of the fruit so they must move from other parts of the plant.

Nicotine is known to be produced only in the roots of the plant, but is found in highest concentration in the leaves, again indicating its movement from the roots to the leaf cells.

Hormones are needed for the growth of the plant, but are not produced by all of the cells that need them.

Many more bits of evidence to support the contention that chemicals do move from cell to cell within the plant could be given, but this should suffice to indicate the importance of this phenomenon and the extent of its occurrence within the plant.

Terms

When discussing the subject of intercellular communication, the cgs or metric system should be used and certain definitions should be followed. It is not always possible to determine just what is meant when an author refers to the rate of movements within the plant, since he may be referring to either the amount of material moving per unit of time or to the distance moved per unit of time. To keep these parameters separate, it is desirable to refer to the distance the solute will move per unit of time as the *velocity*, and to the amount of materials that will be moved per unit of time as the *flux*. The rate will be used only when it is impossible or unnecessary to distinguish between the two.

Diffusion

When one thinks about the movement of chemicals in a biological system, one automatically thinks of that universal phenomenon called diffusion. Indeed, diffusion always plays a role in such movement, but this role is minor in translocation, since net movement of chemicals by diffusion is a very slow process when the distance involved is greater than 1 mm. Certainly, it is much too slow to be the mechanism through which chemicals move great distances in plants. Also, the velocity of such movement is much too great to be explained by diffusion. Therefore, it can be said that although diffusion is occurring in the plant and does play a role in translocation, this role is so minor as to be insignificant, except in the case of volatile, insoluble gases as discussed below. Diffusion is much more important for intracellular transport (*see* Chapter 1) than for translocation between cells.

A unicellular plant or even those consisting of a few cells, particularly if they are filamentous, do not have problems obtaining the chemicals they need from their environment. However, as the plant structure becomes greater in diameter, the procurement of such materials becomes a problem. Diffusion is no longer adequate to transport these materials the distance that is required, and transport through cells is hindered by the cell through which the transport occurs monopolyzing the materials and only transmitting those in excess of its needs. Therefore, long distance transport has necessitated the development within the plant of specialized cells and of mechanisms other than diffusion. Such cells include the vessel elements and sieve tube elements as well as certain parenchyma cells.

Although the translocation of water is an important translocation phenomenon, it was discussed sufficiently in Chapter 4 to make further discussion in this chapter unnecessary. Therefore, that subject will receive little attention here.

Ethylene

Perhaps the simplest form of translocation of organic compounds is expressed by water-insoluble gases, such as the hormone *ethylene* (ethene). Ethylene is volatile and very insoluble in water. Its translocation seems to begin with its loss from the cell by leakage or secretion, and its subsequent diffusion in the gaseous form through the intercellular spaces and cell wall spaces from cell to cell and then out into the atmosphere through the stomates or lenticles of the plant. From there, it can move from one plant to another and from one part of the plant to another part with the atmospheric currents. The volatile essential oils and hydrocarbons may also be translocated in this manner.

Other chemicals can be translocated from cell to cell by moving out of the cell and moving through the cell wall and intercellular spaces. Just what percentage of translocation occurs through this pathway is unknown.

PLASMODESMATA

It is also possible for substances to move from one cell to another without ever passing through the plasma membrane. Perhaps much translocation occurs in this manner. Such movement occurs through either plasmodesmata or through the strands of protoplasm that connect sieve tube elements together — strands that indeed seem to be continuous from sieve tube element to sieve tube element and through these individual cells.

Indeed, every living cell in a multicellular plant is interconnected with every other such cell, either directly or indirectly, by strands of protoplasm. Such strands are called *plasmodesmata*. These plasmodesmata can be seen in Fig. 1.4. They are of small diameter but appear to be continuous with the ground substance of adjacent cells. They are numerous, but vary in number from cell to cell, with estimates of 6,000–24,000 per cell having been reported. They appear to be composed of ground substance, limited by a membrane similar to and continuous with the plasma membrane, and they probably allow a free exchange of ground substance and the included solutes, between these adjacent cells. In some cells, ten or more of these strands are grouped together in what are known as *primary pit fields*, while in others, the strands are randomly spaced around the cell. No doubt plasmodesmata are formed during cell division (cytokineses) and probably persist for the life of the cell. The theory of their origin states that they are formed from spindle fibers, and are remnants of these fibers that formerly penetrated the cell plate during its formation. The large number of plasmodesmata per cell could support this hypothesis, since about 500 spindle fibers appear to extend through each cell plate. Endoplasmic reticulum does not extend through these plasmodesmata, and therefore is not continuous from cell to cell.

The sieve tube elements are found only in the phloem of the vascular tissues of the multicellular plant. They are very specialized cells, oriented to be joined end to end to form long tubes known as sieve tubes. The ends of each of these cells is covered with a plate, known as the *sieve plate*. This sieve plate is porous, and through these pores extend strands of protoplasm which pass on through the sieve plate of the adjacent sieve tube element, and possibly on through other sieve tube elements to form long strands of cytoplasm that may extend the length of the sieve tube.

AUXIN TRANSLOCATION

One type of translocation phenomenon is that illustrated by the polar movement of auxin. Auxin is a plant hormone that is produced in some cells of the plant but not in others, even though it is probably required by all cells. Therefore, it must be translocated from the sites of its synthesis to all cells of the multicellular plant. *Indoleacetic acid*, abbreviated IAA, is the most common auxin of plant cells and is very widely distributed among all of the species of higher plants. Its chemical structure can be seen from Fig. 15.1. It is synthesized in the meristematic cells of the root tips, apical and lateral buds, and in the young developing seeds, leaves, and fruits.

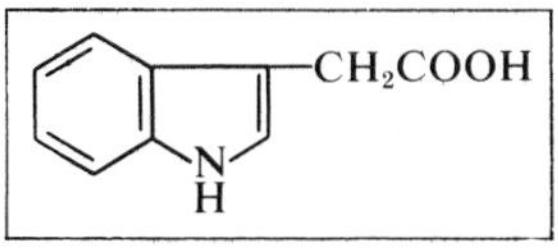

Fig. 15.1   The molecular structure of indoleacetic acid (IAA).

Polar Movement

The translocation of auxin is different from that of most other organic compounds since such translocation has a lower velocity and is polar, meaning that it will travel laterally through plant cells only in one direction, and not in the opposite direction. This polarity can be easily demonstrated by cutting a section of an oat seedling coleoptile just below the tip, placing this section on an agar block, and placing another agar block, but one that contains IAA, on the top of the section. After a period of time, agar will be found in the lower block of agar, having been translocated there from the upper block. However, if one now repeats this by inverting the coleoptile section, no IAA will appear in the lower block even after longer periods of time. This is illustrated in Fig. 15.2.

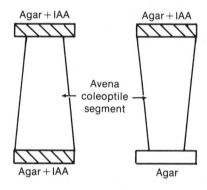

Fig. 15.2   Drawing to illustrate the polarity of IAA movement. Shaded area indicates the presence of IAA. Section is normally-oriented at left, and inverted at right.

Two types of auxin exist in plants, namely, free and bound auxin, and whether or not the auxin is transported depends upon which of these two forms it is in. Only the free auxin is able to be transported, although possibly an equilibrium exists which converts bound auxin to free auxin as the free auxin content of the cell decreases.

The polar transport of auxin must occur through the living parenchyma cells of the plant, since polarity disappears when the cells are killed, and since mature sieve tubes have not yet been formed in the coleoptile sections used for the demonstration in Fig. 15.2.

The velocity of the polar transport of auxin is rather slow, being about 1 cm/hr, which is, however, too fast to be attributed to diffusion. This velocity is also quite temperature independent, but the flux is definitely temperature dependent. Increasing the temperature has an effect on translocation that can be best visualized by equating the transport mechanism to a conveyor belt. Increasing the temperature would not speed up the conveyor belt, but would be synonymous with piling more IAA on the belt so that more IAA is transported per unit time, at the higher temperature, but the velocity of movement is no more rapid. This can be illustrated in Fig. 15.3.

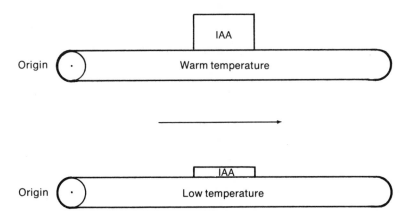

Fig. 15.3   The temperature effects on IAA transport is much like the transport of materials on a conveyor belt, wherein increased temperature does not speed up the conveyor belt but puts more material on it.

However, the flux of auxin transport cannot continue beyond a certain concentration. If progressively higher concentrations of auxin are added to the top agar block in Fig. 15.2, the flux will increase until a certain value is reached, and subsequent higher concentrations of added auxin will have no added effect. Apparently, the transport system becomes saturated at certain auxin concentrations.

Polar transport of auxin is also influenced by the stage of development of the tissue used to study such transport, being less vigorous in older tissues than in younger ones. Perhaps this is correlated with the fact that the rate of metabolism is greater in younger tissues, and since metabolism is needed for polar auxin transport, higher metabolic rate would yield increased auxin transport.

Among the chemicals known to inhibit polar auxin transport, triiodobenzoic acid is the most effective, and therefore is a valuable tool to use in studies of such transport. However, phenoxyacetic, benzoic, and naphthylphthalamic acids are also inhibitory. It is worth noting that some of these are auxins and therefore would be expected to behave similarly to IAA. However, they are inhibitory when their concentrations are high enough, but not at low concentrations.

Perhaps the polar transport of auxin involves movement through the plasmodesmata, but how this could explain polarity is not known. Some evidence also indicates that passage through cell membranes is involved, but this has not been well substantiated.

As important as the polar transport of auxin is to the plant, it must not be assumed that all movement of auxin within the plant occurs by such a mechanism. Indeed, auxin can, and does, move in the vascular tissue also, in association with the sugars and most of the basipetal movement of auxin is mediated through metabolic activity, a small amount is brought about by diffusion. However, such diffusion could not, in itself, explain the movement of most of the auxin in plants due to its inherent slowness.

In some manner entirely unknown, light can cause a lateral movement of auxin. This has been discussed in relation to phototropism in Chapter 7. When one side of a plant apical tip is exposed to unilateral light, auxin moves away from the lighted surface toward the dark side of the stem, as seen in Fig. 7.8. Such movement involves metabolic activity and is related to polar auxin transport.

CARBOHYDRATE TRANSPORT

The type of organic compounds that moves in the greatest quantity within the green plant is the carbohydrates. More carbohydrates are transported over great distances within the plant than any other organic compound. To get some appreciation of the magnitude of this transport, an acre of apple trees will store more than four tons of carbohydrates in the fruit during a growing season, and most of this is transported to the fruit from other parts of the tree. In plants, these carbohydrates move primarily in the form of sucrose, with their ultimate source being the photosynthetic cells located primarily in the leaves, although any green cell, such as those of fruits and stems, can also serve as this source. Sugars can also originate from the hydrolysis of stored carbohydrates in the stems and roots of the plant.

Age

Not only is it important to the plant that chemicals move through the plant, but it is also important where they move, and what their pattern of distribution is. If one applies either radioactive sucrose or the radioactive carbon dioxide from which the plant can make its own sucrose, to a leaf of a plant and then determines the location of the radioactivity within the plant after a period of time, it can be seen that the sucrose moves out of the leaf in large quantities both into the stem and into the young leaves of the plant, but very little of this sucrose goes into the older leaves, especially those that are beyond maturity and those that are senescent. This distribution pattern has been observed many times and seems to be the rule rather than the exception. The stage of development of the leaf definitely plays a role in determining whether or not organic materials will be imported into it. If one considers the leaf as it develops in terms of the destiny of the sucrose it synthesizes, one finds that during the earlier stages of leaf development, until it is about half of full size, it does not export sucrose. This leaf not only uses what it produces but also imports additional quantities from other, more mature leaves, or from storage tissues of the plant. By the time it reaches maturity, it exports most of the sucrose it produces although it also imports a lesser amount. As the leaf becomes senescent, essentially all of the sucrose it synthesizes is exported either to the younger leaves, the flowers, fruits, or the storage tissues of the plant. This distribution pattern is not only true of sucrose, but also of other translocated organic compounds. Young leaves are net importers, and senescent leaves are net exporters.

In the Spring of the year, as the new leaves are developing, there is a net movement of organic compounds from the storage tissues in the roots and stems into the young leaves. As the season progresses, the leaves approach maturity and are then exporting sucrose to the storage tissues of the stems and roots and to growing apical buds, so there is now a net increase in carbohydrates within the plant. As flowers and fruits begin to develop on the plant, the direction of transport is changed so that now most of the sugars produced move into the young flowers and fruits and little into the storage tissue. As Fall approaches, the sugars are translocated from the leaves again in the direction of the storage tissues of the stem and roots, with little going to the apical buds since they are now dormant, and dormant buds import very little sucrose compared with that imported by actively growing buds.

Location

The distribution pattern just described applies to plants whose leaves mature at a uniform rate with each leaf on the plant being in the same stage of development. However, such is rarely the case, since progressively younger leaves are found as one looks up the branch or shoot of the plant. In such cases, the direction of movement of the sucrose synthesized by the leaf depends upon the location of the leaf on the branch. The basal leaves transport nearly all of their exported sucrose into the stem and down into the roots, whereas those higher up have their sucrose largely transported to the young leaves and apical buds near the tips of the branches. Those in between have their sucrose transported in both directions. On the other hand, those leaves near flowers or fruits have their sucrose transported almost entirely into the flowers or fruits, as these structures develop on the plant. So the direction of transport depends upon the location of the source of sucrose in the plant.

However, these distribution patterns can be changed by selectively removing leaves from the plant. For instance, if one removes some of the leaves from a grape vine near a cluster of developing grapes, this has no effect on the amount of sucrose that goes into the grapes because now the leaves farther away from the cluster translocate sucrose into the cluster instead of into the roots as they normally would. Therefore, removing leaves from the plant will alter the distribution pattern of the sucrose translocated through that plant.

Light

Whether the plant is in the dark or the light also seems to have an effect on the pattern of distribution. During the night, carbohydrates, stored in the leaves during the daytime, move downward into the roots. However, when the light is turned on, the direction changes and most of that translocated goes into the growing points of the plant.

Summary

Perhaps the direction of movement of organic solutes can be summarized by saying that movement is from older tissues to younger, growing tissues. There is an upward movement through the phloem when growth resumes in perennial plants—trees and shrubs—in the Spring, from lower leaves and storage tissues into developing flowers and fruits, during seed germination and seedling development when movement is out of the storage tissues of the seed and into the growing seedling, during early stages of shoot growth from bulbs, rhizomes, tuber, etc., during the Winter when leaves are no longer on the woody plant, and during the day when younger tissues are above the source tissues. There is a net downward movement through the phloem during the day and night when photosynthesis is or has been substantial, on the plant before flower and fruit development, and after fruit development has ceased and the fruit has reached maturity or fallen from the plant.

By what mechanism the distribution pattern is determined is one of the very intriguing problems of botany and one which cannot be answered at the present time.

Although there is some lateral movement of translocated materials as they move up or down the stem, most of the movement is vertical and is confined to a rather narrow region of the phloem. As an example, if a tree is growing at the edge of the forest so that one side of the tree receives high light intensity and the other side a low light intensity, the rate of photosynthesis will be higher on the lighted side and more sucrose will be transported down that side, supplying more nutrients for the growth of the cells nearby. Therefore, the annual rings on such a tree may be much wider on the lighted side than on the dark side.

Sucrose is not the only organic compound that is translocated through plants. Many organic compounds and even some inorganic compounds move through the phoem. In all cases, the rate of movement of such is the same as the rate of movement of sucrose. In fact, it is possible to study the rate of sucrose movement by studying the rate of movement of radioactive phosphate through the phloem. Many herbicides also behave similarly, and their translocation will be further discussed in Chapter 18.

NITROGEN TRANSLOCATION

Although, as stated in Chapter 9, each plant cell has the potential to reduce nitrates and synthesize amino acids and therefore the potential to produce the other organic nitrogen compounds needed, there is a translocation of organic nitrogen in the plant. Amino acids synthesized in the roots moves into the shoots. Such movement may be either through the phloem or through the xylem. Perhaps it seems strange that amino acids should move through the xylem, but some alkaloids do also. Nicotine, which is synthesized only in the roots of tobacco plants, moves up the xylem, dissolved in the transpiration stream, to get into the leaves. As the reader may recall, in some species most of the nitrate reduction occurs in the roots.

Amino acids are also synthesized in the leaves. In some species, much of the nitrate reduction occurs in the leaves, but in all species some amino acids do move out of the leaves and into other plant parts. In immature leaves, no such export is observed, but as these leaves reach maturity some export begins, and reaches its peak by the time the leaves have reached senescence, at which time most of the amino acids exported arise from protein hydrolysis rather than nitrate reduction. Particularly is this noticeable in the Fall of the year just prior to leaf abscission, when the nitrogen content of the leaves is greatly reduced due to a net export of nitrogen from the leaves to the more permanent parts of the plant.

Most of the nitrogen that moves from the leaves is in the form of one of the amino acids and such movement is restricted to the phloem sieve tubes. Nevertheless, some nitrogenous growth regulators can also move through the phloem and parenchyma cells, as previously discussed.

RATE OF MOVEMENT

Methods used to determine the rate of movement of materials between the cells of a plant can be illustrated by considering those used to study auxin transport, but will vary somewhat with the type of chemical studied and the tissue used for such study and the means of applying exogenous chemicals. First, one can measure the amount of auxin which accumulates in a receiver and correlate this with the time interval between the time the auxin was applied and the time the receiver was removed. This is illustrated in Fig. 15.2. Second, one can use plant segments of different lengths, and measure the difference in time it takes for each receiver to attain the same auxin concentration and correlate this with the difference in the lengths of the segments. Third, one can apply radioactive auxin to the agar at the top of a segment, and measure the time it takes for the radioactivity to move to the basal agar block by exchanging blocks periodically and assaying them for radioactivity. A commonly-used modification of this is the application of radioactive carbon dioxide to a leaf and the time interval determined when the radioactivity passes a certain point of the plant's stem. Fourth, is a method to study endogenous auxin velocity whereby the source of auxin production within the plant structure is excised and then determinations are made to find out how long it takes to reduce the auxin content of those parts of the plant that would normally obtain their auxin from the source removed. This cannot be used with stable chemicals that are stored in the cells.

Phloem-transport

The movement of organic compounds through the phloem occurs through the sieve tubes — tubes made of sieve tube elements joined end to end. The phloem sieve tube element is a very specialized cell at maturity. In earlier stages of development, it looks much like other parenchyma cells and contains those organelles found in other cells. However, as the cell reaches maturity, cytoplasmic strands develop that run the length of the cell and continue through the cell plate to connect with similar strands of adjacent cells. Other than these strands, the interior of the cell appears to be a large compartment with few organelles. Mitochondria are present but the nucleus disappears as do the plastids and even most of the endoplasmic reticulum. Probably the functions of the nucleus are taken over by adjacent companion cells. Also, the sieve plate, through which the strands of protoplasm penetrate, is composed largely of the chemicals called calose, instead of cellulose.

Movement through the phloem is not polar, since movement occurs both up and down the plant and even in both directions at the same time. The path of such translocation is through the sieve tubes of the phloem. These tubes must be very long in some species, being continuous in plants from the root tips to the leaves. Also, in some plant species, root grafts occur which join adjacent plants together. Thus it is conceivable that sieve tubes may be over 100 feet in length. Even if the sieve tubes are restricted to the individual plant, some of these trees that are over 100 feet in height would have some phloem sieve tubes that would be in excess of 100 feet.

Killing much of the phloem tissues of a plant has little effect on the amount of materials that move within that plant. Perhaps there are two reasons for this phenomenon. First, there appears to be more sieve tubes in each plant than are actually necessary for the survival and well-being of the plant, and second, new phloem cells are soon formed. In fact, phloem cells are rather short-lived anyhow, seldom surviving for more than a season or two.

Movement through the phloem is rarely too slow for maximum plant development. The velocity and flux of translocation are normally adequate to keep the cells of the plant supplied with abundant carbohydrates, as long as these are available, and therefore phloem transport is rarely a limiting factor in plant growth, except at temperature extremes. At high temperatures, transport through the phloem may be slow enough to be limiting, and at very low temperatures, such is also conceivable, but at most temperatures, such transportation occurs at an adequate rate.

TRANSLOCATION MECHANISMS

Transport through the sieve tubes requires metabolic energy. Killing these cells stops such transport, and withholding oxygen or applying respiratory inhibitors slows it down. How this metabolic energy is related to the mechanism of translocation is unknown. Does it alter membrane permeability, or is it required for active movement through the protoplasm?

The mechanism of translocation through the sieve tubes has been studied for many years, and the amount of data that has been accumulated from these studies is tremendous. However, the final word on this subject will not be said for a long time. Currently, there are two main theories to explain the mechanism of phloem transport. One is the massflow or Munch hypothesis, discussed in Chapter 4, and the other involves a mass transport system in which the cytoplasm of the sieve tubes takes an active part as the transporting medium. In both of these theories, the sucrose is considered to move in mass, but the question yet to be resolved is whether the cytoplasm of the sieve tube elements plays an active role in transporting the chemical through the cell, or a passive role, in which case one might say that the solution being transported is being forced through it. The biggest problem we face in trying to resolve these two views is that it has not been possible to accurately define conditions in the living sieve tube elements. Much data are available which describe the dead sieve tube, but observations on living sieve tube elements are lacking. In many respects, this translocation system behaves a great deal like diffusion, but with a much higher velocity. In fact, its velocity is about 30,000 times that of diffusion, or up to 150 cm/hr. This is about 150 times more rapid than the polar transport system previously discussed to explain auxin transport.

Long distance transport of sucrose occurs in the sieve tubes of the phloem tissue. However, the sucrose must get into these sieve tubes before such transport can occur. How the sucrose gets from the cells of its origin to the phloem sieve tubes is unknown, but perhaps this occurs through the plasmodesmata.

Entrance into the sieve tube elements can occur even against a concentration gradient – even if the sucrose content in the sieve tubes is greater than that on the outside of these cells. This, plus the fact that energy and ATP are required, eliminates diffusion as the process. The details of this packing of the sieve tubes with sucrose are not known. However, ATP and metabolic energy are required, and it appears that the sucrose must somehow be first transformed chemically before it can move across the membrane of the sieve tube element. This shows similarities with active uptake discussed in the previous chapter.

Also, little is known about how the chemicals get out of the sieve tubes. No doubt, these tubes are not impermeable to such chemicals and some continuously move out, perhaps by diffusion, as they move down or up the tubes. Such movement laterally is much less extensive than the vertical movement but does occur. Lateral movement gets the chemicals into the ray cells, and other cells surrounding the phloem sieve tubes, and even a small amount ends up in the xylem. If there is some metabolic pump that removes chemicals from the phloem sieve tubes, it is unknown. However, such a mechanism is quite likely.

More will be said about the translocation of substance through the plant in subsequent chapters, and particularly in Chapter 18. The next chapter will begin a discussion of plant development, a process that begins with cell division and ends with death.

Chapter Sixteen

Plant Growth: Cell Division

This chapter begins the study of the physiology of plant development. Plant *development* is the sum total of growth and differentiation (plant development = growth + differentiation). It begins with the division of the cell and ends with the death of the plant.

PLANT GROWTH

Growth is defined as an increase in size, and is due to either cell *division*, which forms more cells, or to cell *enlargement*, which forms larger cells, or more often to both. Fruit increases greatly in size before ripening and this increase is an example of growth. In the case of fruit, most of this growth is due to cell enlargement rather than to cell division, since cell division in fruit is essentially complete before the flower petals wilt and fall off. On the other hand, the growth of the embryo within a seed, from the time of fertilization until the seed reaches maturity, is largely by cell division, with only a limited amount of cell enlargement. Nevertheless, the growth of the multicellular plant, as indicated by an increase both in height and diameter, is a result of both cell division and cell enlargement. Cell division will be considered in this chapter, with cell enlargement being reserved for Chapter 17.

Differentiation refers to specialization of the cells, tissues, or organs. Such specialization may be either morphological specialization or functional specialization, and begins following division. Differentiation is usually complete just after the cell reaches its maximum size, and, will be discussed in Chapter 18.

CELL DIVISION

Cell division is the source of all plant cells. No cells arise other than by division of existing cells. In unicellular plants, this is the method by which new plants are propagated, and in multicellular plants, division results not only in plant propagation, but also in the growth of the plant.

In a population of unicellular plants, such as bacteria, or unicellular algae or fungi, each cell has the ability to divide so each cell represents either the site of cell division or the potential for cell division. As long as the environment is favorable, these unicellular plants will continue to divide periodically, doubling the population each time such a division occurs. How often each division will occur will depend upon several factors, such as the plant species and the environment. Some will divide every half hour, but more often division occurs every 20–24 hours.

Growth of unicellular plants, such as chlorella, can be measured either by counting the number of cells, periodically, in a given volume of water; measuring, periodically, the change in density of the cell suspension with a photometer; or by determining the change in dry weight, fresh weight, or the change in the concentration of such chemicals as proteins, RNA, or DNA.

Fewer methods are available for determining growth by cell division, since cell enlargement and cell division are so closely associated in most plant species that it is impossible to find cells undergoing one without the other. Therefore, such determination must be made either by counting the number of cells per unit area or volume, or, if working with a unicellular plant, by photometrically determining the change in turbidity of a cell suspension. Of course, if one can be sure that growth is limited entirely to cell division, then the other methods listed in the previous paragraph can be used.

LOGARITHMIC GROWTH

Growth by cell division can be visualized as a process by which one cell divides to form two cells, these two cells divide to form four cells, these to form eight cells, and so forth. This type of increase is known as a logarithmic increase and can best be illustrated by plotting the data on semilog paper with time as the arithmetic parameter and growth rate as the logarithmic parameter, as shown in Fig. 16.1. Or this relationship can be expressed mathematically by the following equation:

$$\log N_t = \log N_0 \times kt$$

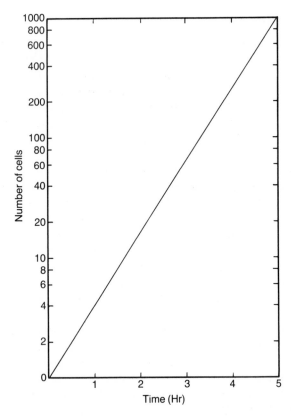

Fig. 16.1 A graph showing the expected rate of division by cells that retain their capacity for continuous division. (Doubling time = ½ hr.)

where N_t = the number of cells after a given period of time
N_0 = the original number of cells
t = duration of growth period
k = the growth constant or the growth rate per unit time

From this equation, or from the graph in Fig. 16.1, the *doubling time* of a cell population can be determined, which is the time necessary for a given population of cells to double in number. If the rate of growth remains constant, both the doubling time and k will be of the same value at all time intervals, and the curve plotted on semilog paper will be a straight line. However, such is not the case. Eventually, the rate slows down. If this were not so, one cell dividing every thirty minutes would, within a few days, produce a cell population whose mass would exceed the mass of the universe.

SIGMOID GROWTH

If the curve showing the relationship between the cell population and time is plotted, invariably the curve will take on an "S" shaped appearance. This type of curve is called a *sigmoid* curve and is expressed by growth generally. Such a curve is shown in Fig. 16.2, and indicates that

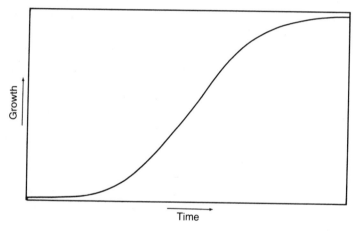

Fig. 16.2 A graph showing the growth rate usually observed. This is called a sigmoid curve.

originally there is a lag period before the rate of growth reaches a maximum. The maximum is then attained and continues until something begins to limit the growth rate and slow it down. Then the growth rate becomes constant and remains so for some time, although it may eventually even decline as the population becomes senescent. The leveling off and subsequent decline of growth rate is due either to a depletion of nutrients, or to an accumulation of toxic materials, or to the influence of growth regulators. Thus, the environment in which the cells are dividing will have a great influence on the rate of division. A population of unicellular plants can be maintained in the exponential growth rate by periodically changing the medium in which the cells are cultured, so that nutrients will be replenished and toxic or inhibitory materials removed. In fact, it is often possible to control the rate of growth by adding, to an *in vitro* culture, just enough of a given nutrient to keep its concentration near the level of deficiency. Thus, the growth rate can also be changed by changing the rate at which nutrients are added to the medium.

SYNCHRONIZED DIVISION

By observing a population of unicellular plants under the microscope, it is usually possible to find all stages of cell division represented. This is so because each cell functions independently in cell division, and divides when it is ready and not until. As a result, one cell will be entering cell division as another is just finishing cell division. In such a population it is difficult to study the physiology and especially the chemistry of cell division. The desirable situation for such studies would be to have a population wherein all of the cells were in the same stage of cell division at the same time so that when one divides, all will divide. Such a situation is referred to as *synchronous* cell division.

Synchronous cell division does occur naturally in rare cases. For instance, in some algal species, such as some diatoms and some green algae, cell division occurs only at night, and only for a short period of time during a 24 hour period can the cells be seen to be dividing. Also, in the pollen mother cells, synchronous division is particularly evident, as indicated in Fig. 16.3. During the process of pollen grain formation, pollen mother cells divide by meiosis to form a tetrad of microspores. Each of these microspores undergoes two subsequent mitotic divisions to form the mature pollen grain. Only during the division of the pollen mother cell is division well synchronized and with subsequent divisions, synchrony is progressively lost. This loss appears to be due to the rupture of protoplasmic strands that join all of the pollen mother cells together but only a few of the microspores together. Perhaps some stimulus that triggers cell division passes from cell to cell through these strands.

The synchrony that occurs in cell division, in unicellular algae, allows studies to be made of the changes that take place in the chemical composition of the cell during various stages of cell division.

THE CELL CYCLE

When cells are active in mitotic division, the cells are said to be in a cell cycle. The cell cycle is a cycle of events which occurs in order and results in repeated mitosis and cytokinesis. Only cells involved in mitotic division are said to be in a cell cycle, not those involved in meiosis.

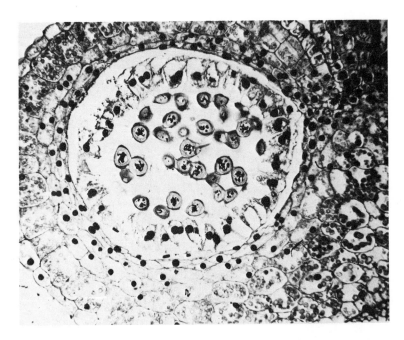

Fig. 16.3 Cross-section of an anther showing synchronous division of the microspore mother cells.

The cell cycle involves interphase, as well as prophase, metaphase, anaphase, and telophase of mitosis. In essence, it involves both nuclear division and cell interphase. However, interphase is said to be made up of three phases of the cell cycle, namely, the G_1, S, and G_2 phases, whereas nuclear division comprises only the M phase. How long each cell remains in each of these phases depends somewhat upon the cell, but is not greatly influenced by the prevailing temperature. Usually the cell will spend about 90% of its cycle in interphase, and 10% in nuclear division. Of the latter, more than half is spent in prophase, with the remaining 4% divided among the other phases, as shown in Fig. 16.4.

Generally, the entire cell cycle occurs about every 20–24 hours, with about $\frac{1}{3}$ in the S phase. The other $\frac{1}{3}$ of the time is shared equally by G_2 and M, although these durations may vary.

G_1

The G_1 phase begins with the completion of cell division. During this phase, the cell is restored to its original size and its nuclear:cytoplasmic

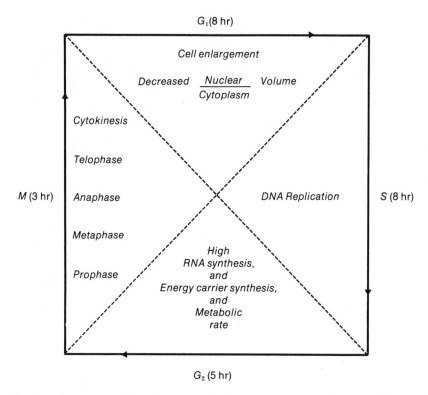

Fig. 16.4   Components of the cell cycle and the duration and activities associated with each.

volume ratio is restored to that of the parent cell before division. Also, the rate of photosynthesis is high during the G_1 phase and low during the M phase, at the time when the nucleus is dividing. This variation in rate may be due to variations in the efficiency of the photosynthesis which would be high during the G_1 phase, and low during nuclear division.

Related to the rate and efficiency of photosynthesis is the observation that an increase in light intensity will increase the number of cells in a population that divide at one time. This is associated with an increase in the ratio of the concentration of carbohydrates to protein. This ratio is also altered by the quality of the light. Red light increases the carbohydrate : protein ratio, whereas blue light lowers it, with white light giving intermediate values, as might be expected. Light quality also alters the relative amounts of the nucleic acids, with red light stimulating DNA synthesis and blue light stimulating RNA synthesis.

During the G_1 phase, there is a duplication of the Golgi apparatus, of mitochondria, and of plastids. Such increases occur during other phases of the cell cycle also, except during the M phase.

During the early stages of the G_1 phase, the chlorophyll content of algae is low but as this phase approaches its end, there is a great increase in chlorophyll content, an increase that is rather short-lived since it soon declines and remains low for the remainder of the cell cycle.

S

During the S phase, no visible changes occur within the cell, but chromosomes and therefore DNA replication — synthesis — is occurring. By the time this phase is over, the cell contains its 4N number of "chromosomes" and the DNA content is twice that of the cell in the G_1 phase.

The replication of the chromosomes, as indicated by a duplication of DNA, marks the first step in cell division. Such replication requires a synthesis of both DNA and of chromosomal protein, since each chromosome consists of one or more DNA molecules to which proteins are attached. Most of this chromosomal protein is histone. It is not passed on from generation to generation and is synthesized even in the absence of DNA replication. As revealed in Chapter 10, the histones are probably repressors for gene activity. Chromosome replication can be revealed by either counting the strands of chromatin, determining the degree of polyploidy, or by a quantitative analysis of the DNA.

Actually, although both DNA and chromosomal protein synthesis occur during the S phase of the cell cycle, it is not proper to speak of this as chromosomal replication, since separate chromosomes are not formed until anaphase. Prior to that time, the replicated structures are called chromatids, with the two chromatids of each chromosome joined together at a *centromere*, separating only during anaphase of nuclear division. As far as evidence is available, it does not appear that the chromatids function as separate chromosomes prior to anaphase so the definitions given above have physiological as well as morphological support.

The replication of DNA involves an unwinding of the two strands of the DNA molecule and the use of each one of these strands as a base to which the other strand forms and becomes attached. Therefore, each of the two DNA molecules formed from one DNA molecule contains one strand of the old DNA and one new strand, as shown in Fig. 16.5.

There is good evidence that the replication of the DNA molecule has certain defined starting points and does not just occur haphazardly. Each

Fig. 16.5 A drawing to illustrate DNA replication. Strands A are from the original DNA molecule, and strands B are being newly-formed.

chromosome does not have just one starting point either, but several may be present on each chromosome.

DNA replication requires the production of appropriate polymerase enzymes as well as the polymerization of DNA-nucleotides. These DNA-nucleotides must be synthesized and made available to the site of DNA synthesis before replication can occur. It is well known that the DNA-nucleotides are not accumulated in the cell at all phases of the cell cycle, but appear just prior to DNA replication. The DNA-nucleotide content remains low until just before DNA synthesis begins, at which time a rapid and great increase occurs. Following DNA synthesis, their content again decreases and remains low, through the remainder of the cell cycle.

What controls this supply of nucleotides? What mode of metabolic regulation is involved here? These answers are not entirely known, but in algae, the synthesis of the DNA-nucleotides appears to occur only in the presence of light. This is not true of all plant cells.

Does this buildup of nucleotides trigger DNA synthesis? In answer to this question, it appears that a deficiency of nucleotides will certainly prevent DNA replication, but just the presence of nucleotides does not guarantee that replication will occur. Something else triggers DNA replication than the supply of nucleotides.

G_2

The G_2 phase of the cell cycle follows chromosomal replication and represents a period when the cell is metabolizing fairly normally. At this time, the rate of respiration is high, with energy carriers being synthesized for use in subsequent nuclear division, and the rate of RNA synthesis is rapid. The lack of either oxygen or the presence of respiratory inhibitors during this phase inhibits subsequent nuclear division.

M

The *M* phase marks the beginning of mitosis, and ends with cytokinesis. During prophase, protein synthesis continues at a high rate, at least until late prophase when the rate declines greatly. After that time, protein synthesis is not necessary for the continued progress of mitosis. The rate of respiration and of RNA synthesis are also high during prophase. These result in an availability of both energy and proteins for subsequent nuclear division, during which time both the rate of respiration and of protein synthesis are low.

Prophase is also marked by the visible appearance of the chromosomes, which form as the chromatin shortens and thickens to form two chromatids, joined together at the centromere, with this entire structure being called a chromosome.

It is essential that if the daughter cells formed from cell division are to carry on the living functions, including eventual reproduction, there must be an eventual synthesis of DNA and chromosomes, and an equal distribution of these chromosomes in the daughter cells. This necessitates a segregation of the chromosomes during cell division so that equal numbers of duplicated chromosomes go to each of the two cells formed as a result of the division. Therefore, the orientation of the chromosomes prior to division and their distribution as they move into the two daughter cells is vital.

During cell division, chromosomes perform several functions. They contract, the chromatids separate, they become oriented at the cell equator, and they exhibit polarized migration. These functions are not the consequences of DNA replication but are due to other factors — factors associated with metabolic regulation.

Chromosomal behavior is always the same during cell division except for two functions. These differences form the basis for the differences between mitosis and meiosis, and include the presence or absence of homologous chromosomal pairing at the equator during metaphase, and the delayed division of the centromere of each chromosome. Even crossingover, which is emphasized during meiosis is not restricted to meiosis but also occurs during mitosis. What causes the chromosomes to pair or not to pair is a question that cannot be answered, but this one act determines whether meiosis or mitosis will occur.

As the cell enters metaphase, RNA synthesis is virtually stopped. It appears that the amount of RNA polymerase does not decrease during nuclear division, so there must be some physical change in the chromosomes during contraction, which reduces RNA synthesis. Details of this are unknown. This may be due, at least in part, to the disappearance of the nucleolus during prophase, since the nucleolus is known to be the site of at least the synthesis of ribosomal RNA.

As the cell enters metaphase, there is a distinct decrease in the rate of respiration. The reason for this cannot be explained.

The visible evidence of metaphase is the lining-up of the chromosomes at the equator of the cell, and the association of these chromosomes with the fibers of the *mitotic spindle*. These *spindle fibers* become evident within the cell during both metaphase and anaphase, and have two major properties, namely, they exhibit polarity, joining at both poles of the cell, and they contract as the chromatids of the homologous chromosomes separate. What controls both of these properties is unknown. It appears now that the spindle fibers are made up of microtubules, but how these become oriented and aggregated to form the fibers, and what causes them to contract to pull the chromosomes apart is unknown. This contraction is somehow initiated after the chromosomes line up at the equator.

An interesting study that needs more emphasis is to determine those factors which control the orientation of the mitotic spindle, since it is such orientation that determines the plane of cell division and therefore the direction of growth. In most plant cells, the plane of division is in the direction perpendicular to the long axis of the elongated cell. But, what about cells that are isodiametric?

Although such division may normally occur in a plane perpendicular to the long axis of the plant, this pattern can be altered, giving two-dimensional growth. To alter this pattern requires blue light of high intensity whereas red light results only in filamentous growth. The intensity of the blue light is directly related to the number of two-dimensional divisions, with the number of such divisions being proportional to the logarithm of the intensity of the radiation. When such cells are transferred to the blue light treatment, cell division in one direction continues for some time but eventually is supplemented by division in two dimensions. The physiological changes associated with this change in cell division pattern are unknown but it appears that prior to two-dimensional division, there is an increase in both RNA and protein contents of the cells.

How the chromosomes are transported to the opposite ends of the cell during anaphase is unknown. The energy requirements must be stored in the cell, since at this time the rate of respiration is low. Chromosome movement is associated with the contraction of the spindle fibers, but how such contraction is brought about — both what causes its initiation and the actual mechanism involved in the contraction of the fibers — is unknown.

PLANT POLYURONIDES

With telophase, after the chromosomes move to the poles of the cell, a nuclear membrane forms around each set of chromosomes, forming, temporarily, a binucleate cell. Then, near the center of the cell, a cell plate begins to form. This cell plate is composed largely of the *polyuronides*, also known as pectic substances.

Polyuronides are carbohydrate derivatives found in all plant cell walls. Actually, these are composed of long chains of galacturonic acid, an organic acid formed from UDP-glucose, although other monosaccharide derivatives are also present. UDP-glucose is formed from the union of a glucose molecule to a uridine diphosphate molecule. Uridine diphosphate is similar to ADP, and is formed by the removal of one phosphate group from uridine triphosphate (UTP), an energy carrier, just as ADP is formed by the removal of one phosphate group from ATP.

Pectic Acid

Three types of polyuronides are found in plant cells. The first to be considered is pectic acid. This is a molecule of about 100 galacturonic acid residues, and, being an acid, forms an ester with calcium or magnesium ions, an ester which is insoluble and is known as *pectate*. Thus, there are calcium pectates and magnesium pectates. These are cell wall components and are present in all cell walls, in addition to the cell plate. Pectic acid is water-soluble, but pectates are not. They are very insoluble.

Protopectins

Protopectins are also polyuronides found in the cell plate as well as in the cell wall. They are made up of much longer chains of galacturonic acid than are the pectates, and many are methylated, with methyl groups attached to the acid residues. These too are insoluble and are sticky materials. They commonly are found in the middle lamellae of plant cell walls.

Pectin

Another polyuronide found in most plant cells, and one that should be mentioned here although it is not a constituent of the cell wall is *pectin*. In contrast to the other pectic substances, pectin is water-soluble, and is located in the protoplasm of the cell. Pectin is also made up of methyl esters of pectic acid, with about 200–300 galacturonic acids in each chain, so the molecule is about twice as large as pectic acid, and differs from it by being methylated. It is, however, a smaller molecule than protopectin. Pectin is well known as a necessary constituent of jelly. Being a large molecule, it forms a colloidal micelle, and therefore imparts colloidal properties to the system, including the property of reversibility of sol-gel formation.

CELL WALL FORMATION

Cell wall formation begins with the formation of the cell plate shortly after the division of the nucleus of the cell during mitosis and meiosis. Sometime during telophase, vesicles appear to line up in the center of the cell between the two new nuclei. These vesicles are formed from the Golgi apparatus and from the endoplasmic reticulum of the cell, and are believed to contain the chemicals from which the new cell wall is constructed. Now the vesicles coalesce, or join together, with their membranes uniting to form the plasma membrane, and their contents form the cell plate. This cell plate is suspended near the center of the cell, but with continued coalescence, the cell plate grows at the ends until contact is made with the mature cell walls of the original cell, as shown in Fig. 16.6. This then gives two cells, and completes cytokinesis. When completely formed, the cell plate is composed of polyuronides, hemicellulose, and a

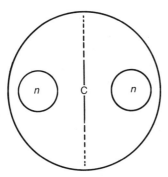

Fig. 16.6   Cell plate formation beginning in the center and lengthening until it extends from wall to wall.

loose meshwork of cellulose, and becomes the middle lamella between the two new cells as the primary cell walls are deposited on both sides of the cell plate.

This type of cell plate formation is a distinguishing feature of plant cells, in comparison with animals cells, since in animal cells, cytokinesis involves a pinching in of the old plasma membrane – animal cells do not have cell walls – until the two sides join, effectively separating the cell into two new cells.

Although each cell of the unicellular plant usually continues to divide periodically, this is not true of most of the cells of the multicellular plant. Most of the cells of the multicellular plant do not divide after they are formed, but increase in size and become so differentiated that subsequent division does not occur. Of course, some of these retain the potential for division, but some become so specialized that they even lose this potential.

PLANT MERISTEMS

A multicellular plant does not "grow all over," but rather most cell division is restricted to certain regions, regions known as meristems. *Meristems* may then be defined as areas of cell division in the multicellular plant. They represent regions of the plant that contain meristematic cells – cells that maintain the potential for both nuclear division and cytokinesis. Such division is by mitosis, rather than by meiosis; with meiosis occurring only in relation to gamete formation, not for the growth of the plant.

The meristems are of three types. First, there are the *apical meristems* found in the buds of branches and in the root tips. These are found in the root tips of all multicellular plants and in the terminal and lateral buds of the shoots. In these tips and buds, the apical meristem is located just under the epidermal cells. The second type of meristem is the *intercalary meristem*. This meristem is found principally in the leaves of monocots, located near the base of the leaf. However, in some monocots, such meristems are also located in internodal areas of the stems. In either case, intercalary meristems are surrounded on both sides with much mature tissue, whereas apical meristems have mature tissue just on one side. However, cell division in both the apical and the intercalary meristems results in an increase in length of the plant part, either the stem or the leaf, as the case may be.

The third type of meristem is the *cambium*. This meristem is located in the stems or roots of perennial plants, particularly dicot trees and shrubs, and its activity gives an increase in the diameter of the stem or root instead of adding to its length. Actually, there are two types of cambiums, namely, the vascular cambium which produces secondary xylem and secondary phloem, and the cork cambium which produces the cork cells of the bark. All multicellular plants have apical meristems, but not all have cambiums.

Since most monocots are not perennial or woody, they usually do not have a cambium but rarely, an *interfacicular meristem* will be formed between the vascular bundles and will increase the diameter of the plant. Palm trees are good examples of plants showing such a phenomenon.

Mitotic Index

Just because the cells of the meristems retain their capacity for cell division does not mean that they are continuously undergoing such division. Even when the meristems are active, only a small proportion of these cells can be seen to be in the state of nuclear or cell division at any one time, as seen in Fig. 16.7. This proportion of cells actually undergoing division, at a given time, compared with the number of adjacent meristematic cells not dividing, is expressed as a ratio and is known as the *mitotic index*. When the meristems are dormant, such as in the case of bud dormancy or seed dormancy, the mitotic index is very low or even zero, whereas when the meristem is active, the mitotic index may approach 100%, although much lower values are more common.

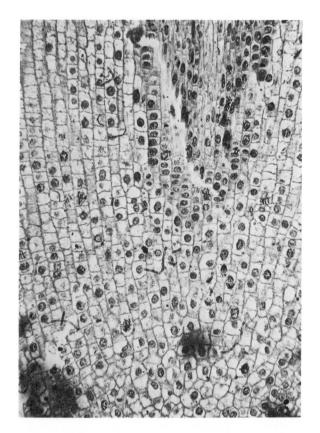

Fig. 16.7 A longitudinal section through a root tip showing the meristematic region with some cells undergoing mitosis.

Location

At the time of floral development, a limited number of flower cells will become meristematic and start to divide. These are called sporogenous cells, and their division by meiosis results in the formation of the embryo sac and the pollen grains. Although such division adds little to the growth of the plant, it is important to the plant.

Following fertilization, a zygote is formed as is an endosperm cell. Both of these then divide by mitosis to form the endosperm and the beginning of the formation of the embryo. However, the embryo soon develops apical meristems which then take over the function of subsequent cell divisions within the embryo.

After the cells have been formed by the meristem, they usually cease dividing. However, some of them do not differentiate so much that they cannot, at some time in the future, again undergo cell division. As a result, meristematic cells may be formed in roots, from mature tissue, or even callus or tumors may develop.

The locations of cell division within the multicellular plant are illustrated in Table 16.1.

Table 16.1   Sites of cell division in multicellular higher plants.

1. Embryo sac mother cell
2. Endosperm cell
3. Zygote
4. Apical meristems
5. Cambial meristems
6. Interfascicular meristems
7. Microspore mother cells
8. Microspores
9. Pollen grains
10. Intercalary meristems
11. Root initiation
12. Tissue culture
13. Callus formation
14. Tumor formation

The typical meristematic cell is a small cell that is isodiametric, and has thin walls as well as all of the organelles found in the cell shown in Fig. 1.4, except the vacuole does not occupy such a large volume of the cell. When present, the cell contains numerous very small vacuoles instead of one large one. Most of the interior of these cells is composed of the nucleus and the cytoplasm, not the vacuole.

CONTROL OF CELL DIVISION

Temperature

Since the rate of cell division in plant meristems is very important in many aspects of plant growth, factors which alter this rate have been studied and do represent some of the most active areas of research at the present time. Temperature is one of the principal external factors that affects cell division in plants. Indeed, frost resistance results in a cessation of cell division some time before Winter sets in. As the temperature approaches 0°, cell division ceases. If such an approach occurs slowly, the cell stops dividing and enters into an apparent state of dormancy. Such loss of activity results in an increase in the frost resistance of the cell.

Hormones

Particularly active is research on the effects of plant hormones on cell division, and within the past few years much of the research work done in plant physiology is of this nature. Phytochrome seems to be somehow involved in this, since both cell division and cell enlargement are inhibited by red light and stimulated by far red light. Of course, this may or may not be a direct effect, but certainly it is not directly essential for cell division since cell division does occur in root tips that are in the dark. In fact, red light is inhibitory to root initiation, whereas far red stimulates it.

Four types of plant hormones have been studied very intensively since they do at least stimulate cell division and in some cases seem to be essential for it. These four are auxin, gibberellin, cytokinin, and ethylene hormones that were previously discussed in Chapter 12.

Studies of root initiation and cambium initiation reveal that cell division requires the presence of auxin and often of cytokinin, gibberellin, and other hormones. Cambium activity begins in the Spring of the year in woody plants, and cell division in the cambium begins first near the top of the tree just after the buds begin to swell and open. At this time the buds are producing much auxin and some of this auxin moves down the stem and into the cambium where it stimulates cell division. Cytokinins may also be either so transported or the auxin stimulates their synthesis in the cambium. Nevertheless, cell division in the cambium is resumed progressively downward, until the entire cambium for the length of the stem is active. As the season progresses, the buds become dormant. With such dormancy, auxin is no longer produced so the cambium near the top of the tree becomes inactive, with cell division ceasing, and this inactivity progresses down the stem until by Winter, cell division has ceased in the entire cambium.

Callus tissue cultures are used very often to study the effects of hormones and other chemicals on cell division. Such studies reveal that cell division is influenced by auxins, cytokinins, and gibberellins. However, cytokinins, such as zeatin, act only in the presence of auxins. In some cases, auxins seem to stimulate the synthesis of cytokinins, and cytokinins seem to stimulate the synthesis of auxins. At least one of the principal effects of cytokinin is to stimulate cell division, which it normally does in conjunction with auxin. However, it is not known which act of cell division is so stimulated or if these are actually essential for division, or if they merely stimulate the division already in progress.

Among the effects of gibberellins on cell division and growth is that associated with the formation of the flowering stalk in plants. There are two patterns of growth by non-perennial plant species, namely, the *caulescent* where a stem is the obvious part of the shoot, and the *rosette* type where a whirl of leaves is formed the first year, and then the next year the plant bolts and forms a flowering stalk. There seems to be two locations of cell division in these two patterns. In the caulescent type, cell division occurs near the tip of the terminal bud and continues down from the tip for a centimeter or more. In the rosette type, cell division normally occurs only near the tip, but with the initiation of bolting, the region of cell division becomes longer. In the latter plants, the tip seems to form the flowers and the subtip meristem forms the flowering stalk. One question of interest to physiologists is what stimulates the flower stalk to form in the rosette type plants. Of course, low temperature treatment causes its initiation, as discussed in a previous chapter, but this initiation seems to stimulate gibberellin production which actually causes the cells below the normal meristem to begin dividing to form the flowering stalk. Gibberellin added exogenously will cause the plant to bolt without vernalization but has no effect on the flowering of caulescent species. Apparently, they produce adequate amounts of gibberellin for their needs.

Although unicellular plants have cells which apparently synthesize all of the chemicals needed for cell division, this is apparently not so in the higher plants. The cells of the apical meristems cannot divide unless they receive certain chemicals from other cells of the plant. In some cases these chemicals travel some distance within the plant and thereby correlate growth, and in other cases, these chemicals are produced by cells that are adjacent to the meristematic cells of the apical meristem. Gibberellins are in the latter class.

At times, cell division in the apical meristems is inhibited by hormones, rather than stimulated by them. For instance, the axillary buds of cotyledons are often dormant, with no cell division occurring. However, these buds undergo active cell division if the tip of the seedling is removed. A similar phenomenon can be observed in the case of lateral buds of plants. The lateral buds remain dormant as long as the terminal bud of a branch or stem is actively growing. However, if the terminal bud is removed, the lateral bud meristems become very active with intense cell division. However, if the terminal bud is removed, and auxin put in its place, then the meristems of the lateral buds remain dormant. In this case, it appears that auxin produced in the terminal bud is moving down the stem and inhibiting cell division in the lateral buds.

Auxin is not the only chemical inhibitor of cell division in plants. Abscissic acid is a very good inhibitor of this type and has been associated with a number of different kinds of dormancies in the meristems of plants. Generally, it reverses the action of auxin, gibberellins, and cytokinins and can be reversed by higher concentrations of these. It seems to function at the level of enzyme production, inhibiting the production of those enzymes normally controlled by the three hormones listed above, especially hydrolytic enzymes. As such, it plays a role in bud dormancy, seed dormancy, abscission, senescence, and flowering as will be further explored in subsequent chapters.

Cell division is initiated, or at least greatly stimulated or inhibited, by plant hormones. These hormones include a group of cell division inhibitors, such as abscissic acid and related compounds, and a group of stimulators, such as auxins, gibberellins, cytokinins, ethylene, and wound hormones. Indeed, there is good evidence that the rate of cell division in many plant meristems is very much controlled by the relative concentrations of these stimulators and inhibitors. Dormancy is due to an excess of inhibitors and the resumption of bud growth to an excess of stimulating hormones. More will be said about these in subsequent chapters. Just how they function is not known, but as revealed in Chapter 11, they do appear to control the synthesis and activity of certain enzymes. Just how these enzymes are related to cell division or what stage of the cell cycle is affected are unknown.

Cell division is also controlled by relative concentrations of the stimulating hormones. For instance, auxin excess will inhibit division, as will a high ratio of auxin: cytokinin. But, as the ratio decreases, either due to a decrease in auxin or to an increase in cytokinin, mitotic activity increases. This ratio seems to be important in determining root initiation, and the formation of callus and tumors.

Indeed, it is through cell division that new cells are formed, and only these new cells can elongate to give the increase in size that is associated with growth. However, cell division has little direct effect on the growth of higher plants, since most of the elongation or increase in size is due to subsequent enlargement of the newly-formed cells, a subject to be covered in the next chapter.

Chapter	Plant Growth: Cell
Seventeen	Enlargement

SIGMOID GROWTH CURVE

Although growth of unicellular plants, such as unicellular algae, fungi, and bacteria, can be expressed as a logarithmic relationship, as seen in the last chapter, this relationship is true only when such growth is the result of cell division, and is not true of growth by cell enlargement. Growth by cell enlargement shows a different relationship, as can be seen in Fig. 17.1, where the growth of a pollen tube is plotted against time after placing the pollen grain in a germination medium. Pollen germination is neither dependent upon nor associated with cell division, so it is useful for studies of cell enlargement. As seen in Fig. 17.1, the rate of growth starts rather slowly, then increases at a uniform rate for a time, and finally decreases. In essence, its curve is *sigmoid*. During the second growth phase, when the growth rate is rapid and uniform, and its curve is linear, the growth rate can be expressed by the following equation:

$$\text{Growth rate} = \frac{N_0 - N_t}{t}$$

Here N_0 is the initial length of the cell, N_t is the length of the cell after time t, and t is the length of time which expires during the growth of the cell or between measurements of cell length.

CELL ENLARGEMENT

A curve such as this is also given if the growth rate of the growing region of a multicellular plant is plotted against time, and is, in fact,

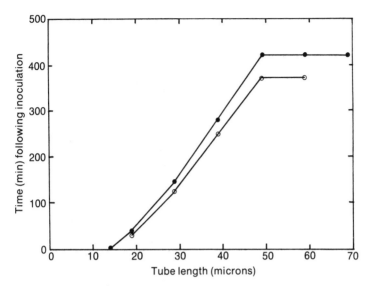

Fig. 17.1 Pollen tube elongation with time. The white and black circles represent data from different experiments.

usually associated with growth. For instance, if the growth rate of the root of a seedling is measured at various time intervals, such a curve may be plotted from the results. This seems to implicate cell enlargement as being a more important contributor to growth in the multicellular plant, than is cell division. The reason for this is that most of the increase in size is due to cell enlargement rather than to cell division. This can be seen by measuring a cell that has recently been formed from cell division, a cell which will have dimensions near 7×7 microns, and by measuring the same or a similar cell after cell enlargement has ceased, at which time the cell may be about 20×200 microns in size. This represents a tremendous increase in cell size, and particularly in its length, as is shown in Fig. 17.2, and also explains why growth may be determined more by cell enlargement than by cell division. Of course, this does not detract from the importance of cell division in growth, since a cell cannot enlarge until it has been formed through cell division, and since cell division is the origin of all cells.

What is responsible for this unequal rate of growth in two dimensions of the cell? In the plant cell wall, the cellulose molecules, which give the wall its physical strength, are oriented in such a way as to allow the cell to enlarge more in one dimension than in another. Also, the unequal

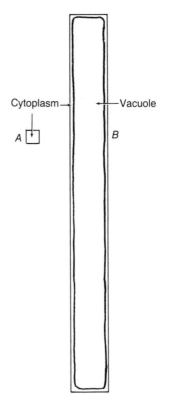

Fig. 17.2 Illustration to scale, of the size changes often associated with cell enlargement by plant cells. Cell *A* is the cell size shortly after division; cell *B* is the size of the cell when fully enlarged.

pressures from surrounding cells may have some effect in this directional control. The interaction of these two factors controls this unequal growth of the cell. Were this not so, the plant would appear much differently. How would an annual such as sweet pea or bean plant look if its stems, branches, or twigs grew about ten times as wide as they do?

Enlargement of the plant cell is the result of the cell wall yielding to pressure produced in the enclosed vacuole. This involves four conditions. First, turgor pressure within the cell must be great enough to supply the pressure needed for this expansion. Second, the cell wall must be softened enough to allow the wall to expand. Third, as the wall expands, new cell wall materials must be added to the wall to prevent its becoming so thin that it ruptures, and fourth, there must be a high metabolic rate within the growing cell.

WATER AND CELL ENLARGEMENT

In contrast to animal cells, where enlargement occurs chiefly through protein synthesis, enlargement in plant cells occurs largely through water uptake. Turgor pressure is essential for this enlargement. In fact, the statement has been made that with no turgor pressure there is no expansion. This is evident from the observation that has been made many times which shows that if water deficiencies develop within a plant, growth stops and will not resume until the deficiency has been alleviated. Summer drought will cause such growth cessation for this reason, and no doubt growth is often limited by sub-optimal water content of the plant. Turgor pressure will be produced within the vacuole as a result of concentration differences of the solutes within the vacuole as compared with the solute concentration outside the cell, as shown in previous chapters. Only when the solute concentration outside the cell is lower than that inside the cell, will a turgor pressure be developed. On the other hand, such concentration outside the cell cannot be too much lower or the pressure produced within the vacuole will be so great that the cell will expand too rapidly and burst. When pollen grains are placed in nutrient media of different sucrose concentrations, the amount of tube growth is related to the concentration of the sucrose in the medium. At high sucrose concentration, no growth occurs, and at low concentration, the tubes rupture, so no growth occurs. This relationship between the solute concentration of the medium and the growth of a pollen tube can be seen in Fig. 17.3, which indicates that the optimum concentration for enlargement in these structures is 12% sucrose, which has an osmotic pressure of about 9 bars. No doubt, each cell has an optimum pressure which will give its best enlargement, but the magnitude of this value will be quite variable. Perhaps cells rarely experience the optimum osmotic pressure of their environment at the time when they are expanding. Expansion does not wait until optimum pressure conditions prevail, but will occur over a range of osmotic pressure values. However, this range is not without limits.

CELL WALL POLYSACCHARIDES

Plant cell walls are rather rigid, and therefore resist expansion. This rigidity is due both to properties of the cellulose molecules and to their

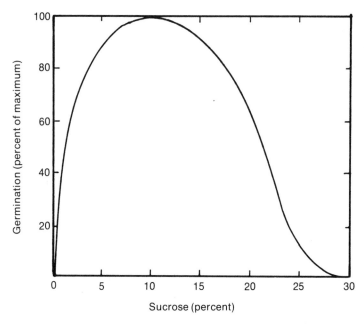

Fig. 17.3 Percent of pollen grains germinating in a medium of varying sucrose concentrations. (After Goss, *Transactions of the Kansas Academy of Science*, 1962, **65** (3): 310–317.)

orientation within the wall. Cellulose microfibrils are commonly oriented perpendicular to the long axis of the primary cell wall, and therefore the cell wall resists enlargement, but with less resistance being offered in one direction than in another. This allows the cell to elongate more in length than in width. In other words, the microfibrils function much like hoops around a wooden barrel, or like the well-known girdles worn by some members of the human race.

Cell Wall Chemical Composition

Before considering further the physical properties of the cell wall, it would be well to consider its chemical composition. The plant cell wall is composed largely of polysaccharides, as contrasted with the cytoplasm which contains largely proteins. These polysaccharides are much more abundant in plants, than are the polysaccharides such as starch and inulin that were previously discussed, and include cellulose, hemicellulose, and other large molecules that are polymers of various monosaccharides. These shall first be studied from the standpoint of their structure, composition, and synthesis, and then from the standpoint of how they are incorporated into the cell wall.

Cellulose

Cellulose has been stated to be the most abundant organic compound on earth. It is a constituent of all plant cell walls, making it an important component of wood. Structurally, it is a large molecule but of variable length and weight. It is formed by the union of 1400–10,000 glucose molecules uniting, by pulling out a water molecule between each pair of glucose molecules much like the peptide bond formed during protein synthesis. These form a long, unbranched chain to become the cellulose molecule.

Unlike starch or inulin, the cellulose molecule is very stable, once it is formed, and cannot be reutilized, i.e., it cannot be broken down to release the sugar molecules, by the cell that synthesized it. However, there are a few species of fungi, known as the brown-rot fungi, that can break down the cellulose molecule and use the released glucose molecules as food. Decaying logs often are brown in color due to the action of fungi that destroy the white cellulose molecules leaving the brown lignin to give the log its color.

Cellulose is synthesized from glucose-1-phosphate by the reaction of glucose-1-phosphate molecules with a nucleotide, guanidine triphosphate, to form guanidine diphosphate glucose. These molecules then join together releasing the guanidine diphosphate and forming the cellulose molecule. This is illustrated in Fig. 17.4.

Cellulose gives the physical properties to wood, as well as to all plant cell walls. Cotton, which is used in the manufacture of cloth, paper, rayon, plastics, and explosives is composed of fibers of nearly pure cellulose.

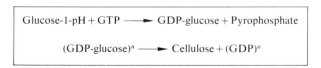

$$\text{Glucose-1-pH} + \text{GTP} \longrightarrow \text{GDP-glucose} + \text{Pyrophosphate}$$

$$(\text{GDP-glucose})^n \longrightarrow \text{Cellulose} + (\text{GDP})^n$$

Fig. 17.4   Pathway for cellulose biosynthesis.

Hemicellulose

Another ubiquitous constituent of the plant cell wall is the substance known as hemicellulose. This is really not a chemical molecule, but rather a mixture of polysaccharide molecules, including mannans, galactans, xylans, and arabans. Since these are not always present in the same proportions, there can be, and are, many types of hemicellulose.

Mannans

Mannans are polysaccharides similar to cellulose except they are composed of mannose molecules. They have been found in many plant cell walls, especially in the cell walls of the straw of many grasses, in the wood of both hardwood and conifer trees, in some leaves, and in the cell walls of many seeds. In these walls, they are combined with glucose, often at a ratio of one glucose to two mannose, to form a type of hemicellulose found in hardwoods, or combined with glucose and galactose at a ratio of one glucose and one galactose to three mannose molecules to form a hemicellulose common to softwoods such as conifers.

Galactans

Galactans are also found in the secondary cell walls of grass straw, wood, and seeds. They too are similar in structure to cellulose except the molecule is composed of galactose molecules. However, like mannans, the galactose molecules often join with other monosaccharide molecules to form long-chained hemicellulose. One example is given above whereby glucose, mannose, and galactose make up a form of hemicellulose.

Xylans

Xylans are polysaccharides composed of xylose molecules. However, xylose molecules usually form polysaccharides with other monosaccharides and these polysaccharides are considered to be forms of hemicellulose. Xylose is, in fact, a component of the most abundant hemicelluloses in plants, especially in wood, apricot seed shells, and grass straw. Xylose is usually associated with glucose or galactose in the hemicellulose molecule.

Arabans

Arabans are polysaccharides composed of many arabanose molecules. These arabans are not hemicellulose, but are closely associated with the polyuronides in the primary cell wall. They are components of the gums of cherry, peach, and plum, are water-soluble, and are very hygroscopic. This means that they are able to absorb large amounts of water and have been said to thereby aid in drought resistance.

From the above discussion, it can be seen that the polysaccharides of the cell wall consist of cellulose (composed of glucose molecules), hemi-cellulose (composed of glucose, galactose, mannose, and xylose mole-cules), and the arabans (composed of arabinose molecules). Cellulose is the basic structural material of cell walls and cannot be reutilized by the cells producing it, hemicellulose is a substance that fills in among the cellulose molecules and can be reutilized, forming a source of carbon and energy for many trees and other plant species, especially in the early Spring, and during seed germination.

POLYSACCHARIDE DERIVATIVES

Lignin

In addition to the polysaccharides, the plant cell wall contains numerous polysaccharide derivatives. One such derivative often found in large quantities, especially in the mature cell walls, is *lignin*. Lignin is a plastic material and is deposited among the strands of cellulose. Its structure is not known. It apparently is a very complex molecule that does contain some sugar but is perhaps primarily a polyphenol. It is very resistant chemically and is not reutilized by the cells producing it, but can be uti-lized and broken down by white-rot fungi which use lignin as a source of food. Since it is not easily destroyed, lignin is important in the formation of soil humus and thereby contributes greatly to the organic matter con-tent of the soil. It is also an important industrial waste that although it sometimes is destroyed by burning, is too often washed into streams causing pollution which kills fish. This is particularly a common waste material from paper pulp mills. Synthetic vanilla is formed from waste lignin utilizing the vanillin molecule, a phenolic compound of importance in food flavoring.

Gums

Various types of *gums* are found associated with plant cell walls. These appear on the surface of tree trunks as a result of gummosis, as is often observed during the Christmas holidays as the Christmas tree is being placed in a stand or as such a tree is being transported from the store to the house. Gummosis is caused by injury to the plant which results in a breakdown of the cell wall and other cell contents. Gums on the surface of conifer and cherry tree trunks are good examples. Chemically, these gums are complex organic substances related to the polyuronides, but their exact chemical structures are not known, and they may consist of mixtures of molecules of different types rather than being a pure chemical compound.

Mucilages

Mucilages are also complex chemical substances related to the poly-uronides. They are found in the outer cell layers of many aquatic species and in the coats of many seeds. They also represent the slimy substance on the inner bark of slippery elm, and the sticky substance in the seed pods of carob and honey locust trees. Mucilages are very hygroscopic and swell in water due to imbibition of the water molecules.

And so, the cell wall, which is characteristic of plants and distinguishes them from animals, is composed of many large organic molecules. The functions of this wall seem to be to protect the cell from mechanical damage, either from forces originating from the outside, or from excessive internal forces, primarily turgor pressure. The wall surrounds the proto-plasm of each cell, and the walls of adjacent cells are held together by a sticky substance, the middle lamella, which is composed of polyuronides.

CELL WALL STRUCTURE

Primary Wall

The plant cell wall is often composed of two layers, the primary wall and the secondary wall. The primary wall is formed following cell divi-sion, as vesicles move to the surface of the cell plate or middle lamella, and rupture, uniting their membranes to form the plasma membranes and extruding their contents to contact the middle lamella and to deposit primary cell wall components on it. The primary wall, however, has a different composition from that of the middle lamella, being lower in polyuronide content and higher in hemicellulose and cellulose, and with some lignin later deposited as the primary cell wall increases in thickness and matures. Here, the cellulose becomes the structural material of the cell wall, comprising about 30% of its dry weight, and the polyuronides, hemicelluloses, and lignin are deposited among the meshwork of cellulose molecules. The cellulose molecules are long chains of glucose. These long chains form a helix with a diameter of 3.5 nm, and are called *fibrils* or *elementary fibrils*. These may or may not join to form thicker strands called *microfibrils*. These fibrils or microfibrils form a screen-like mesh-work with many holes among them. Some of these holes are filled with the other cell wall materials, but the strength of the cell wall is largely due to the cellulose.

Secondary Wall

As the primary wall matures, but even before the cell stops enlarging, another cell wall—the *secondary cell wall*—is laid down on the inside surface of the primary wall. This wall is composed largely of cellulose and lignin. Whereas the fibrils of cellulose are interwoven and randomly arranged in the primary wall, the fibrils of cellulose are definitely oriented in the secondary wall. In some cases, they form spiral patterns, in some cases circular patterns, and in other cases a variety of patterns within the wall of a given cell. The secondary wall is particularly prominent in the vessel elements, in fibers, and in stone cells, whereas it is lacking in parenchyma cells. As yet, what causes the vesicles to change their internal composition to form middle lamella, primary, and secondary walls is not known, nor is it understood how the cellulose fibrils can be oriented in the directions they are oriented in the secondary walls. However, by the time the secondary wall is formed, cell growth has stopped and the cell is mature.

Cells with only the middle lamella and with the primary wall, such as parenchyma cells, are capable of later division to form new cells. However, once the secondary wall is formed, division becomes less likely, due to the mechanical strength of the secondary wall, and these cells can no longer divide.

Wall Protein

In addition to the polysaccharides and their derivatives, plant cell walls contain protein. This protein is peculiar since it has a high hydroxyproline content. In fact, essentially all of the hydroxyproline in the cell is located either in the cell wall or in the vesicles that aid in cell wall formation. The function of these wall proteins is probably enzymatic, whereby they function in cell wall synthesis and in the softening of the wall during cell enlargement.

MECHANISM OF WALL EXPANSION

It should be remembered that as the cell wall is being formed, the cell is growing by expansion. Due to the turgor pressure, the cell is exerting a pressure on the wall just like the pressure exerted on a balloon when it is blown up, or like the pressure exerted on the wall of an automobile tire as air is forced into it. This causes the cell wall to stretch. If additional cell wall material was not subsequently formed, the cell wall would become so thin, as the cell continued to expand, that the wall would rupture from the turgor pressure exerted against it. Therefore, it is necessary that new cell wall material be formed and deposited in the wall. How this is coordinated is unknown, but it is essential that a high metabolic rate be occurring while enlargement is occurring and one reason for this is the dependence of the synthesis of new wall materials on it.

Cell enlargement is accompanied by a high rate of metabolism. Inhibiting the rate of respiration by a lack of oxygen or of carbohydrates or by the presence of respiratory inhibitors will reduce cell enlargement. Just how these two are tied together is not entirely known, but certainly metabolism is needed for cell wall softening, for the synthesis of new cell wall materials, and for the synthesis of cytoplasm generally.

Growth by cell enlargement is not accompanied by an equivalent amount of protein synthesis, but by a considerable increase in the amount of cell wall materials formed. As a result, the cell wall does not decrease in thickness, but the cytoplasm, which originally completely filled the cell, becomes restricted to a thin layer just inside the plasma membrane, with by far the greatest volume of the cell becoming occupied by the vacuole, and with the cytoplasm occupying but little of the cell volume.

WALL HORMONES

The osmotic pressure of the plant cell does not increase just prior to cell enlargement, so this is not the mechanism by which enlargement is initiated. Rather, such initiation is caused by the cell's ability to control the plasticity of its wall. As indicated, cell enlargement causes the cell wall to stretch. This stretching causes the cellulose fibrils to become oriented somewhat in the direction of the length of the cell. However, cellulose molecules are often very resistant to stretching, and would not allow the cell to increase in size if they were not softened first. This softening occurs only in the presence of certain plant growth hormones. In fact, it has been said that without these hormones there is no growth. Since growth is defined as an increase in size, and since the increase in the size of the plant is due largely to cell enlargement, and since the cell cannot enlarge without hormones being present to soften the wall, it is easy to see the importance of such hormones in plant growth. Growth hormones represent a functionally-related group of organic compounds.

These include the auxins, gibberellins (of which over twenty kinds are known), the cytokinins (three are known and the presence of more kinds is suspected), ethylene, and abscissic acid, as well as others. These all work individually or in various combinations to stimulate, if not initiate, or to inhibit cell enlargement. In the following discussion, consideration will first be given to how the hormones work individually, and second to how they interact to bring about cell enlargement. Although these hormones do seem to have the capability, as will be shown, to act as individuals in cell enlargement, perhaps they never do so, but rather they act together in various combinations, at least in most plant tissues.

Gibberellins

The most striking example of the effect of a hormone on plant growth may be seen by the action of gibberellin on growth. A dwarf corn plant, that would normally reach only about one-half meter in height at maturity, will develop into a normal corn plant over three meters in height merely by adding a solution of gibberellin to the apical meristem. Or a bush bean plant, what would normally grow to a height of about one-third meter, will turn into a pole bean of five meters in length by the application of gibberellin. Similar effects on pea plants are shown in Fig. 17.5. Most of this extra growth is due to cell enlargement.

Gibberellins have a tremendous stimulatory effect on cell enlargement. They do this by one, two, or three mechanisms. First, gibberellins have been shown to weaken cell walls, and such weakening would stimulate cell enlargement. Second, gibberellins induce the formation of proteolytic enzymes, an action which would increase the breakdown of proteins, releasing tryptophan which could then be used for the synthesis of additional IAA. Indeed, IAA concentration does increase in the presence of GA. Such IAA would then be able to soften the cell wall, resulting in a stimulation of cell enlargement. Finally, GA is known to bring about the hydrolysis of starch, a phenomenon that occurs during cell enlargement. Such hydrolysis would release sugars and thereby increase the solute concentration and osmotic pressure of the cell, increasing the turgor pressure and stimulating cell enlargement. It appears that GA functions in some of these activities by forming m-RNA molecules which then interact someway with IAA to stimulate cell enlargement. Considering all of these possible mechanisms by which GA could stimulate cell enlargement, it is no wonder that this hormone gives such striking effects on growth.

Cytokinins

Cytokinins, well known for their control of cell division, can also control cell enlargement. For instance, fresh lettuce seeds will not germinate in the dark, but will germinate in the light or will germinate in the dark if previously treated with cytokinins. Also, it is well known that the

Fig. 17.5 Growth alterations of pea plants by applications of GA. A = normal plant. B = plant similar to A but treated with GA. (Experimental plants of Steven Spiker, Kansas State University.)

expansion of young leaves requires the presence of cytokinins, and this expansion is primarily due to cell enlargement. Such responses occur even at very low concentrations—concentrations as low as 10^{-11} M. Such concentrations are too low for conventional chemical analyses and must be determined either by bioassay methods or by use of a mass spectrometer.

As indicated in Chapter 12, cytokinins are components of some of the t-RNA molecules, and therefore might be expected to function by control of protein synthesis. However, how this relates to the observed stimulation of cell enlargement by cytokinins is not known.

Auxins

Auxins have long been known to stimulate cell enlargement, as previously discussed. However, some of this cell enlargement previously attributed to auxins is now known to be due to ethylene produced as a result of the presence of auxin. Also, other plant growth hormones are known to stimulate auxin production.

Gibberellins, auxins, cytokinins, and other hormones stimulate cell enlargement when present in appropriate concentrations, but inhibit it at higher concentrations. This is not to say that all growth hormones function alike in cell enlargement. Indeed, GA is unable to stimulate cell enlargement if DNA synthesis is inhibited, but such DNA inhibition has no effect on stimulation by IAA. It would appear strange that inhibition of DNA synthesis should affect cell enlargement, since such synthesis is normally associated with cell division, not with cell enlargement. However, it appears that what is being inhibited is the DNA synthesis that occurs in the mitochondria, not that in the nucleus. Cytoplasmic DNA synthesis must be necessary for the stimulation of cell enlargement by GA either because of its need for mitochondrial replication or for a continuation of mitochondrial function.

Inhibitors

On the other hand, there are certain inhibitors of growth by cell enlargement too, especially the inhibitor known as abscissic acid (ABA).

Abscissic acid is a growth inhibitor, which inhibits the growth stimulating activities of auxins, gibberellins, and cytokinins.

In addition to ABA, there are other compounds found in plants that similarly inhibit growth, but to a lesser extent. There is, however, some question whether or not these act as hormones or even inhibit growth *in vivo*.

Just how growth hormones stimulate or inhibit cell enlargement is unknown but the evidence is ample to indicate that this action occurs through the control of protein synthesis and function. The next question that needs answering then is how this alteration of protein synthesis affects cell enlargement. It could do so either by altering the osmotic pressure of the cell, by altering the availability of energy for metabolism, or by altering the cell wall structure.

RATES OF CELL ENLARGEMENT

Sometimes the rate of growth by cell enlargement is very rapid and sometimes it is very slow. For instance, growth rate may be so slow as to be imperceptible, particularly during dormancy or as seen in the growth of roots in Winter, whereas at other times it may be very rapid. The filaments of the stamens of wheat flowers will triple their length in just 3 min at the time of anthesis, and the stems of bamboo plants have been reported to grow about 30 cm per day. Rapid growth is also seen in the closure of the hoops of nematode-capturing fungi, where the cell volume may triple in a 0.1 sec interval as the nematode comes in contact with the cell. However, the growth of most plant species seldom exceed a small fraction of this rate, namely, about 3 cm per day.

TIME OF CELL ENLARGEMENT

The time when growth by cell enlargement occurs will also vary considerably. Some plants grow at equal rates both day and night, whereas others grow more rapidly at night than in the daytime. (Some people even claim they can hear corn growing at night.) No doubt the effects of temperature and growth inhibition by light are factors which control this. Cell enlargement is known to be stimulated, at least in some plant cells, by red light, and this stimulation is far red reversible, implicating the phytochrome system. However, as shown in earlier chapters, light can also reduce cell enlargement.

PATTERNS OF CELL ENLARGEMENT

There are two patterns of cell enlargement in plants. In one case, such enlargement occurs by adding new cell wall materials only at the tip of the cell and in the other type, these materials are deposited more or less evenly over the entire area of the inside of the cell wall. What controls or determines which type of enlargement will occur is unknown. Enlargement by the addition of materials only or primarily at the tip occurs in the growth of the pollen tube and in the growth of certain filamentous algae and fungi.

CELL ENLARGEMENT IN SEEDS

Plant seeds represent a good structure to use to study cell enlargement. The reason for this is that during the development of the seed, until it reaches maturity, most of its growth is by cell division. From maturity to seed germination, growth is almost entirely by cell enlargement. Therefore, the study of seed germination is largely a study of cell enlargement.

Seeds are amazing structures. They are produced by all gymnosperms and angiosperms, they are compact, efficient, and represent a resting stage in plant development by which the plant can be propagated without being killed or badly damaged by adverse environmental conditions, especially by the cold weather of Winter.

Seed Structure

If a seed, such as a bean seed, is soaked for several hours in water to soften it, the seed can be dissected into three main parts, namely, the brown seed coat on the outside of the seed, the white cotyledons which make up the largest part of the bean seed, and the small cream-colored embryo inside. These structures can be seen in Fig. 17.6.

The seed is always limited on the outside by the seed coat, composed of two or more layers of tissue. The outer layer is hard and woody, with cells packed closely together and covered with mucilage and with a waxy cuticle. The inner layer or layers are thin and membranous. Together, the

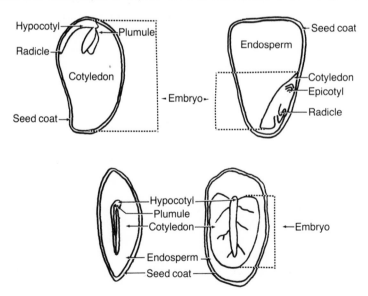

Fig. 17.6 The structures of three types of seeds. Upper left is lima bean, upper right is corn, and the two lower seeds are castor bean.

layers of the seed coat protect the seed from mechanical injury, from pathogens, from excessive desiccation, and often from germinating at a time when the survival of the seedling would be jeopardized due to dry or cold weather.

The embryo, which is usually embedded within the storage tissue, is composed of a radicle, hypocotyl, and plumule, as seen in Fig. 17.6.

Seed Chemicals

The chemicals found in seeds can be grouped into three categories. First, are those normal constituents found in all plant tissues and considered to be essential to the life of the seed. Second, are those found in much higher concentrations in the seeds than in other plant tissues, and considered to be storage products. Third, are the so-called secondary plant products. Those chemicals comprising the first and third groups have been discussed in other chapters so this chapter will consider mostly those in the second group, namely, the storage products.

The storage products are found within the storage tissues of the seed, namely, in the endosperm or cotyledons. There may be one cotyledon, as in the monocots, two cotyledons, as in the dicots, or many cotyledons, as in the conifers. These cotyledons may be filled with cells containing much stored food materials or they may not function in food storage, in which case they would be thin and nearly empty sacs; in which case also, the endosperm outside the cotyledons would contain the stored food.

Storage tissue occupies the largest volume of the seed, and is composed of cells with high contents of stored foods such as starch or fats, storage proteins, RNA, and mineral nutrients. These storage products will be used by the embryo and seedling during and following germination. They are also of great economic value to man, since most of man's food supply comes from plant seeds, as does many of his industrial products such as vegetable oils.

Table 17.1 reveals that the seeds store more fats or carbohydrates than any other chemicals. Some seeds, such as the cereals, store mostly starch, and some mostly fats. Actually, the seeds of most plant species store more fats than starch, but this is not true of those most common to man, seeds such as the cereals. Generally, the larger seeds store starch while the smaller seeds store fats. Perhaps this is because more than twice as much energy is stored in 1 gram of fat than in 1 gram of starch. Seeds high in fat are also high in protein, whereas those high in starch

Table 17.1   Some chemical components of selected plant seeds.*

Species	Water	Protein	Fat (%)	Carbohydrate	Ash
Corn (*Zea mays*)	13	9	4	73	1
Sorghum (*Sorghum vulgare*)	10	11	4	74	2
Wheat (*Triticum esculentum*)	12	12	2	72	2
Pea (*Pisum sativum*)	12	24	1	60	3
Soybean (*Glycine soja*)	8	35	18	35	5
Hemp (*Cannabis sativa*)	7	28	37	–	–
Pumpkin (*Cucurbita pepo*)	2	23	32	13	4
Sunflower (*Helianthus annuus*)	5	18	28	–	3
Ginkgo (*Ginkgo biloba*)	7	7	2	41	2
Northern Red Oak (*Quercus rubra*)	49	7	22	34	3

*From *Handbook of Biological Data*, W. S. Spector (Ed.), 1956. W. B. Saunders Co., Philadelphia, Pennsylvania. © Federation of American Societies for Experimental Biology. (Reprinted with permission.)

are low in protein. Therefore, larger seeds are low in protein and smaller seeds high in protein, on a dry weight basis. These relationships are shown in Table 17.2.

Seed fats are composed of glycerides of fatty acids. These fatty acids are largely of the unsaturated type, with oleic, linoleic, and linolenic acids being the most common. However, a large variety of saturated fatty acids are also present, but in lesser amounts. Soybean seed fats are composed of about 15% saturated fatty acids and 85% unsaturated fatty acids (Table 17.3).

Seeds contain a great deal of protein, as seen in Table 17.1. Soybeans have been reported to produce more than 400 pounds of protein per acre per year, and wheat more than 200 pounds per acre per year. Some of this

Table 17.2   Relationships among seed size, and starch, fat, and protein content.

Seed size	Starch	Fats	Protein
Large	high	low	low
Small	low	high	high

Table 17.3   Principal fatty acids of some seed fats.*

Species	Saturated			Unsaturated		
	palmitic	stearic	(%)	eleic	linoleic	Linolenic
Corn (Zea mays)	10	3		50	34	—
Wheat (Triticum esculatum)	14	1		30	49	6
Soybean (Glycine soja)	15			27	52	6
Hemp (Cannabis sativa)	10			16	46	28
Sunflower (Helianthus annuus)	11			30	60	—

*From *Handbook of Biological Data*, W. S. Spector (Ed.), 1956. W. B. Saunders Co., Philadelphia, Pennsylvania.© Federation of American Societies for Experimental Biology. (Reprinted with permission.)

protein in seeds is in the form of cell structures, being structural protein, some is present as enzymes or enzyme constituents, but most of the seed protein is storage protein. That this protein is high in food value is evident from the fact that seeds furnish most of the protein for the human diet. Unfortunately, these proteins are low in lysine, tryptophane, and methionine. This low methionine content is not conducive to using plant proteins as an only source of protein for human nutrition unless supplemented with methionine, lysine, and tryptophane.

Within the storage tissues, storage proteins are located in aleurone grains. These aleurone grains are limited by a membrane, and consist largely of crystals of proteins of a type known as globulins. Globulins represent most of the storage protein in seeds, except for cereals where the stored protein is either prolamins or glutelins or both. The globulins are characterized by high molecular weights, high contents of proline, arginine, glutamic acid, aspartic acid, and glutamine and/or asparagine (Table 17.4). They are also soluble in dilute neutral salt solution, such as sodium chloride or ammonium sulfate, but insoluble in distilled water. Actually many types of proteins are globulins, and some of these have been given names, such as the edestin from *Cannabis sativa*, glycinin from soybean, *Glycine max*, legumin and vicilin from peas, *Pisum sativum*, and phaseolin from beans, *Phaseolus vulgaris*. In grasses or cereals, the storage proteins are not principally globulins, but rather prolamins— which are soluble in alcohol—and glutelins, which are only soluble in alkali or acid. The major storage protein of wheat seeds is gluten, which consists of two fractions, namely, gliadin and glutenin, and the next most

Table 17.4 Principal amino acids in some plant seeds (% of seed weight).*

Species	Arginine	Histidine	Isoleucine	Leucine	Lysine	Methionine	Phenylalanine	Threonine	Tryptophan	Valine
Corn (*Zea mays*)	0.45	0.24	0.36	1.1	0.29	0.21	0.46	0.34	0.08	0.5
Wheat (*Triticum esculentum*)	0.63	0.31	0.58	0.91	0.35	0.22	0.7	0.38	0.19	0.64
Pea (*Pisum sativum*)	2.6	0.4	1.2	1.9	1.5	0.1	1.4	1.2	0.2	1.2
Soybean (*Glycine soja*)	4.1	1.3	2.6	4.3	3.1	0.5	3.1	2.3	0.7	2.7
Sunflower (*Helianthus annuus*)	5.46	1.43	2.78	3.71	1.45	1.61	2.39	1.64	1.14	2.7

*From *Handbook of Biological Data*, W. S. Spector (Ed.), 1956. W. B. Saunders Co., Philadelphia, Pennsylvania. © Federation of American Societies for Experimental Biology. (Reprinted with permission.)

important wheat seed protein is gliadin. In corn, zein is the most abundant storage protein, whereas in grain sorghum, it is kaferin.

Seed proteins function as potential sources of organic nitrogen during seed germination and seedling development. With the initiation of germination, they are hydrolyzed to release their amino acids, which are then translocated from the cells of the storage tissue to the embryo to be resynthesized into enzymatic or structural proteins.

With this introduction to the chemical structure of seeds, it is now appropriate to consider seed germination and dormancy, since these involve cell enlargement and its inhibition, respectively.

SEED DORMANCY

The mature seed is a compact, independent biological entity with the capacity of germinating to form a new plant if conditions become favorable. However, if freshly-harvested seeds are sown, they usually will not germinate. Some crop seeds are exceptions to this rule, since selective breeding has reduced their dormancies, but by far the majority of species produce seeds that will not germinate when mature. They are dormant at this time. *Dormancy* refers to a delay in seed germination after the seed is mature or produced. Sometimes this dormancy is due to a lack of a favorable environment, such as the seeds in a packet in the seed store, but usually it is due to other causes.

In the field, if seeds germinated shortly after they were produced, the chances of the young plant surviving would be very slim. Such a seed would germinate in the late Summer or Fall, and the young plant would be killed by the cold weather of Winter. However, if some mechanism functioned to delay germination until Spring, then the chances for survival of the young plant would be much greater. This is the value of dormancy. It delays germination until the chances for survival of the seedling are good.

Evolution has developed many ways to prevent the germination of the mature seed. Not only are there many types of seed dormancy, but there are usually more than one type of dormancy in any one seed. If one fails, germination is still prevented. What are these types of dormancy?

Seed Coat Dormancy

One group of seed dormancies is known as seed coat dormancy. These dormancies reside in the seed coat. The first type of such dormancy is represented by the seed coat being impermeable to water. This type is very common in legumes and water lilies. Water is essential for seed germination, and water uptake by the seed must occur before any other physiological activity associated with germination. This is not unexpected, since enzyme activity requires that the enzymes be hydrated, and since water is necessary for the development of turgor pressure before cell enlargement can occur. If the seed cannot increase its water content, it cannot germinate. Impermeability can be detected by weighing a sample of seeds, placing the seeds in water overnight, and weighing the seeds again after removing the surface water. If the two weights are similar, the seed coat is impermeable to water. If the seed weighs considerably more, then such impermeability is not present. If the seed of honey locust is selected, the seeds will be found to be impermeable to water. In fact, seeds of this species can be left in water for 15 years and still no water will be taken up. Only after the seed coat becomes damaged, is dormancy broken. Other examples of plant species producing seeds with such dormancies include members of the *Malvaceae, Leguminoseae, Nymphaceae, Cristaceae*, and *Nelumboaceae*, which include such species as red bud, water lotus, cotton, okra, locust, and *Symphoricarpus* sp.

Another type of seed coat dormancy is due to the impermeability of the seed coat to gases, especially to carbon dioxide or oxygen. This type of dormancy is common in the grasses and in *Xanthium* sp. Gas exchange with the environment is necessary because the rate of respiration is low in mature seeds and must greatly increase prior to germination to supply the energy needed for cell enlargement. Such an increase is prevented if the seed cannot obtain adequate amounts of oxygen from the atmosphere or cannot get rid of the carbon dioxide which accumulates during respiration. This dormancy is also broken by damage to the seed coat.

Germination involves growth of the embryo and therefore an increase in its size. Such an increase cannot occur if the seed coat is too tough, as it functions much as a straight-jacket. Dormancy due to the mechanical resistance of the seed coat is found in walnut, *Prunus* sp., smooth sumac (*Rhus glabra*), pigweed, water plantain, and mustard. This dormancy too must be broken by mechanical damage to the seed coat.

The fourth and final type of seed coat dormancy is due to chemical inhibitors in the seed coat. All white-kernel wheat seeds sprout at maturity, but red-kernel varieties exhibit varying degrees of dormancy due to the presence of inhibitors in the seed coat. Only after the inhibitor has been removed, will the seed germinate.

Seed coat dormancies of the first three types, namely, seed coats impermeable to water, impermeable to gases, or hard seed coats that offer insurmountable mechanical resistance are broken by *scarification* of the seed coat. Scarification refers to treatments that damage the seed coat. These treatments may be natural or artificial but the result is the same, namely, the seed coat is broken and the seed coat dormancy is therefore broken.

Much of the scarification in the field is biological, with the seed coat being destroyed by microorganisms that use the seed coat for food. Much scarification, both in the field and in the laboratory, is chemical in action. In the laboratory, seed coat dormancy is often broken by soaking the seed for a period of time in chemicals such as alcohol, sulfuric acid, or even hot water. Many locust and *Sporobolus* sp. seeds will germinate if hot water is poured on them, although many seeds require stronger chemicals, with *Rhus glabra* seeds requiring soaking for 1–2 hours in concentrated sulfuric acid. This is rough treatment, and indicates the resistance of some of these seed coats. In the field, chemical action to break dormancy is seen in the action of carbonic acid in the water, or of other chemicals produced by living organisms, chemicals that corrode the seed coat and thereby break dormancy. Scarification may be applied under natural conditions by alternating freezing and warming temperatures to which the seeds are exposed during Fall, Winter, and Spring, or by wetting and drying of the seed. Both treatments cause strain on the seed coat and eventually result in its rupture. Finally, the seed coat may be ruptured by mechanical scarification such as breaking dormancy in the laboratory by cutting or filing a nick in the seed coat or by damaging the seed coat with an abrasive such as sand paper or sand blasting. *Dalea spinosa* is a plant that grows in the dry washes of the deserts. Its seeds fall to the ground and mix with the sand beneath this small tree. When there is a good rainstorm in the area, the dry wash becomes a roaring river, carrying the seeds and sand with it. The sand erodes the seed coat and when the water stops flowing, the seed can germinate. This guarantees that the young plant will have adequate amounts of water for its growth and survival.

Yes, plants have several kinds of seed coat dormancy, and nature has found several ways to break these dormancies. However, natural breakage of the dormancies takes time, time to allow the seed to survive the cold weather of Winter before it germinates.

Dormancy Due to Inhibitors

Another common type of dormancy is due to chemical inhibitors that are produced in the fruit and move into the seed preventing it from germinating. Such dormancies are found in several of our desert plants, as well as in the fruit of tomato, apples, pears, some citrus fruits, etc. Would it not be unusual to find a young apple tree growing out of the apple purchased at the grocery store? It is not difficult to demonstrate the presence of such inhibitors. All that is necessary is to squeeze the juice out of the fruit, onto some filter or blotting paper, place this in a petri dish or covered jar, spread some seeds that are not dormant on the paper and wait for them to germinate. Usually even the seed from the same species of plant need not be used. For instance, the juice from the fruit of *Symphoricarpus* will inhibit the germination of wheat seeds. This type of inhibition can be overcome by destruction of the chemical inhibitor in the fruit, as occurs during rotting. In desert plants that show this type of dormancy, the dormancy is broken by leaching or washing the inhibitor, which is soluble in water, out of the fruit. To do so requires a substantial rain, and therefore guarantees that adequate water will be present in the soil to allow the plant a good chance for survival. Abscissic acid is the inhibitor in some fruits.

An interesting type of dormancy that has been studied in some detail, lately, is dormancy originating in the embryo or endosperm. This type of dormancy is due to the presence of a chemical inhibitor or to the deficiency of a metabolite. One such type of dormancy requires exposure to light to break the dormancy. Many grasses, lettuce, *Pinus* sp., *Lepidium virginicum*, *Mimulus ringens*, *Oenothera* sp., *Epilobium*, some varieties of tobacco, and *Ranunculus sceleratus*, are said to be examples. In these seeds, not only is light required, but the correct color or wavelength is important. Red light, with wavelength near 660 nm will break this dormancy whereas far red light, with a wavelength near 730 nm will not. Not only will far red light not break this dormancy, but it will reverse the action of red light and not allow seeds exposed to red light to germinate. For example, lettuce seeds when freshly harvested may not germinate unless exposed to red light after which they will germinate. However, if exposed to far red light after the red light exposure, the seeds will not germinate. One can repeat alternating red and far red light treatments many times, but whether or not the seeds germinate will depend upon what color of light was used for the last treatment. If red light, the seeds will germinate, if far red, they will not germinate. Incidently, gibberellin will take the place of red light. If the seeds are soaked in a solution of gibberellin, they will germinate even in the dark.

How phytochrome controls germination is not known, but this type of dormancy is due to the inhibition of cell enlargement, so phytochrome must control cell enlargement in these seeds, either by the control of inhibitors or essential metabolites.

Whereas some seeds require light for germination, some will not germinate in the presence of light, including seeds from most of the *Liliaceae*, *Datura stramonium*, many *Hydrophyllaceae*, and California poppy. The mechanism for this dormancy may involve phytochrome also, but the details are unknown. This dormancy is also associated with the embryo or endosperm.

Another common type of dormancy originating in the embryo or endosperm is that requiring low temperature (*stratification*) to break. Such seeds are rendered germinable by storage in moist soil or peat moss for 2–3 months at 4°C. Such dormancy is broken in the field during the low temperatures of Winter as long as the temperatures are low enough and of long enough period of time. If either the temperature is not low enough or does not remain low enough for the required period of time, the dormancy will not be broken so the seed will not germinate. Dormancy of this type is common in apple, peach, juniper, linden, dogwood, iris, ash, rose, *Xanthium* sp. and *Pinus* sp. as well as others. Breaking of such dormancy by stratification can be reversed by oxygen deficiency, excess water, or high temperature. The cause of such dormancy is unknown.

Immature Embryos

Another type of dormancy associated with the embryo but rather uncommon, is the presence of immature embryos in the mature seeds. By the time the seed is mature and ready for harvest, the embryo is not mature enough to germinate, so germination is delayed until the embryo matures. Many orchids, holly, European ash, *Gnetum* sp., and *Ginkgo biloba* seeds are examples.

Chemical Deficiencies

The witchweed plant, found in the Carolinas, is an example of an unusual type of dormancy associated with the embryo. In this case, dormancy is due to the lack of an adequate amount of a purine needed by the embryo for germination and the seed will not germinate until the embryo gets an adequate supply. The seed gets this supply only from grass plants, on which the witchweed is a parasite. The seed will lie in the soil unable to germinate until grass plants grow nearby. From these grass plants the needed purine moves out and into the soil, eventually finding its way into the witchweed seed, allowing the seed to germinate, and the young plant can then live parasitically off the nearby grass plant. This is another type of dormancy that delays germination until the young plant has a good chance for survival.

Soil Inhibitors

Another type of dormancy is that originating from chemical inhibitors in the soil. For instance, several inhibitors of seed germination can be extracted from seeds, and from other plant structures. These inhibitors move through the soil and into the seeds preventing their germination, and thereby eliminating or reducing competition for light, water, and mineral nutrients.

Studies of the dormancy of seeds is often frustrating because the seed will have two or three types of dormancy. For instance red bud seeds have impermeable seed coats and a low temperature requirement. One dormancy may be broken but the seed will still remain dormant. This may be frustrating to the scientist, but it is beneficial to the seed, allowing it more time and better guarantees of survival for the seedling and propagation of the species.

SEED GERMINATION

A seed cannot survive forever. Eventually, it must either die or germinate. Its purpose in life is to someday germinate to form a new individual plant and in this manner propagate the species.

For germination, the seed must be viable, non-dormant, and placed in a favorable environment. Such an environment must have adequate amounts of available water, a favorable temperature, and a sufficient supply of oxygen for germination.

Water

The water content of the mature seed is very low, and this must increase before germination can occur. In fact, seed germination is usually limited by the lack of water once dormancy has been broken.

To obtain the needed water, seeds are planted in the soil and irrigated. The irrigation increases the soil water content to the value necessary to furnish the water needs of the seed. To furnish these needs, the water content of the soil need not be high. Some seeds can germinate even in an atmosphere near water saturation, some when the soil is saturated with water. However, most seeds will not germinate under such extremes but germinate best when the soil moisture content is near field capacity (*see* Chapter 2).

When seeds are planted in the soil, it is customary to press the soil down around the seeds. The reason for this is to give better contact with the seed. Otherwise, if the soil is loose, this contact may not be sufficient to allow adequate amounts of water to enter the seed, so fewer seeds will germinate.

Temperature

A favorable temperature is also required for germination. Just what temperature is required varies with the species of plant producing the seed. Those species endogenous to the tropics require higher temperatures than those of the temperate or arctic-alpine regions. Cotton seeds require a higher temperature for germination than corn seeds, which require a higher temperature than wheat seeds.

Even within species of a given region, requirements differ. In the early Spring, germination is inhibited by low temperature, although some seeds even germinate on ice. As the temperature increases, the seeds of some species germinate early in the season whereas others germinate later, due to their requirement for higher temperatures for germination. This is one reason why some plant species appear in the early Spring, later Spring, early Summer, and late Summer, rather than all growing in one season.

Seeds are not adapted to best growth at a constant temperature. In their natural environment, the temperature fluctuates over each 24 hour period, being cooler at night and warmer during the daytime. Yes, the soil temperature is more constant than air temperature and shows less fluctuation, but it too fluctuates considerably. Therefore, we expect seeds to germinate better when the temperature fluctuates over the 24 hour period, than at constant temperature. At least this is true for many seeds, especially grass, flower, and vegetable seeds. For example, bluegrass seed germinates best at temperatures of 15°C for 16 hours and 25°C for 8 hours of each day.

Oxygen

As previously mentioned, the seeds of most plant species will not germinate well in soil that is saturated with water. This is not true of species that normally inhabit swampy areas, such as cattail (*Typha latifolia*), or of aquatic plants, but is true of most other species. Perhaps the main reason for this is lack of adequate amounts of oxygen in such soil. In Chapter 2, it was shown that the pore spaces of the soil are filled with atmospheric gases or water, and as the water content increases the gas content decreases. If the pore spaces are filled with water, there is no space for gases so oxygen will be deficient. Germination requires a rapid rate of respiration, and respiration requires oxygen, so if oxygen is deficient, respiration rate is slow and germination does not occur. Under such conditions, fermentation may occur. But fermentation will soon kill most seeds due to the accumulation of toxic substances.

Seeds that get buried too deeply in the soil do not germinate due to inadequate amounts of oxygen in the soil, but as soon as they are uncovered or moved closer to the soil surface they germinate. This is one reason why so many weeds grow after a farmer turns over or plows a field.

PHYSIOLOGY OF SEED GERMINATION

Once the seed loses its dormancy, if it is in an environment that furnishes adequate amounts of water, a favorable temperature, and sufficient oxygen for its needs, the seed will begin the physiological activities that lead to cell enlargement and germination.

Germination involves the resumption of growth by the embryo of the seed and is said to end when the radicle of the embryo can be observed to penetrate through the seed coat. Such germination occurs when a viable, non-dormant seed is placed in a favorable temperature, and a sufficient supply of oxygen. From the time the seed is placed in such an environment until the radicle penetrates through the seed coat a number of invisible physiological activities occur in the seed, beginning with an uptake of water.

The water content of the mature seed is low. As shown in Chapter 4, the water content of the growing tissue is often about 85% of the fresh weight of the tissue, whereas the water content of mature seeds is often less than 10% of the fresh weight. However, this value will vary with the relative humidity of the environment in which the seeds are kept. Corn seed, for example, may have a water content of about 6% of the fresh weight when kept in an atmosphere with a relative humidity of 15%, but may have a water content of about 20% if stored at a relative humidity of 90%. Nevertheless, the water content of the mature seed is too low for germination, at the time the seed is planted in the soil, so first its water content must increase. This is done by osmosis since its water potential (*see* Chapter 2) is greater than that of the soil. Mature seeds have a very high water potential. A look at the cells of a dry, mature seed, reveals that these cells are small, with small, shrunken vacuoles, and the cells are also plasmolyzed. This condition is conducive to a high solute concentration, due to the low water content, and therefore to a high osmotic pressure. In addition, the colloidal materials of the cell are not fully hydrated. Colloidal materials have the ability to adsorb water molecules and thereby become hydrated. This is one mechanism for binding water within the cell. These adsorbed water molecules are oriented or arranged in an orderly manner around the colloid, and bound tightly to it. This ability of the colloid to adsorb water molecules, strongly attracts water molecules in the vicinity of the colloid. This attraction is measured as the *imbibitional pressure* (IP) and adds to the water potential of the cell just as osmotic pressure does. However, the IP is greater in mature seeds than is the OP, so as soon as the seed is placed in a favorable environment, water moves into the seed in response to this strong attraction. Since the water molecules are oriented around the colloid as they become adsorbed to it, this increases the size of the colloid-water complex causing it to swell. As a result of this swelling, the seed increases in volume. Place a dry seed in water and observe the increase in the size of the seed the next day. Swelling causes the seed coat to rupture, increasing the amount of water taken up by the seed, the amount of oxygen available to it, and the rate at which carbon dioxide can be lost from it. In short, such swelling speeds up cell enlargement and therefore the germination process.

As the colloids become saturated with water, the IP value decreases rapidly and soon approaches zero. By now, the water potential of the seed is due to OP rather than IP, but is nevertheless great, so water continues to move into the seed, increasing its water content and causing the cells to become turgid. When the water content of the seeds reaches 40–70% of the fresh weight, metabolic activities of the seed begin to increase rapidly. First, a rapid increase in the rate of respiration occurs. This is due in part to the rupturing of the seed coat, making the seed more permeable to oxygen, but also to an activation of respiratory enzymes by the increased water content. It also appears to be due to an increase in the number of mitochondria within the cells. This increase in respiration rate occurs in both the embryo and the storage tissues, perhaps increasing mostly in the embryo. However, since the embryo only occupies a small volume of most seeds, the increase in respiration of the seed must be due mostly to an increase in the respiration rate of the cells of the storage tissues.

Following closely behind this increase in respiration rate, is a marked increase of hydrolytic activity within the storage tissues of the seeds, which results in the breakdown of large food molecules, such as stored protein, starch, fats, and RNA to form smaller molecules, such as amino acids, sugars, and nucleotides. These small molecules are translocated to the embryo where they are used for the resumption of embryo growth. Hydrolysis of stored food materials in the storage tissues results in a continued decrease in the amount of these materials and is eventually followed by death of the cells of the storage tissues. Hydrolysis of fats results first in the conversion of fats to fatty acids and glycerol, from which the fats are formed. These fatty acids are then broken down to form acetyl CoA, which is used to form sugars by reversing glycolysis. Glycerol, from fats, is also converted to sugars by being converted to dihydroxyacetone phosphate and then into sugars. Starch is also hydrolyzed to sugars. These sugars formed from fats or starch, are transported to the embryo to be used in its growth.

FRUIT GROWTH

Another example of growth primarily by cell enlargement, is the growth of fruit. By the time the flower reaches the stage of anthesis, the parts of the flower that will form the fruit are composed of small cells, but contain nearly all of the cells that will be found in the mature fruit. Only for a few weeks after anthesis, does cell division continue in the fruit, and most of the fruit growth is due to subsequent cell enlargement. Fruit growth, following pollination, is due almost entirely to cell enlargement, at least in many plant species.

Fruit growth is slow prior to pollination, but shortly thereafter there is a tremendous increase in this growth as cell enlargement is greatly stimulated. This results in larger cells and in an increase in the volume of intercellular space.

Growth of fruits can be determined either as changes in the fruit volume or as an increase in fresh weight. Volume measurements are usually made by measuring the diameter or length of the fruit, or by measuring fruit displacement of water or some other liquid.

Plotting growth measurements of fruit against time gives a growth curve that is either sigmoid, as in Fig. 17.1, or double-sigmoid, as in Fig. 17.7. Dry fruits, as well as fruits of tomato, apple, pear, and certain others, exhibit a sigmoid curve. In these, the growth rate is slow at the time of fertilization, increases rapidly shortly thereafter until the fruit approaches its maximum size, and levels off as the fruit ripens. And yet,

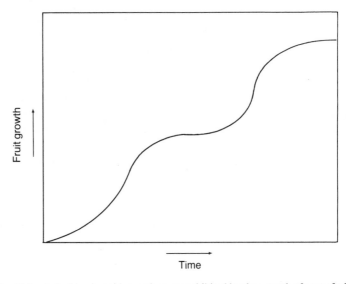

Fig. 17.7   A double-sigmoid growth curve exhibited by the growth of stone fruits.

if fertilization does not occur, this rapid increase in growth rate is not observed, but rather growth is slow for a time and the flower soon abscises.

The double-sigmoid curve is shown by stone fruits, such as apricot, peach, cherry, or grape. In this case, the sigmoid pattern is followed by another rapid increase in growth rate followed by a leveling off as ripening begins. The first rapid growth rate is due to the growth of the fruit. The first plateau in the curve occurs at the time the seed is growing. When the seed stops growing, the fruit again resumes growth. However, if these fruits are seedless, such as with seedless grapes, the curve is sigmoid rather than double-sigmoid.

Fruit growth can be brought about in non-fertilized flowers by the application of growth hormones. This is best illustrated with strawberry fruits where the acenes containing the seeds control the growth of the fleshy fruit, as illustrated in Fig. 17.8. Here it can be seen that the role of the seeds in fruit growth can be largely replaced by auxins. Indeed, it is often possible to produce fruit of several species without fertilization by applying hormones to the non-pollinated flowers. Usually though, a mixture of auxin and gibberellin will work better than either one alone.

Thus, it appears that the role of seeds in fruit growth is to furnish hormones to the surrounding cells so that cell enlargement can occur in

A   B   C

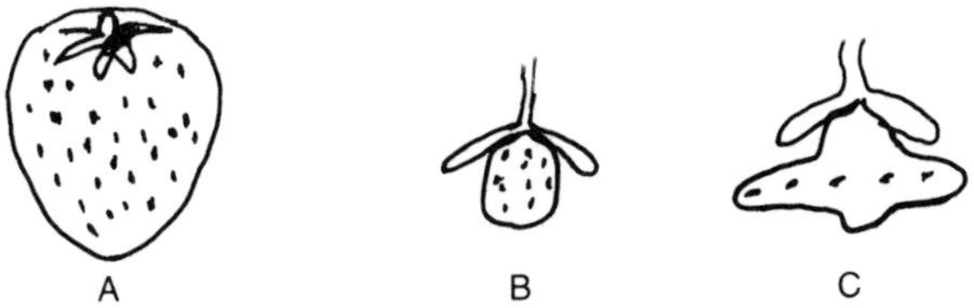

Fig. 17.8 Development of strawberry fruit. A = fruit as it normally develops. B = green, immature fruit. C = fruit as it would develop if all achenes were removed at the immature stage, except those near the equator of the fruit.

the fruit. However, seedless fruit often has a different chemical composition than seeded fruit, so seeds also alter fruit chemical composition.

In addition to growth hormones, fruit requires water, carbohydrates, and mineral nutrients for growth, and the availability of these to the fruit will alter fruit size. These factors will be discussed in Chapter 18.

In addition to those examples used, there are other aspects of plant development that are dependent upon cell enlargement. However, it is not always possible to separate cell enlargement from cell division. These are usually very closely related so that what affects one will also affect the other. On the other hand, cell enlargement is also closely related to cell differentiation, a subject to be covered in the next chapter.

<table>
<tr><td>*Chapter*
Eighteen</td><td># Differentiation and
Correlative Development</td></tr>
</table>

If all cells of a multicellular plant were the same, the plant would be little more than a sphere of identical cells, and would change only as growth and death occurred. That such is not the case, is due to differentiation and correlative development.

CONSTANCY OF DNA

Indeed, since cell division results in an equal distribution of the chromosome and gene components between the two daughter cells, it would be easy to explain why all cells should be the same. Even the constancy of DNA would support this observation. It was formerly believed that differences among cells were due to differences in the amount of genetic material, material which is now referred to as DNA. However, such differences cannot be demonstrated since, with few exceptions, all cells of the individual have the same amount of DNA. These exceptions are first, the gametes, which have one half as much DNA as the resting somatic cells, and second, the cells that are actually undergoing division and which therefore have twice the normal amount of DNA during the latter part of the S phase, and during the G_2 and M phases prior to cytokinesis. However, such deviations from the normal amount could not explain the many differences which exist from cell to cell within the same individual plant.

Further evidence that all cells of a multicellular plant are basically similar comes from the fact that cells that do appear to differ greatly from one another within the same individual can be treated so that they de-

384

differentiate to become meristematic and form identical plants. For example, an injured epidermal cell of the flax plant can develop into a new flax plant, or a moss plant can form from a single moss leaf cell, or an entire carrot plant can form from a single cell of a carrot root. Thus, even though cells from an individual plant differ greatly from one another, they all have the same amount of DNA and they all retain the same potential.

INTERCELLULAR DIFFERENCES

Nevertheless, different cells of the same individual do differ both as to their chemical composition and to their physical structure. Urease, an enzyme, is particularly high in concentration only in the cotyledons of jackbean; α-amylase is found in highest concentrations in the aleurone layer of barley seed; polyphenylase and lipoxidase are found in detectable amounts in the root endodermal cells, but not in adjacent cortical cells; grass root epidermal cells alternately are long and short in size, and the short cells contain acid phosphatase even before they form root hairs, whereas the adjacent long cells do not.

No, it is not difficult to observe evidence of differentiation in plants. Such evidence is abundant. What is difficult is to determine the mechanism by which such differences are brought about and what initiates such changes.

Specialization in cells results from conditions imposed upon them by their environment as well as by their enzymes. The latter may be the result of the presence or absence of the potential to produce certain enzymes, which will function to cause differences among individual plants, or the result of metabolic regulation changes often brought about by the turning on or turning off of the genes of the DNA of the cell. Each cell of the individual plant will have the same genetic composition, but only certain genes in each cell will be turned on, and which genes are actually functional will vary from cell to cell. The study of differentiation is the study of differences in metabolic regulation, and an important aspect of this is what causes certain genes to be masked or repressed.

DIFFERENTIATION IN THE ZYGOTE

Perhaps the best place to start on a discussion of differentiation in plants is with the zygote, and to follow differentiation in the cells that result from its division and from subsequent divisions.

Many of the aspects of embryo development have already been determined by the time the zygote is ready to divide. It might be expected that the division of the zygote would result in two cells that are identical, but this is not so. Even such aspects as in what plane the first division will occur is among those very important in determining many aspects of the future development of the embryo, and this is determined by conditions within the zygote prior to the time for its division.

The zygote may divide because it reaches a certain size that is so large that division is essential for the survival of the cell, but in what plane it will divide and what differences will be evident in the daughter cells will be determined largely by the polarity of the zygote. Polarity refers to differences which exist within the cell. This polarity seems to be caused by, or associated with, differences in the location of metabolites and/or organelles within the cell. Shortly after fertilization occurs, for some unknown reason, the cell experiences a redistribution of its metabolites, with opposite sides of the cell ending up with different amounts of each metabolite. In multicellular plants, surrounding cells aid in this redistribution or polarity, but in free zygotes of lower plants, polarity appears to be controlled by the environment. This latter observation can be seen in fucus zygotes which divide to form a rhizoid on one side and a thallus on the other side. The side on which the rhizoid is initiated is always on the side away from the light, and on the side toward a positive pole of an electrical circuit, towards the side of increasing pH, and toward the side with the highest temperature.

In zygotes of higher plant species, either before or after fertilization, the distribution of metabolites in the egg becomes heterogenous, with accumulations of different metabolites occurring in two opposite regions, thus establishing polarity, with an apical and a basal site. The apical region will become the principal site of protein synthesis, of growth, and of morphogenesis, and is the site at which the new plant will form, as seen in Fig. 18.1. The basal region becomes vacuolated and often distended and accumulates osmotically-active substances. This region is destined to become the suspensor and to temporarily aid in the transfer of food materials from the parent cells to the newly-developing embryo, as seen in Fig. 18.1.

As the zygote increases in size and in heterogeneity of metabolic activity, division occurs with a cell wall being laid down at right angles to the cell axis, separating the regions of maximum heterogeneity. This is the position which apparently gives maximum equilibrium to the system and gives daughter cells that are internally most homogeneous. However,

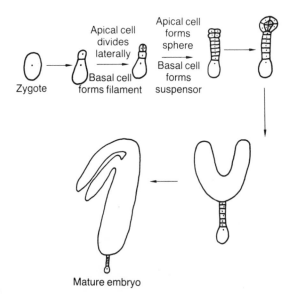

Zygote

Apical cell
divides
laterally

Basal cell
forms filament

Apical cell
forms
sphere
—————
Basal cell
forms
suspensor

Mature embryo

Fig. 18.1   Stages in plant embryo development.

such a division also results in the formation of two cells that are most unlike each other, so already differentiation is evident.

EMBRYO DEVELOPMENT

Structural Changes

Subsequent development of the embryo occurs largely through cell division, with little enlargement of the cells being involved. The basal cell divides, for a time, to form a filamentous, multicellular structure that remains attached, at one end, to the mother tissue, and attached at the other end to the developing embryo. This filamentous structure is called a suspensor, and it functions in the absorption and translocation of water and nutrients to the apical cells. However, not all absorption and translocation of nutrients takes place through the suspensor, since much of these nutrients can be absorbed by the apical cells directly from the adjacent cells of the mother tissues.

Subsequent divisions of the apical cell formed from the first division of the zygote, result in a sphere remaining smaller than those of the suspensor, and remaining meristematic for extended periods of time.

Originally, there is little differential gene action in the development of the embryo, as indicated by the fact that the young embryos of many plant species are very similar in appearance. However, eventually, both the environment and the genetic composition of the cells become active and the embryos become more species-specific, taking on more and more characteristics that are associated only with the species of plant being developed.

Nutritional Requirements

As the embryo develops, it not only takes on different forms, but it also has different nutrient requirements. In the very young stages, its nutritional requirements must be very extensive and complex, which would indicate that its cells are not capable of synthesizing all of the metabolites required. This is indicated by the fact that it is not possible to remove very small embryos (proembryos) from the plant and culture them using tissue culture techniques. Occasionally some growth is evident in a complex nutrient medium, containing endosperm extracts, but such growth results only in the formation of a callus, not in the formation of a mature embryo. By the time the embryo has taken on the heart-shaped appearance, it can be cultured in a complex medium, containing endosperm extract, and a mature embryo obtained, indicating that it has now developed the enzyme systems needed for the synthesis of additionally required metabolites. Have these additional enzyme systems been formed due to the action of the surrounding cells, or does this represent derepression of gene action within the embryo cells themselves? The answer is not known. More mature embryos can be cultured to maturity in the absence of an endosperm extract, if water, sugar, mineral nutrients, and occasionally certain vitamins and amino acids are supplied to them, and by the time the embryo is mature, it will develop into a new plant if only water, sugars, and mineral nutrients are present. Thus, embryo development involves not only changes in the form of the embryo but changes in its metabolism as well, with the early stages of embryo development being entirely dependent upon surrounding cells for many factors and this dependence decreasing as development continues.

When considering the products of the activity of meristematic cells, with respect to differentiation, it should be remembered that the cell can be considered to be in a cell cycle which repeats itself periodically. For a newly-formed cell to differentiate, this cycle must be interrupted, and the preparation for this interruption can be considered to be the first step on the road to differentiation. However, what causes this interruption is not always known.

Differentiation in Meristems

Perhaps the simplest system to consider in relation to cell differentiation is that of two cells formed from cell division, wherein one cell remains meristematic and the other ceases further division and becomes differentiated. How do these two cells differ and to what are these differences due? This, of course, is the situation which exists in the meristems of plants. Meristematic cells divide and one of the two cells remains meristematic while the other differentiates. Much of the work carried out on differentiation has been done using root tips since these allow serial sections to be made to study cells in progressively-later stages of maturation. These cells also show polarity due to different locations of metabolites and organelles within the cell. They also show that the location of the cell within the root tissue has a definite relationship to the ultimate direction of differentiation of that cell. Although two cells of a division may be of different sizes and would become different through differentiation if they remain in the position in which they normally are found following division, if these two cells are separated following their formation, they may still be of different sizes, but they no longer assume different functions. They behave similarly to each other when removed from their normal environment, and placed in a similar environment. Apparently, even adjacent cells within plant tissues are subject to different environmental conditions, conditions which differ enough to cause adjacent cells to become very different after differentiation has occurred. Thus, their orientation with respect to one another in the multicellular plant must be important in determining their natural destiny. Whether or not the cell is to remain meristematic must also be determined by some correlative influence that passes from one cell, producing or transporting this influence, to an adjacent cell in which the influence prevents further mitosis, allowing differentiation to occur. The nature of this influence is unknown.

DIFFERENTIATION IN CALLUS TISSUE

However, nutrition also has an influence on differentiation, as can be shown with callus tissue. Callus tissue is nothing more than a mass of undifferentiated parenchyma cells. Such tissue can be treated in such a way that xylem and phloem cells will form within it whereas without this treatment such differentiation does not occur. Such differentiation can be brought about by grafting a bud to the surface of the tissue, in which case the xylem and phloem tissues develop below the graft. This can also be done now in the absence of a bud merely by supplying the tissue with appropriate concentrations of auxins and sucrose. The ratio of the concentrations of these two chemicals will determine the relative amounts of xylem and phloem cells formed. However, in the absence of the bud, the xylem and phloem cells are randomly distributed through the callus and lack the orientation normally present in the intact plant.

Nutrition

It is also possible to furnish callus tissues with an appropriate nutrient medium and induce the formation of buds and/or roots, as seen in Fig. 18.2, where a new tobacco plant is developing in a flask from a piece of callus tissue cultured on an agar medium. It is even possible to separate individual cells from a piece of plant material, such as from a section cut

Fig. 18.2   A new tobacco plant forming from the differentiation of callus tissue. (Courtesy of Dr. Jerry Weis, Kansas State University.)

from the root of a carrot, and to treat these isolated cells so that some of them become meristematic and eventually form individual plants that are similar to the original plant from which the section was taken. However, not all of the cells which are so isolated will behave this way. Only a few will, so this presents a problem especially since the cells that will so form a new plant cannot be identified beforehand. Also, the medium for such development is rather complex and often must contain some natural mixture of chemicals, many of which are of unknown composition, such as coconut milk. Rarely can these natural foods be replaced with a pure chemical solution but it has been possible to supply water, mineral nutrients, carbohydrate, a hormone of the auxin type, an amino acid as a source of nitrogen and a sugar alcohol, and obtain some success in this type of culture. No doubt, the specific nutrient requirements for each species would be different, and these differences still prohibit extensive cultures of this type. However, enough success has been achieved to indicate that many cells of the multicellular plant retain *totipency* (ability to form a new plant) even after some differentiation has occurred.

CONTROL OF GROWTH DIMENSION

Associated with some forms of differentiation is the change in the direction of cell division to give a change in growth dimension. Cells dividing in just one plane will form a filament. By appropriate treatments, division can be made to occur in two planes, giving a plate-like structure. Further treatments cause the division to occur in three planes, giving the three-dimensional structure common in multicellular plants. Some treatments that can cause these changes in the plane of cell division, but what causes the switch from one plane to another or what turns on multiplanar divisions within the living plant are still unknown and remain as challenging studies for future plant physiologists.

CORRELATIVE DEVELOPMENT

Within the multicellular plant, differentiation is often the result of correlative development. Such correlative development results in the differentiation of the cells of the plant to form tissues, and in the positioning of these organs to form the individual plant.

A multicellular plant in which the correlation of growth and differentiation was not active would be a strange plant indeed. If such a plant could possibly have organs such as leaves, roots, etc., the chances are good that the leaves would outgrow the roots, or the flowers would be misshapen and large or small where compared with their normal size, and fruits may form anywhere on the plant and be of any size or shape. Indeed, without a correlation of development in the multicellular plant, the plant world as is known today, would not exist and no doubt the animal world would not either.

Correlative development refers to the control of plant development at one location within the plant by another part of the same plant. Thus, it involves a location at which some message or influence is produced, the translocation of that influence from the site of production, and a site at which the influence acts to alter plant development. The site of production of the influence must be away from the site of its utilization. The distance may be as great as many feet, or as short as between one cell and its neighbor. *Lonicera*, honeysuckle, flowers are in pairs on opposite sides of the stem, and yet, one flower of each pair has an influence on the other.

Shoot:Root Ratio

One type of correlative development is shown by deviations in the so-called shoot:root ratio. This ratio is determined by weighing the shoot and the root of a plant and expressing the results as a ratio. The concept behind this is that the growth of the root can and does influence the growth of the shoot and vice versa. This ratio will be increased by any factor that stimulates shoot growth more than root growth and decreased by any factor that stimulates root growth more than shoot growth. For instance, the availability of abnormally high amounts of nitrogen in the soil will increase the ratio. This increase is explained by the high nitrogen resulting in good shoot growth, and this good shoot growth utilizes the photosynthate, which would reduce the amount of carbohydrates normally transported to the roots, so root growth is less. Thus, when high concentrations of nitrogen are available, the ratio would increase due both to increased shoot growth and to decreased root growth. An increase in the amount of water available for plants or a decrease in soil aeration will also increase this ratio. Why?

Generally, if mineral nutrients or water are deficient, the roots will use what they need and the shoots will suffer most from the deficiency — not the roots. This would reduce the shoot:root ratio. On the other hand, if the rate of photosynthesis is reduced, the shoots will use the photosynthate available and the roots will be deprived of their needs for this nutrient, thus increasing the shoot:root ratio. A plant growing in the shade would be expected to have a greater shoot:root ratio than one of the same species that was growing in a habitat where it was exposed to full sunlight.

Nutrient Control of Organ Development

Perhaps one of the most common and most important types of correlative development is that associated with the distribution of nutrients within the plant. To study this phenomenon, it is necessary to know what nutrients are required by the different plant organs, and which of these nutrients are available to it and in what quantity. For the determination of what nutrients are required, a tissue or organ culture technique is commonly used. For example, the tips of a growing root may be excised and transferred to a flask containing water and various nutrients. If the root continues its development at a rate similar to that on the intact plant, it is assumed that the nutrient solution in the flask contains the nutrients needed by the root. If such development is not observed, then one or more nutrients is assumed to be absent. Such a study reveals that root tips need water, a source of carbohydrates, mineral nutrients, and certain vitamins, especially thiamine, pyridoxine, and nicotinic acid. In the intact plant, the root tips would normally obtain the water and mineral nutrients from the soil, the carbohydrates from the leaves and vitamins from the young, developing leaves. Thus, the leaves do control the rate of root growth by the amounts of carbohydrates and vitamins that they supply to the roots.

On the other hand, if a very young leaf is transferred to a flask, then water, mineral nutrients, carbohydrates, adenine, kinetin, and red light are required for their continued normal development. In the intact plant, the water and mineral nutrients would be obtained from the roots — so the growth of the leaves would be controlled by the roots — and the carbohydrates would be obtained from more mature leaves or from cotyledons or other storage areas, the adenine from mature leaves and roots, and the red light from sunlight. Here again, the possibilities of the older leaves or the roots controlling leaf development can be seen.

Stem development can also be studied through tissue cultures, wherein it is found that stem apices require water, mineral nutrients, carbohydrates, and a growth hormone such as IAA or GA. Thus the roots and leaves can control stem growth by supplying or not supplying adequate amounts of nutrients, and the young leaves and buds can control stem growth by the supply of hormones furnished. Thus it can be seen how one organ of the plant can and does control growth and differentiation in another.

Fruit development is often very dependent upon the nutrient status of the plant and is therefore correlated by roots and leaves. The biennial habit of bearing fruit by trees is well known, whereby these trees bear heavy crops one year and very light crops the next. Perhaps this is due to the heavy demand for nutrients during the heavy bearing years which depletes the storage of these materials, and also such heavy demand would monopolize nutrients at the time when the flower buds of the next year's crop are forming, reducing the number of such buds formed and thereby reducing the next year's crop.

An interesting aspect of correlative development is associated with photosynthesis. The rate of photosynthesis in one leaf can be greatly influenced by treatment of leaves or other organs some distance from the site of photosynthesis. This indicates that there is some influence that controls the rate of photosynthesis but it is produced by the plant at locations somewhat removed from the sites of photosynthesis. The nature of this influence is not known.

Another example of a similar correlative phenomenon is associated with protein stability. When a plant organ, such as a leaf or cotyledon, is excised from the plant, there soon occurs a net breakdown of the protein structures of these organs. Such net hydrolysis can be prevented by supplying the excised organ with kinetin. Does this mean that other parts of the individual plant furnish kinetin or related hormones, which normally prevent a net breakdown of the proteins in the intact plant?

Bud Development

As considered in a previous chapter, Winter dormancy in buds is also a correlative phenomenon. Perhaps this would be a good time to point out some of the factors that can and do stop or reduce the growth of plant buds. The apical buds of plants, which contain meristems, would appear to have the potentialities for continuing growth indefinitely. However, such growth does not continue throughout the year, since there are times when the plants appear to cease their growth even when conditions for good growth seem to prevail. There are four reasons for this interruption. First, the apical bud may change into a flower bud, with a terminal florescence being formed. This occurs in many annual plants, including pepper and hibiscus plants, and results from floral induction. Once such a transformation has been made, it is not reversible. The flower bud does not change back into a vegetative bud so growth of the shoot permanently stops. Second, the buds may become senescent and die, as occurs in lilac, euonymus, and *Rhus glabra*. When this happens, apical dominance is broken, and the lateral buds start growing with continued growth of the shoot. Such a phenomenon may occur several times during the growing season in an individual plant. More will be said about this in the next chapter. Third, growth may cease due to unfavorable environmental conditions for growth. Often, water deficits cause a cessation or near cessation of growth. This is usually temporary, and growth will resume when the water supply becomes adequate. Also, the temperature may be too high or too low to support growth. Fourth, the buds may become dormant, a phenomenon observed in many trees, shrubs, and herbaceous plants, including *Salix*, *Populus*, and *Quercus*. This is referred to as seasonal bud dormancy.

Two types of seasonal bud dormancy exist, depending upon the time of the year when the dormancy occurs, namely, *Summer dormancy* or *Winter dormancy*. Summer dormancy occurs during the later Spring or early Summer, at a time when environmental conditions appear to be best for growth. At this time, the buds of oak will become dormant and will remain dormant until next Spring. Note that the shoots of these plants only grow for about two months or less each year, with growth beginning about April and ending in June. No wonder they are slow-growing trees. In some species, Summer dormancy is broken the same Summer and the same bud may enter and break dormancy several times during the growing season. The cause of Summer dormancy is due to an accumulation of inhibitors, such as abscissic acid, or to an increase in the inhibitor: growth hormone ratio within the bud. High night temperature does seem to be related to this type of dormancy in some way, perhaps by stimulating the accumulation of the growth inhibitor, or by reducing the accumulation of growth hormones.

Winter dormancy is shown by the buds of many biennial and perennial plants, but not by annual plants. Growth of the buds continues during the Summer but stops before frost sets in due to the appearance of bud dormancy in the Fall. This type of dormancy is particularly beneficial to plants in the temperate, alpine, or arctic regions, because growing tissues are particularly sensitive to low temperature injury whereas dormant tissues are much more resistant, allowing dormant buds to overwinter without frost injury, whereas growing tissues could not.

In the temperate regions of this country, early in the Spring even before the last traces of snow have left, certain flowering plants, such as the crocus, begin to grow vigorously. The amount of vegetative growth which the crocus produces is not great, but is sufficient to support the formation of conspicuous flowers and inconspicuous, subterranean corms. Soon the flowers have served their purpose, and the shoots become senescent and die. During the remainder of the Spring and the Summer, conditions for again initiating growth seem to prevail. The day and night temperatures are warm, moisture is usually sufficient, and aeration is probably favorable. However, the crocus corm will not sprout. It has become dormant, and will sprout again only after this dormancy has been broken by a period of cold temperature.

The peach tree produces flowers and leaves in the Spring. Apical meristems are active in bringing about stem and root extension and cambial meristems are increasing the diameter of the stems. By midsummer, the flower buds are formed — flower buds which will develop next year. However, even though these buds are formed by mid-summer, and conditions appear favorable for their rapid development, they enter a state of dormancy. This dormancy will not be broken until these buds have been exposed to a temperature below 8°C for about 1000 hours. This cold temperature treatment is generally acquired during the Winter, and allows floral development to proceed with the advent of favorable environmental conditions in the Spring. In the subtropical areas, many varieties of peach trees cannot be commercially grown due to the lack of sufficiently low treatments during Winter and consequently few blossoms are formed in the Spring.

All perennials do not exhibit this phenomenon of bud dormancy. In the temperate region probably 50% of the trees and shrubs have buds which do not grow during the Winter due to a lack of favorable environmental conditions for growth, but if they are transferred to a warm greenhouse, growth will continue.

The intriguing question is, what causes Winter bud dormancy? Why do these buds stop growing when they should be under very favorable conditions for growth? In 1914, Klebs reported that the growth of beech, ash, oak, and hornbean could be maintained in Winter under continuous illumination. Later, it was shown that continuous illumination was not necessary as long as the day length or light period was of sufficient duration. If the day length is short, the buds became dormant, if long the dormancy is broken or does not develop at all. So short days initiate Winter bud dormancy, in many perennials. However, plants differ in their response to short days. Some, such as the maples, birches, catalpa, dogwoods, beech, *Ficus*, *Liriodendron*, sycamore, poplars, *Rhus*, black locust, willows, hemlock, elms, *Weigelia*, and *Virburnum* grow continuously when the day length is long, whereas others will grow in flushes under long day conditions and stop growing when the days are short. In other words, growth is not continuous under long day conditions but the plant will grow for a few weeks, rest for a few weeks, and then resume growth again. The cause of this rest period is still unknown, but is probably due to an accumulation of growth inhibitors. Examples of plants of the second type are the pines, oaks, and rhododendrons.

Winter dormancy is caused by short day lengths, and can often be prevented by exposing the plant to longer days, for example by turning a light on when the sun goes down. Plants kept under long day lengths or whose nights are interrupted with light do not exhibit Winter dormancy. Even a brief flash of light is able to prevent Winter bud dormancy. To cause this dormancy one can expose either the leaves or the buds to short days. This indicates that perhaps a growth inhibitor is produced in the leaves or buds, and moves from the leaves to the bud to stop its growth — an example of correlative development. One such inhibitor has been found to be abscissic acid.

Winter dormancy is broken by long day lengths, if the leaves are on the plant. If the leaves have fallen off, as usually happens to deciduous plants in the temperate regions during Autumn leaf fall, then the bud dormancy can be broken either by the application of growth regulator called GA to the bud or by exposing the bud to a certain period of low temperatures. The latter is the way bud dormancy is broken in the field, and was discussed in Chapter 15.

With the breaking of bud dormancy, the concentration of inhibitors in the bud decreases and the concentration of auxin increases. Associated with this increase in auxin production is the resumption of cambial activity in the stems of woody plants.

The control of cambial activity, as discussed in Chapter 16, is another example of correlative development in plants. In this case, hormones produced in the apical buds during their growth, move down the stem of the plant and cause cambial cells to resume meristematic activity. Then, as the growing season nears its end, such hormones are not produced, so cambial activity ceases.

Getting back to the subject of bud dormancy, the Irish potato tuber is an example of the complications which must be expected relative to bud dormancy. This tuber is an underground stem, and therefore it possesses buds, as do most stems. In this case, the eyes contain the buds. Since the tuber is a stem and does possess buds, apical dominance might be expected to be present, and indeed it is. As long as the apical bud remains healthy, sprouting is delayed, but if the apical bud is removed or destroyed, sprouting is initiated quite rapidly.

At the time of harvest, potato tubers exhibit a dormancy of the buds even when the apical dominance is eliminated. Tubers which are cut so as to remove all buds except one, show this dormancy, which may be evident up to about three months after harvest. In contrast to the many examples of breaking bud dormancy by cold treatment, this dormancy cannot be so eliminated, but only at high temperatures (20–35°C). Also, this dormancy can be broken by low oxygen supplies. No doubt some inhibitor must be associated with the potato peel, and this inhibitor must be inactivated before bud dormancy can be broken. Under some conditions, it is desirable to break this dormancy so that the tubers can be planted soon after the harvest. As was shown to be the case with some other types of bud dormancies, ethylene chlorohydrin can be used for this purpose. It is not toxic to the tubers at the concentrations generally employed, and it does not adversely affect crop yield.

It might also be suspected that high concentrations of auxins applied to the buds could delay sprouting. This is so, and is the basis for a successful method of lengthening the storage life of potatoes. Potatoes sometimes present a problem when they are stored, because after a period of time, their dormancies are broken and they sprout, with a resulting increase in transpiration, and soon wilt and shrivel up. Sprouting can be delayed by low temperatures, but this also results in a conversion of starch to sugar with a concomitant decrease in the quality of the tuber as food. However, there is available now a treatment which is used quite extensively.

Certain growth regulators will inhibit bud development and, therefore, sprouting. They are applied by mixing some shredded paper, which has been impregnated with the growth regulator, with the potatoes; by mixing them with talc or other inert dusts and dusting the tubers with this mixture; or by spraying with a liquid solution.

The tuber, such as that of the potato, is an underground stem that forms as a result of a cessation of stem elongation, and is associated with considerable cell division and growth of the pith, pericycle, and perimeuclary zones. This is followed by a tremendous accumulation of food materials, primarily starch, in these newly-formed cells. Just what brings about these changes is not known, but there is good evidence that in response to short days and lower temperature, there is a hormone produced in the shoot of the potato plant which moves down into the stem and induces tuber formation. The nature of this hormone is unknown.

Reproductive Control of Vegetative Development

Another interesting and important type of correlative development is that exhibited by the control of reproductive development over vegetative development. Flowering and fruit formation is often harmful to vegetative development. Many annual and perennial plants will continue to grow vegetatively until they flower and set fruit, after which vegetative growth ceases. For instance, it is well known that if one wishes sweet pea plants to flower profusely and long, it is necessary to remove the flowers shortly after they form. And the century plant of the southwest deserts will grow vegetatively for decades until it flowers, but following flowering and fruit development, it will die. Perennials seem to be immune to this lethality of sexual reproduction, but this phenomenon will be considered in more detail in the next chapter, since it delves into the intricacies of senescence.

No plant lives forever. Some live for centuries and some for only a few days, but all eventually die, and if a method for the propagation of new individuals was not available, the species would soon become extinct. Vegetative propagation is one method through which the species is continued. However, even vegetative propagation cannot usually continue forever, since eventually the species loses vigor unless sexual reproduction occurs. Also, many species (the annuals) cannot survive the Winter in the vegetative stage, but survive only in the form of seeds. Yes, sexual reproduction is necessary for the continued survival of most plant species and it, along with vegetative propagation, allows the species to increase in the number of individuals.

Sexual reproduction in the gymnosperms and in the angiosperms, begins with floral induction and proceeds through floral initiation, flower bud development, anthesis, pollination, fertilization, and fruit and seed development. It results in one plant producing numerous seeds each of which has the potentialities for developing into a new plant.

Floral Induction

Floral induction is a result of differentiation and is often also a result of correlative development. It includes all of the invisible changes that take place in a plant and which will result in the initiation of flower buds. After the seed begins to germinate, the seedling is formed. This seedling does not ordinarily flower. It is too mature. However, as it continues to grow, eventually it reaches a stage in its development wherein flowering can occur, along with its associated sexual reproduction. This might be likened to the adolescent stage in humans and, in plants, this stage at which flowering can occur is known as the *ripeness-to-flower stage* of development. Each individual plant must attain this stage of development, before it can flower. Just what changes occur in the individual to result in this potential for flowering are unknown, and just how long after the seed germinates before this stage is reached is extremely variable. Each plant must produce at least a few leaves even though the leaves need not be mature. Younger plants cannot be induced to flower, but the amount of vegetative growth needed to attain this stage is often very small. For instance, cocklebur (*Xanthium*) seedlings are capable of floral induction after just one week from seed. Some varieties of tomato plants reach this stage after 13 leaves are produced, rye plants must have 7 leaves, and corn requires about 11 leaves. However, annuals do flower the same year that the seed germinates, whereas this is not true for all biennials and perennials. Most biennials do not reach the ripeness-to-flower stage until the second year, and for perennials, such as trees, this stage may not be reached for many years following germination of the seed. However, among all three groups, the species vary greatly in the extent of their vegetative development required to attain the ripeness-to-flower stage.

When the ripeness-to-flower stage is reached, some plants go ahead and flower irrespective of the environment, while others must wait for specific environmental conditions which must first be present. For example, a tomato plant may form flower buds as soon as the plant develops 13 nodes, whereas Winter wheat plants will not flower until after being exposed to low temperature and long days.

In many plants, floral induction is controlled by the environment and especially by the day length (*photoperiod*) and/or the temperature. In the temperate regions of the world, the season of the year during which an herbaceous plant flowers is determined largely by the day length. Some species flower only when the day is long and others only when the day is short. Since the day length varies with the latitude, it might be expected that the geographical distribution of plants is controlled in part by the prevailing day length, and indeed this is true. However, even within a given species it has frequently been observed that strains differ in their photoperiodic requirements. For example, sideoats grama grass, *Bouteloua curtipendula* in Kansas can be found with long or with short photoperiodic requirements or some strains even show no photoperiodic requirements for flowering.

Flowering begins with certain biochemical changes that take place within the plant, which ultimately will result in the conversion of a vegetative bud into a flowering bud. These changes are known as *floral induction*. Just what these biochemical changes are is not known, but must involve a change in metabolic regulation of some sort. Is this change quantitative or qualitative? Is it associated with the appearance of new enzyme systems or just with changes in the activity of pre-existing ones? These answers are not known. However, it is known that in many cases a flowering hormone—*florigen*—is involved and the synthesis of both DNA and RNA are necessary for floral induction to occur. Such induction is also followed by a rapid increase in the mitotic index of the cells of the plant apex, and in a rapid increase in the RNA content of the meristems of the buds, whereas the ratio of histone, a supposed gene repressor, to DNA decreases. This indicates a derepression of DNA activity that may be necessary for the synthesis of enzymes needed for flowering.

In many plant species, floral induction will occur only after the plant has been exposed to favorable day lengths. Such induction is known as *photoperiodic induction*. Some species require that the day length be short, and are known as *short-day* plants, whereas others require long day lengths and are known as *long-day* plants. In either case, there is a certain day length above or below which the plant will not flower. This is known as the *critical day length*, abbreviated CDL.

The number of hours in the CDL will vary somewhat from one species to another, but generally for short-day plants, the CDL is 13 hours, and for long-day species, the CDL is 12 hours. This means that at day lengths less than 13 hours, short-day plants will flower, whereas at longer day lengths, they will not, but rather will continue their active vegetative growth.

The times of the year when plants will flower can be predicted by considering their CDL and the day length of their habitat throughout the year. This day length is not constant but varies from day to day and month to month. These variations can be seen by referring to Fig. 18.3, which gives the day length throughout the year at Manhattan, Kansas. However, even before sunrise and after sunset, there is some light available to the plant, so the light intensity necessary for photoperiodic induction must also be known. Much lower intensities are required than are required to saturate photosynthesis and for vigorous vegetative growth. In some plants, intensities of only 0.1 ft candle are required, although more consistent flowering is observed if the intensity is 10 ft candles or greater. Others require 100 ft candles or more. As previously indicated, day lengths can be altered by using electric lights to extend the day length, or light shade to shorten it, and these are frequently done in greenhouses. In some of the coffee growing regions of the world, other trees are planted in the coffee groves among the coffee trees to furnish shade, which effectively reduces the day length and allows these coffee trees to flower.

In some respects, the term, critical day length is a misnomer, since it is

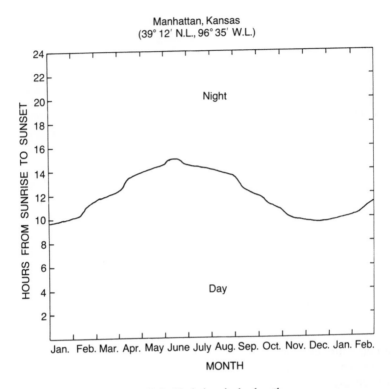

Fig. 18.3   Variations in day length.

actually the length of the dark period that determines whether a plant will flower or not. A short-day plant flowers only when the dark period— night—is long, and a long-day plant only when the length of the dark period is short. This can be demonstrated by growing a short-day plant under conditions of a long night period, and interrupting the night period near the middle with a brief flash of light. Such an interruption will prevent the plant from flowering even though the total length of the dark period was longer than that required for the plant to flower normally. This practice is used to delay flowering of sugar cane in Hawaii. This phenomenon seems to be associated with the phytochrome system since only red light will nullify the long dark period, and this action can be reversed by a subsequent exposure of the plant to far red light. Alternating red and far red light flashes will result in floral induction or not depending upon which quality of light the plant was last exposed to.

Just how many light and dark cycles are necessary for floral induction will vary with the plant species. Most species require a number of favorable cycles for induction to occur, but a few species initiate flowering with just one such cycle. Such species include the short-day plants *Xanthium pensylvanicum*, *Chenopodium rubrum*, and *Pharbitis nil*, and the long-day species *Lolium temulentum*. However, even in these, an increase in the number of days on treatment will often increase the number of flowers formed, and shorten the time required to reach anthesis. The grass, *Lolium temulentum*, is a long-day plant requiring short nights for flowering. It is very sensitive to the photoperiodic treatment since exposing just one of its leaves to a single long-day period is sufficient for floral induction.

Since light is involved in photoperiodic induction, a pigment must be involved which participates in the photochemical reaction. If the wavelength or color of the light necessary to bring about photoperiodic induction is known, the color of the pigment involved will be known, since pigments are of a complementary color to the light which they absorb. Also, the nature of the structure of the pigment molecule will be known. The color of light required for this reaction can be determined by exposing a plant to light of one color and finding which color is most efficient in photoperiodic induction. Such studies have been conducted and reveal that maximum efficiency occurs when the light has a wavelength of about 660 nm. This is red light, so the pigment must be blue or green in color. No such pigment can be seen in plants, so it must be present in very minute concentrations. Recently, this pigment has been detected, with very sensitive instruments. It appears to be a protein, and is called phytochrome. This is the same pigment that plays such an important role in seed dormancy, leaf enlargement, and many other physiological phenomena in plants, as previously studied. Therefore, this photoperiodic effect should be reversible in the presence of far red light (730 nm). It is. If a short-day plant has its long dark period interrupted by light, the plant will not flower. However, if shortly following this exposure — within less than an hour thereafter — the plant is exposed to far red light, the plant will flower just as though its dark period had not been interrupted. Yes, evidence indicates that at least the phytochrome system is involved, but the functions of other unknown pigments are also suspected.

Floral induction can be inhibited, in an environment favorable for its formation, without inhibiting vegetative growth, by applying either 2-thiouracil or ethionine to the leaf. How these act is also unknown.

In addition to the pigments discussed above, floral induction involves a type of clock-like mechanism, referred to as a *biological clock* or circadian rhythm. This clock is implicated in many physiological activities although essentially nothing is known about its mechanism. It is indeed very strange, but seems to function much like an oscillator, such as the pendulum on a grandfather's clock, or the balance mechanism on more modern timepieces. Its mechanism must be largely physical, rather than biochemical, since its Q_{10} is only about 1.2–1.3, values that indicate physical not biochemical activity. This clock is largely independent of the environment, in its activities, but can be set or reset by light and/or temperature changes in the environment.

In addition to its role in photoperiodism, the biological clock also functions in such widely diverse physiological activities as rhythmic sporulation patterns in algae and fungi, alternate branching–non-branching in fungal mycelia, alternating luminescence, alterations in photosynthetic capacity, cell division, carbon dioxide evolution, banding of starch grains, the change of shape of plastids with the time of day, changes in the volume of the nucleus, and in night movements, anthesis, guttation, and others. Indeed, such diversity makes one wonder if there are different biological clocks for different control phenomena, and if there are different mechanisms for biological clocks in different cells.

The biological clock functions to measure time intervals for the cell or plant. However, this is not the only means by which such intervals are measured in the living cell. Photochemical reactions, such as the phytochrome system, measure diurnal intervals, and longer or shorter intervals can be measured by the accumulation of metabolites or by the temperature control of metabolism, respectively.

Where is the pigment, or pigments, located in the plant? This can be determined by exposing just a part of the plant to favorable day lengths for flowering, and leaving the remainder of the plant under unfavorable day length conditions. This has been done, and has shown that photoperiodic induction occurs only when the leaves are exposed to favorable photoperiods. No other plant organ, not even the buds, can function in this capacity. Not only are the leaves the receptive organs, but not any leaf will do. It must be recently matured, not immature or old or senescent. So the pigments are located in recently mature leaves.

When such a leaf is stimulated by a favorable day length, a hormone is produced by the pigments involved. This flowering hormone is called *florigen*. Florigen has never been isolated nor has its chemical structure been determined. In fact, it appears that the name florigen may refer to a number of chemically unrelated substances that are synthesized in the leaves in response to favorable day lengths, but are physiologically related because they induce a similar response in the buds.

After the florigen has been synthesized in the leaves, it moves out through the living cells of the petiole. Such translocation is non-polar and does not occur through the xylem. This is known because low temperatures hinder it. However, the velocity of movement is quite slow, with rates of about 2 cm/hr being reported. About four hours are necessary for the synthesis and movement of sufficient florigen to initiate induction after exposure to favorable day lengths. However, this time interval will vary with the distance the florigen has to travel.

Once the florigen arrives at its site of action in the bud, it can only act if the cells of the meristem are in a state of active cell division with a reasonably high mitotic index. Dormant meristems cannot be induced to flower by the normal inductive treatments. Chemical treatments which cause bud dormancy or inhibit cell division will interfere with floral induction. Such chemicals, include 5-fluorouracil, 5-fluorodeoxyuridine, 2-thiouracil, 6-azauracil, and the growth retardant CCC, as well as by 8-azaguanine, and maleic hydrazide.

Also, photoperiodic induction requires that the plant be exposed to high light intensity prior to the exposure to a favorable light period. Without this treatment, induction cannot occur irrespective of the subsequent photoperiod. Perhaps this high light intensity is required for adequate metabolic substrate, but a high intensity photochemical reaction is also a possible requirement.

Photoperiodic induction correlative phenomena are illustrated in Fig. 18.4.

FLORAL INITIATION

As the florigen moves into the terminal or lateral buds, certain changes take place. At this time the entire nature of these changes is not known, but appears to involve RNA metabolism, since chemicals that interfere with normal RNA metabolism also interfere with initiation. Two or three days after the florigen moves out of the leaf, marked changes occur in the appearance of the bud. Where the vegetative bud was round and rather smooth in appearance on its upper surface, it now begins to flatten, swell, and become very irregular. Small protuberances appear on its surface, and these later become the appendages of the flower, namely, the sepals, petals, stamens, etc. This visible change in the appearance of

Fig. 18.4   Illustration of mechanisms involved in floral induction.

the vegetative bud as it is transformed into a flower bud is known as *floral initiation*.

Environmental Control

Floral initiation is also controlled by the environment, although the environmental requirements for initiation may not be and often are not the same as for induction. For instance, it has been shown that for floral induction in orchard grass or in Kentucky bluegrass, short days and a low temperature are required (5–10°C). Floral initiation requires longer days and warmer temperatures whereas growth of the flower requires long days and warmer temperatures. Are these not the conditions that normally prevail when the grass is growing in the field?

Long-day plants are usually biennials. They produce leaves which form a rosette pattern, growing closely to the ground and spreading out in a circular pattern. In effect, this is due to a lack of elongation of the internodes of these plants. When the day length becomes long, the internodes elongate and a long stalk is produced which ultimately bears the flowers. The lack of internodal elongation and stalk development seems to be associated with the leaves. If the leaves are removed or even cooled, flowering results. Therefore, it is believed that a hormone must be continuously synthesized, within the leaves under short days, which is inhibitory to flowering, and long days stops its synthesis and allows flowering to occur. Gibberellins can replace the long-day conditions in these plants and cause them to flower under short days. However, GA cannot cause short-day plants to flower under long days.

Competing Organs

Floral initiation and subsequent flower development are greatly influenced by competitive organs for growth substances, and cannot strongly compete with other organs for these materials. This is one cause of biennial variation in the fruit yield of many trees. These trees, for instance apple, initiate floral buds the growing season before the buds open. Peaches, dogwood, and many forest trees initiate flower buds in late Summer, while apples initiate them in early Summer. In any respect, floral initiation for the following year's crop is taking place while fruits are forming on the tree. Fruit formation on any plant requires a great deal of growth materials. Therefore, the fruits tend to monopolize the sugars and other growth materials produced by the plant. If the fruit crop is heavy, so much of these substances will be translocated to the fruit that sufficient amounts will not be available for floral initiation. Therefore, the next year, fewer flowers will be present. Even if initiation may be successful, the flowers may not form due to nutrient deficiencies.

Nutrition

The relative amounts of carbohydrates and nitrogenous compounds in the plant also have an effect on the efficiency of initiation. It has long been known that if a plant is heavily fertilized with nitrogenous fertilizers it will flower neither early nor abundantly. The carbohydrate level must be relatively high. If excess available nitrogen is present, the carbohydrates are used up in amino acid and protein synthesis instead of being used for floral initiation. Nasturtium plants will flower much more abundantly on a poor soil than on a well-fertilized one. It is true that nitrogen is needed by plants for floral initiation, and optimum nitrogen content gives maximum flowering, but an excess is detrimental. If it is desired that flowers be abundant on plants, these plants should be fertilized well early in the season, but fertilizer should not be applied as the season progresses.

Light

Floral initiation is also very much affected by light intensity. It has been known for many years that those plants that are growing in the sun produce many more flowers than similar plants growing in the shade. Apparently, high light intensity increases floral initiation but how it does this is not known. Opening a stand of forest trees also increases seed production so trees must be similarly affected.

Hormones

Following initiation, profound changes occur in the hormone content of the plant. These changes are both qualitative and quantitative. New hormones are produced and the hormones initially present are produced more abundantly. Many of these hormones have not been identified at the present time and more research must be done before we can explain their significance to flower development.

Many plant species produce perfect flowers, meaning that both pistillate and staminate organs are found in the same flower. However, some do not. Some are monoecious. Each individual will produce flowers that possess the staminate structures and flowers that possess the pistillate structures. Often both staminate and pistillate flowers are formed on the same stem or branch. One very interesting problem in biology is what causes some flowers to be pistillate and others to be staminate. Indeed, it is now possible to alter the ratio of the number of staminate to pistillate flowers on a plant. This can be done in several ways. Pistillate flower formation is increased by exposing the plant to low temperatures, to above optimum concentrations of nitrogen, to an external supply to auxin or by keeping short-day plants, such as corn, under short-day conditions. On the other hand, staminate flower formation is favored by exposing the plant to high temperatures, to less than optimum amount of available nitrogen, by applying gibberellins to the plant or by keeping short-day plants under long-day conditions following induction. Physiologists generally agree that the optimum level of auxin for flower initiation is less than that for vegetative growth and it is also possible that auxin levels for the production of staminate flowers are lower than that for pistillate flowers.

BUD DORMANCY

After initiation, the flower bud forms its young structures such as sepals (that protect it by forming a cover over it), and petals, stamens, pistils, etc. The cells of these may then enlarge to form the mature flower, which is the case in annual or biennial plants or they may become dormant as is usually the case in perennial plants. This bud dormancy is the same as that previously discussed and is usually broken by a certain characteristic number of hours of exposure to low temperatures. The exposure normally occurs during the low temperatures of Winter, and if the Winters are not cold enough for a sufficient length of time, the buds may remain dormant the following growing season and not form fruits and flowers or the plant will be less productive. This dormancy protects the bud during the cold weather of Winter.

ANTHESIS

Eventually, the cells of the bud do enlarge and the flower bud develops toward maturity. Flower buds reach maturity at anthesis. Anthesis is that stage of floral development when the flower bud opens. It is the stage of flower development that people are most familiar with. In many plant species, the opening of the flower bud is controlled by the environment. In most species, such as species of *Campanula, Geranium, Cistus*, and *Ipomoea*, the buds open in the morning. This opening is influenced by the dark period since the buds fail to open if the night is interrupted by light. The light must be red and have a wavelength of about 660 nm. This may also be reversible by far red light, and controlled by phytochrome, although very little research has been carried out on this subject.

As is well known, some species have flower buds which open only in the evening. For instance, the buds of evening primrose (*Oenothera* sp.) open at about 6:00 p.m. However, they will open at 6:00 a.m. if the plants are exposed to darkness from 6:00 a.m. to 6:00 p.m. and to light from 6:00 p.m. to 6:00 a.m. In other words, if the normal light-dark cycle is reversed. The buds always open 12 hours after a dark-light transition or normally, about 12 hours after sunrise. The timing mechanism is not known in detail, but appears to be located in the buds, as each bud must be exposed to this dark-light transition before it will open. This too has not been studied as well as it should have been. Perhaps the biological clock is also functioning here.

As might be expected, there are also some plants whose buds open at rather irregular times. For instance, it has been reported that the coffee tree has flower buds whose opening is controlled by the climate. The buds open when rain follows a long dry spell. Of course, in this plant, as well as in all others, the bud must first reach a certain stage of development before it will open irrespective of the environmental conditions to which it is exposed.

Nectar

Shortly after anthesis, the anthers rupture, releasing the pollen. This rupturing of the anthers is followed closely by a secretion of nectar by the flowers. This nectar is secreted by nectaries – specialized cells at or near the base of the petals. Secretion is an activity that is dependent upon metabolic activity of the flower, and especially of the nectary cells, and anything that reduces the rate of respiration reduces the secretion of nectar. This secretion then must be dependent upon energy derived from respiration. However, the water status of the plant is important too, because the amount of nectar produced is decreased by increasing the soil moisture stress, or by saturating the soil with water. Nectar is primarily an aqueous solution of sugar, with sugars comprising about 99% of its dry weight. Therefore, the amount of nectar produced also depends upon the availability of sugars. Nectar secretion increases when the plant is exposed to a high light intensity the day previous to secretion. Also a high nitrogen availability to the plant reduces the amount of nectar produced, probably because increased amounts of available nitrogen utilize more of the sugars for amino acid synthesis, reducing the sugar content of the plant.

As your nose no doubt has already revealed, nectar, also contains certain odorous, volatile substances, particularly amines and essential oils.

Nectar secretion continues at variable rates as long as fertilization does not occur or the flower petals do not senesce. If either of these happen, nectar production ceases.

Abscission

Flowers vary, according to the species, in the amount of time they remain open. Following anthesis, the flowers usually remain open either until fertilization is complete or until pollination and pollen tube growth occur. Orchid flowers will remain open for two months or so awaiting for fertilization to occur and therefore are rather long-lived. On the other hand, there are species of plants whose flowers close or abscise (fall off) one day after anthesis, irrespective of whether these flowers are fertilized or not. For insect-pollinated flowers, if the insects are not present the day the flowers are open, no fruits are produced. Some of our common flowers come in this category, namely, *Oxalis* sp., day lily, and watermelon. The mechanisms for closing or abscissing are not clearly understood, but will be discussed in Chapter 19.

Irrespective of the length of time the flower remains open, eventually it will close, and after this happens, nectar secretion stops and the petals and stamens will wilt and abscise. To conserve the nutrients in these petals and stamens, now that the plant has no further need of them, various nutrients including nitrogen and phosphorus are translocated to the ovary, where they are used for the growth of the seed and fruit, that now begin to form. This translocation is associated with a rapid breakdown of protein in the petals and stamens, and possibly the sepals, following the closing of the flower.

Perhaps one of the most interesting phenomena associated with both differentiation and correlative development is fertilization and the subsequent development of the embryo. No doubt the influence which attracts the pollen tube to the ovule is a correlative phenomenon, and the influence transferred from the sperm to the egg cell that initiates the formation of the embryo is a fantastic phenomenon associated with differentiation. Without this influence, the egg cell does not divide and eventually dies, but with it division occurs and the new plant develops.

VEGETATIVE REPRODUCTION

Differentiation and correlative development are also evident in the phenomenon known as vegetative reproduction, a very important and widespread form of reproduction in plants.

Some plants can grow for many years or centuries vegetatively, but eventually all plants die. Therefore if a means of reproduction was not available the plant life on this earth would soon disappear and with it all life would vanish. Reproduction is necessary for the continuation of the species. When reproduction is mentioned, sexual reproduction is thought of as the only type of reproduction of which higher animals are capable. However, much of the propagation of new individuals in the plant kingdom is vegetative and does not involve the sexual cells.

Vegetative reproduction involves either the differentiation of meristematic cells that are capable of producing a new species or the development of dormant meristems that are so capable. Some examples of vegetative reproduction in plants follow.

Many plants reproduce by rhizomes which are underground, but horizontal, stems. These rhizomes form nodes just as do the normal aerial stems, and buds and roots can be formed at these nodes. In this way, new plants are potentially able to be developed at each node. Examples of plants that are capable of reproduction in this manner are most of our perennial grasses, buckbrush, ferns, cattails, and sedges, poison ivy, many mints, wild morning glory and a beautiful but pesty weed, the water hyacinth. This is quite an impressive list of plants and does indicate the wide distribution of this method of propagation. This is also a very efficient and rapid method of reproduction. For instance, it has been observed that water hyacinth plants in a pond can produce over 650,000 new plants in just one growing season. Yes, plants can be quite prolific by the use of this method of vegetative reproduction. Many weedy plants, such as buckbrush and white clover, as well as strawberry plants, reproduce readily by runners or stolons. These are aerial stems that lie on the soil surface and grow out from the plant in many directions. As in the case of rhizomes, roots and buds can be and are produced at each node where the node comes in contact with the soil surface. Thus, a new plant can potentially be produced at each node. By what mechanism does contact with the soil cause the new plant to develop?

Another common method of vegetative reproduction is by horizontal roots which grow through the soil and new shoots grow upward from various points on these roots. Some examples of such plants are common to us and include bindweed, milkweed, canada thistle, osage orange, sumac, beech, elderberry, and lilac.

Suckers are often formed on many plants and permit reproduction.

There are also some plants that are rather unusual in that they form new plants vegetatively on the vegetative organs of the mature plant. Some species of *Liliaceae* produce bulbils on their bulbs and these are capable of developing into new plants. Some species of *Kalenchoe* form new, young, plants on their aerial leaf margins, as seen in Fig. 18.5. These young plants later fall from the mature plant, take root in the soil, and develop into new plants. This is another very prolific method of vegetative reproduction.

These previously-discussed methods are often used commercially for propagation purposes, along with a rather artificial method known as the rooting of cuttings which shall be discussed in Chapter 20.

Yes, there are many examples of differentiation and correlative develop-

Fig. 18.5   Plantlets which originate at the margins of the leaves of some species, by the differentiation of leaf cells.

ment in plants, and these are necessary for the normal development of the individual. As this development continues, eventually, the plant leaves its mature state and becomes senescent, which eventually results in the final stage of plant development, namely, death. These will be considered in the next chapter.

Chapter Nineteen

Longevity, Senescence,
and Death

LONGEVITY

Plant development begins with cytokinesis, and continues until the cell either divides or the plant dies. The time interval between these two events is known as *longevity*. It is proper to speak of the longevity of the plant, or of a cell, or of a seed, etc.

Longevity, in plants, varies greatly, with some cells living but a few minutes, after which they either divide or die, whereas some multicellular plants may live for several centuries.

Age is respected, if for no other reason than the fact that to survive beyond the normal lifespan is an exceptional achievement. This is true irrespective of the biological entity concerned. The author has often looked, with respect, upon a tree that has survived for thousands of years, and never ceased to be amazed at its persistence. An example of such a specimen is given in Fig. 19.1. This specimen is a *Juniperus scopulorum* tree growing in Cache National Forest in Utah. Usually such old specimens are found in a desert environment due to the fewer diseases present in such an environment, and also because the environment is not conducive to rapid plant development. This tree is reported to be over 3000 years old. Indeed, it was an old tree when Christ was on the earth.

It is true that the tree survived for centuries, but what about its cells? Are they that old too? A close observation of Fig. 19.1 will show that a much smaller percentage of that tree is composed of living leaves and branches than is a mature or young tree of the same species. In fact, most of the mass of that tree is composed of dead materials, the remnants of

Fig. 19.1 An old juniper tree, over three thousand years of age, illustrating the sparsity of leaves and the preponderance of woody tissues.

415

cells that ceased living many years ago. Any old tree is composed largely of dead cells, and of the remnants of dead cells. Yes, only a very small percentage of its mass is actually composed of living cells, and although these are now alive, how long have they been so? Is a meristematic cell in the tip of one of the branches of this tree immortal, or does it have but a short existence? How much of the living protoplasm of the original zygote that started the development of this tree is now actually present in its living cells? Perhaps the answer to these questions is none. The tree is old because it has retained some growth activity for many years, but its living cells are new since they were recently formed. Longevity in the perennial plant is more a matter of replacing cells that die, than of maintaining the cells it has for long periods of time.

Control of Longevity

Longevity is determined by the length of time spent in each of the stages of development, and can be increased or shortened by either extending or shortening one or more of these stages. Perhaps one of the most interesting examples of longevity would be in the case of plant parts such as seeds, spores, or pollen grains. Longevity of these can be extended by storage under conditions of low moisture, low temperature, or in an atmosphere with a low oxygen content. It is possible to store seeds at a given temperature and relative humidity, but if the temperature is raised, the relative humidity must be lowered to give the same longevity, or vice versa. It is not the actual age of the seed which is important, but rather the conditions under which the seed has been stored. This indicates that longevity is determined by factors of the environment which alter the structure of the plant part in some way to bring about the death of the cells. What these changes are is unknown, but the extent of oxidation is a factor.

It is true that temperature, relative humidity, and oxygen supply will alter longevity, but it is also true that longevity differs from one plant to another. What can this be attributed to? First, as the size of the individual increases, the longevity often decreases. Since more time is involved to produce a large plant, it might be expected that the larger its size, the greater its longevity, and this is true.

The more rapid the rate of respiration, the shorter its longevity. This might be expected, in view of the fact that a high oxygen concentration of the air shortens the longevity and therefore the storage life of seeds, spores, etc.

In those plants where the staminate and pistillate flowers are on separate individuals, those with female flowers survive longer than those with male flowers. This is similar to the situation in man, where the woman has a greater lifespan than does the man.

A correlation also exists between the rate of transpiration and longevity. Those plants with a higher rate of transpiration have a shorter longevity. This may be due to the increased uptake of minerals by these plants with a more rapid accumulation of those that may become toxic.

Talking about toxic substances, these also could shorten the lifespan. Many substances are produced by plants that have toxic properties. If these accumulate to toxic concentrations the lifespan would be shortened. No doubt this is a factor which determines differential longevity between individuals.

The rate of food used during reproduction or food monopolization by some parts of a plant at the expense of others will also alter longevity, as will be seen later. Indeed, once an annual or biennial plant flowers, it is short-lived.

Radiation also has a detrimental effect on longevity, perhaps through its effects on chromosomal aberrations.

SENESCENCE

So far in this book, the earlier stages of plant development have been considered rather thoroughly. It is now time to consider the later stages, namely, senescence and death. Senescence and death are just as normal as stages of development as are cytokinesis and germination and juvenility and maturity.

Senescence can be defined as the stage of development between maturity and death, and it is often referred to as old age. Senescence begins when the plant ages beyond maturity, and ends with the death of the plant. It is a stage in the life cycle represented more by degradative and hydrolytic processes than by processes of synthesis.

Decreased Weight

Senescence is, in some respects, a reversal of growth since during senescence there is a continuous decrease in the fresh weight of the plant or plant part. This decrease in fresh weight is due both to a decrease in the water content of the cells and to a decrease in the dry weight of the plant. This decrease in dry weight is associated with both a decrease in the size of the cells, so that senescent tissues have smaller cells than do mature tissues, and either a decrease in the rate of photosynthesis or a decline in the efficiency of the photosynthetic process that does occur. Also, as will be discussed later, senescence is associated with fewer leaves on the plant since some leaves abscise, which would also result in less photosynthetic activity by the plant.

This decrease in the dry weight of the cells is not associated with a significant breakdown of the cell wall materials — materials which impart most of the dry weight to the plant. These materials such as cellulose and lignin, cannot be broken down by the plant. On the other hand, there is a decrease in the amount of protoplasm which more than accounts for the decrease in dry weight. In fact, the amount of protoplasm decreases more than does the dry weight, in the senescent structure.

Respiration

As might be expected, this decrease in the amount of protoplasm is associated with a decrease in the rate of respiration, at least during later stages of senescence. This is due to both a decrease in the amount of sugars available for respiration and to a decrease in the activity of the mitochondria, and in some cases, even to a decrease in the number of mitochondria per cell.

Protein

This fact, plus the decrease in the amount of protoplasm, is also due to, in part, or caused by a decrease in protein content of the cells during senescence. A mature plant shows a stable protein level, with the rate of synthesis equalling the rate of protein breakdown. But, with senescence, the protein content decreases due to a more rapid rate of protein hydrolysis than protein synthesis. This is due largely to an interference with the mechanism of protein synthesis, due in part at least, to a decrease in the number of ribosomes per cell, and to the presence of fewer polysomes. Ribosomes degenerate during senescence and polysomes disappear. This may be associated with the reported increase in RNA-ase and DNA-ase activities, which reduce protein synthesis. With a decrease in the rate of respiration, less energy is also available for protein synthesis.

Pigments

Protein synthesis is not the only synthetic activity which declines during senescence. Perhaps the most obvious is that of the synthesis of chlorophyll. Senescent leaves are usually yellowish in color. Much lighter in color than young or mature leaves. This is due to a decrease in chlorophyll content, which is associated with senescence. This decrease is due more to interference with chlorophyll synthesis than with stimulation of chlorophyll destruction. If the chlorophyll level becomes low enough, this may also interfere with photosynthesis, reducing its rate. Indeed, advanced stages of senescence in leaves are frequently associated with leaves of yellow, rather than green, coloration. This is especially evident with certain Autumn leaves. Changes in the environment which bring about Autumn leaf coloration will be discussed later in this chapter.

Just because there is a decrease in the concentration of chlorophyll with senescence, does not mean that there is a decrease in the content of all pigments. Indeed, some even increase in concentration with senescence. Anthocyanins and other flavanoid pigments are noticeable in this respect. Autumn leaves often turn red or purple due to this increase in their concentrations of these pigments during senescence, and ripe fruit has lost its green color, due to chlorophyll breakdown, and taken on red or purple coloration due to increased anthocyanins. Carotenoids also behave similarly, giving some fruit a yellow or orange color when ripe. Actually, ripe fruit is senescent fruit, and fruit ripening is associated with the fruit becoming senescent.

Ash

The ash content of the plant increases with senescence. This may seem strange, since leaching might be expected to reduce the ash content, but mineral nutrients continue to be taken up and dry weight decreases, resulting in the increased ash content. However, there is another factor to consider, in this respect, in plant organs. During senescence, there is a net translocation of certain mineral nutrients, such as nitrogen, phosphorus, potassium, and magnesium, out of senescent leaves, flower petals, etc. into other parts of the plant. This would decrease the ash content, but apparently there is a greater increase in the immobile nutrients, such as calcium, sodium, and silicon, and the micronutrients which more than compensates for the loss.

Viscosity

There is also a change in the viscosity of the cell cytoplasm with senescence, with the viscosity increasing. Although cell viscosity changes often and quite rapidly throughout the life of the plant, the average viscosity actually increases with senescence.

Senescence is also associated with a significant change in the types and quantities of hormones produced by the plant or its organs, as will be discussed later in this chapter.

Intracellular Changes

Changes associated with senescence in the multicellular plant are due to senescent changes on the cell level plus correlative influences. What changes occur in the individual cell as it becomes senescent? If these changes were known, the subject of senescence on the multicellular plants would be better understood. As a cell becomes senescent, changes occur both in its organelles, and in its chemistry. Ribosomes seem to degenerate, and protein and RNA contents decline. Chloroplasts also degenerate, which might be due to the reduced protein content, and with them, the chlorophyll concentration decreases. Some other pigments increase in concentration, while some mineral nutrients decrease, especially nitrogen, phorphorus, potassium, and magnesium, and other inorganic elements increase in concentration. Changes which occur in respiration have already been mentioned, with an eventual gradual decline in the rate, and an eventual change in the respiratory pathway, with fermentation becoming more active. Auxin concentrations decrease while ABA concentration increases.

As these, and other changes, progress, eventually the cell becomes so degenerate that it cannot supply the energy needed for its own maintenance so its organization breaks down, with the membranes losing their differential permeability and respiration ceasing. All of these changes could be brought on by alterations in RNA and/or DNA metabolism or functions.

FRUIT RIPENING

Changes which take place during senescence can be better illustrated by considering the ripening process of fruit, since this has many of the characteristics of senescence.

During the development of the fruit, it eventually reaches its approximate maximum size. At this time it is mature and green in color, due to its chlorophyll content, and is quite hard. This state of maturity is required before ripening can occur, and is soon followed by initiation of senescence.

The fruit becomes ripe and subsequently continues to degenerate until its cells die and the fruit rots, releasing the seeds so that a new generation can appear.

If a mature apple fruit is removed from the tree, little change can be noted in its physiology for a couple of days. Then there is a tremendous increase in the amount of ethylene produced within the fruit. This ethylene is volatile and some of it escapes into the surrounding atmosphere, where it will initiate ripening in other fruit. An example of this is the rapidity with which bananas will ripen while being transported in a ship if only one bunch of the bananas ripen prior to arrival at their destination. Once one bunch ripens, the escaping ethylene will soon cause all other bananas in the ship to ripen, often resulting in the loss of most of the fruit before it arrives at its destination.

Shortly after the ethylene content increases, there is a very marked increase in the respiration rate. This increase is known as the climacteric rise, and continues until the maximum rate of respiration is observed, a rate known as the climacteric. After this, the rate continues to decline and will do so until death occurs. Such changes in the rates of respiration have been reported in many fruits, such as apple, apricot, banana, peach, pear, plum and tomato, in flower petals, and in other senescent tissues, and therefore seem to be quite common. Theories to explain the mechanism of the climacteric rise were discussed in Chapter 8.

As might be expected, the climacteric rise is associated with a decrease in the oxygen content of the fruit and an increase in its carbon dioxide content. It is also associated with other changes in the regulation of metabolism, changes which result in alterations in the chemical composition of the fruit.

Green fruit is sour, whereas ripe fruit is sweet. There are two reasons for this. First, green fruit contains a high content of certain organic acids which give the fruit its sour taste. During ripening, these acids are used in the synthesis of sugars, which decreases the acid content and increases the sugar content. Also, ripening is associated with a breakdown of the starch, which is present in very high concentrations in green fruit, with the subsequent formation of sugars. This also adds to the sweet taste of the fruit.

In addition to being sour, green fruit is hard. During the ripening process this changes so that the ripe fruit is soft. This softening is a result of changes in the polyuronides of the fruit, changes whereby pectates and protopectins are changed into soluble pectins. This, in essence, dissolves the middle lamella which allows adjacent cells to separate, softening the fruit.

Green fruit is green because of its high chlorophyll content. With ripening, this chlorophyll content greatly decreases, and the fruit changes color, becoming either yellow, orange, reddish, or purple as the carotenoids are unmasked and their synthesis, plus the synthesis of anthocyanins, is greatly increased.

There is also a noted change in the aroma of fruit during ripening which is due to changes in the chemical composition of the fruit. During the ripening process, certain volatile substances, such as esters of formic, acetic, caproic, and caprylic acids are formed, and these affect both the taste and the odor of the fruit.

By the time the climacteric is reached, the fruit is ripe. Beyond that stage, senescence continues with other marked changes in the metabolism of the fruit. For one thing, there is an increase in the rate of anaerobic respiration, with the subsequent production of ethanol and acetaldehyde, which affect the taste of the fruit. Also, softening continues until the fruit literally falls apart. Have you tried to pick up an overripe fruit, such as an overripe tomato, and had it squash in your hand?

ABSCISSION

Senescence is also often accompanied by the abscission of the senescent structure. As a result, the fruit or leaf falls from the plant when it becomes senescent. However, abscission precedes death and the plant part is not dead when it abscises.

In the temperate regions, it is a common observation that each Autumn, the leaves of deciduous trees turn a variety of colors and fall to the ground. In fact, this periodic falling of the leaves is probably the origin of the term *Fall*, which is synonymous with Autumn. But, leaves are not the only plant parts that fall off. Petals fall from flowers, after the flowers have been fertilized, and as fruit matures, it too falls from the plant. This shedding of plant organs is known as *abscission*, and when the plant part falls, it is said to have abscised.

Value

Although the abscission of plant parts has numerous agricultural applications, it is also of value to the plant under natural conditions. When fruits abscise, they and their seeds become dispersed by the wind, by water, by animals, or through other means so that the new offspring may have a better chance of survival. When leaves abscise, water loss is greatly reduced so that plants may better survive drought and Winter conditions. When fruits abscise before maturing, the crop is naturally thinned so that the fruits which remain on the tree will contain more nutrients to enable the offspring to have a better chance for survival. In the tropical rainforest, bark abscission frees a tree of epiphytes. Finally, the abscission of leaves, bark, etc. acts as a method of excretion whereby the plant loses many chemical substances that may have accumulated and which may become toxic if allowed to accumulate indefinitely. It is also conceivable that their abscission makes available to the actively-growing parts of the plant, some of the mineral nutrients that have become bound in older tissues. In fact, in the tropical rainforest, new vegetation could not survive if abscission and leaching of leaves did not occur.

Parts Abscising

Leaves, leaflets, stems, bud scales, bark, flower parts, cotyledons, stipules, inflorescences, flowers, and fruits have been observed to abscise. Most deciduous trees lose their leaves in the Fall, and evergreens too lose their leaves periodically. However, in the case of evergreens, all of the leaves do not abscise from a plant at one time, so that the shoot is never devoid of all of its leaves. In some trees, such as cherry, linden, elm, and birch, the leafy stem tips abscise at the end of the Spring growing season, so that small leafy branches can be observed to fall from the tree in early Summer. In such cases, growth is resumed the following year, by the lateral bud immediately below the abscission zone becoming dominant and taking over the functions of the terminal bud. In certain cases, even the large branches of woody plants abscise.

The leaves of all plants do not abscise, and especially is this so with many of the herbaceous and annual species. Therefore, abscission is not an essential phase of plant development. With the exceptions of a few herbaceous plants – coleus, fuchsia, begonia, and others – annual plants die with their leaves remaining intact on the stem, and these leaves remain intact until they decay or are broken off. Certain of the woody species also exhibit this lack of abscission under certain conditions.

The leaves of some conifers do not abscise individually, but rather fall off in bundles by the abscission of the dwarf branches. Coast redwood and bald cypress trees shed branches with leaves rather than individual leaves.

Environmental Influence

Leaf abscission is brought about by certain environmental factors. Many plants lose their leaves when drought develops. The extreme examples of this can be seen in some desert plants, such as *Fouquieria* sp. and *Euphorbia splendens* which form leaves during the brief rainy seasons in the desert, and lose them when drought begins. This results in their spending much of the year without leaves. Many mesophytes also lose some of their leaves during periods of drought.

Low, but not freezing, temperatures also stimulate leaf abscission. Indeed, this is considered to be one of the factors associated with leaf fall in the Autumn. A reduced light intensity can also result in leaf abscission and this may be one of the chief reasons why the inner regions of the canopy of trees are devoid of leaves.

Another factor which results in abscission, is the change in the day length or photoperiod. Short days and cool temperatures are responsible for Autumn leaf coloration and leaf fall.

A healthy plant part does not abscise, so anything which results in injury to the plant, such as spraying with toxic chemicals or the activity of certain pathogens, may cause abscission.

Mechanism of Abscission

To understand the mechanism by which abscission is brought about, it is desirable to know something about the morphological alterations of the plant that are associated with abscission. Early in the development of the plant, the part that will ultimately abscise forms a so-called *abscission zone* near the base. This zone can be detected since it is often narrower than the adjacent areas, and the cells are smaller. It is the weakest area of the organ with the cells having cell walls that are weak, thin, and which lack both lignin and suberin. It is in this zone that abscission will eventually occur.

As the plant part becomes senescent, certain changes take place in a very narrow segment of the abscission zone. This may occur in a matter of days or of weeks before abscission. This segment is a few to several cell layers in thickness, and before abscission occurs, the middle lamella and cell walls in this layer swell, become gelatinous, and finally break down and dissolve or at least the cells separate with the dissolution of the middle lamellas. These middle lamellas are destroyed as the insoluble pectates and protopectins are changed into soluble pectin, due to an increase in the activity of several enzymes, including pectin methylesterase. This narrow segment of the abscission zone, where the cells are damaged, is called the separation layer. The formation of the separation layer leaves the plant part attached only by thin vascular elements, so mechanical pressure, such as raindrops hitting on the leaf or wind blowing against it, will break off the part, and abscission is said to have occurred.

The real problem for the plant physiologist is to learn what happens between the initiation of the stimulus which causes abscission, and the formation of the separation layer. What alterations of metabolic regulation occur within the plant which cause the cells to separate and to be destroyed to form the separation layer? Most of the studies carried out to find an answer to this problem have been done on leaves.

As mentioned earlier, as the leaf becomes senescent, there are changes which occur in the hormone composition of the leaf. Auxins production by the leaf declines greatly and ABA synthesis increases. Thus one might suspect that such changes could bring about abscission. Indeed, if auxin is applied to the blade of a senescent leaf, abscission does not occur, as shown in Fig. 19.2. Or, if the leaf blade is cut off, and auxin is applied to the cut end of the petiole, the petiole does not abscise. However, if auxin is placed on the petiole between the stem and the abscission zone, abscission will occur and will do so more rapidly than if such treatment was not given, as shown in Fig. 19.2. Thus it appears that it is not the absolute amount of auxin produced by the leaf which controls the formation of the separation layer, but rather the auxin gradient across the abscission zone.

However, in addition to these known effects of auxin, ABA has been shown to increase in concentration in senescent tissues and cells. Also, abscission can be brought about by increasing the amount of ABA in the

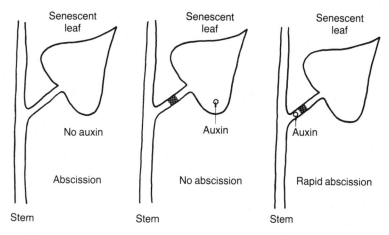

Fig. 19.2   Illustration of the relationship between location of exogenous auxin application and abscission. Hatched area represents location of abscission zone on the petiole.

leaf. This would indicate that it is the presence of ABA which is responsible for abscission. On the other hand, such observations are also compatable with the theory that it is the concentration ratio of the auxin to ABA which controls abscission. When the ratio is high, abscission is prevented; when the ratio is low, the separation layer forms, and with it abscission. Again, this is an example of correlative development which seems to be controlled by the relative concentrations of growth stimulants and growth inhibitors, a phenomenon which is so common in the control of plant development.

The initiation of abscission lies in how senescence is able to alter the hormone balance of the plant part to inhibit or bring about the formation of the separation layer.

After the plant part abscises, cork forms below the separation layer due to an increase or initiation of cell division activity in that area. This cork prevents the entrance of pathogens into the plant, and reduces the water loss from the broken ends of the vascular tissues.

DELAY OF SENESCENCE

It might be well to indicate some ways in which senescence can be delayed. These will be elaborated on in the last chapter, but include the removal of adverse conditions for development, removal of conditions needed for rapid development, and the prevention of flower and fruit formation. An example of the first way to extend longevity or reduce senescence is shown by the many species of plant which are annual plants in the temperate region, but become perennial plants in the tropics. *Ricinis* sp. and *Melilotus officinalis* are examples of these.

Removal of conditions needed for rapid development is illustrated by storing a seed in a dry environment where its development is delayed until the water content of the environment is increased greatly.

Perhaps the most famous example of the effects of flowering is shown by *Agave americana* plants of the desert. These plants normally flower and die after 8–10 years, but may be kept alive for 100 years if flowering is delayed by some means such as an unfavorable environment. Many annual plants become perennials if kept under conditions of day length which will prevent their flowering.

It should not be construed that the entire multicellular plant becomes senescent all over at a given time. It is common knowledge that a deciduous tree or shrub does not become senescent when it loses its leaves in the Fall, and the flower petals of many plants become senescent while other parts of the same plant may be just undergoing early development, and the lower leaves of annual plants may be senescent and abscise while leaves further up the stem are just expanding. Senescence is usually occurring at different rates in different parts of the same plant.

INTERNAL CAUSES OF SENESCENCE

This brings up the next question of interest: just what are the internal causes of senescence? Why do plants not remain in the mature stage of development indefinitely? Why do they become senescent? A unicellular plant does become senescent, but this senescence may be completely reversed by cell division. What is there about the process of cell division that is so able to overcome this senescence condition and return the cell from old age to youth? Is this due to a dilution of toxic materials or senescence hormones, or to a reduction of adverse chromosomal aberrations which have accumulated during the growth of the cell? Any of these answers may be correct. Perhaps there are a number of internal causes of senescence even at the cell level.

Correlative Phenomenon

In the multicellular plant, senescence is often a correlation phenomenon. This is evident in the case where a senescent, yellow, tobacco leaf can have its senescence reversed, by removing all of the leaves from the plant except the senescent leaf. This then would indicate that senescence, in this case, is brought on by either the movement of a senescence influence from the other leaves to that which will become senescent, or the monopolization by the other leaves of materials needed to prevent senescence in the specific leaf. The monopolization of nutrients by younger plant parts, leaving the older parts deficient, has been discussed often in this book. Supplying a senescent leaf with cytokinin will also reverse senescence, at least partially, but this does not imply that senescence is due to a lack of cytokinins. Applying cytokinins to a leaf causes many organic chemical substances to be attracted and to increase in concentration in the leaf. This includes such materials as proteins, amino acids, phosphorus, sulfur, sugars, and auxins.

Nutrient Deficiency

That a deficiency of nutrients will cause senescence is evident from the fact that fruit formation often leads to senescence and in some cases, this is known to be due to the fruit monopolizing the supply of available nutrients so that the meristems and mature vegetative cells suffer from their deficiency. In such cases, any treatment which will increase the availability of nutrients will increase longevity by delaying senescence.

Hormones

Hormones may also be involved in senescence in multicellular plants. They may do this either by the appearance of hormones which cause senescence or by the disappearance of hormones needed for normal growth and for the maintenance of maturity. The evidence for senescence hormones is shown by spinach plants, which are of two types. Some plants are staminate, producing only staminate flowers, whereas some are pistillate, producing only pistillate flowers. If the staminate flowers remain on the plants, these plants become senescent. If the flowers are removed shortly after they are formed, the plants continue their vegetative growth and do not become senescent. One might suspect that this senescence is due to the flowers monopolizing nutrients, but this cannot be so, since the flowers are very small so the amount of nutrients they monopolized would be insignificant. Therefore, these flowers must be producing a hormone which causes the plant to become senescent. After all, after the flowers have served their purpose, there is no need for the plant to continue to survive. This introduces the interesting question, though, as to why do the cells which produce the senescent hormone not become senescent themselves? How are they protected from the action of such hormones? Or, is it a matter of the extent of accumulation, with the cells producing the hormones, having the hormones present in much lower concentrations than the cells in which senescence occurs? If so, what causes the hormone to accumulate in some cells and not in others? These are all good questions for which answers seem not to be available.

In some cases, senescence is due to a decline in hormone production. For example, if pea plants have their flowers removed as soon as they are formed, the plant still becomes senescent. However, if the apical bud is removed and a young bud is grafted in its place, the plant does not become senescent. Apparently, with an increase in age, a hormone that delays senescence is no longer produced in adequate amounts. What this hormone is, is not known. However, the application of kinetin and GA will delay senescence, so these may be involved. Or, since kinetin is known to stimulate protein synthesis, perhaps senescence is due to the turning-off of RNA or protein production. It will be valuable to learn what causes these to turn off, and to learn how this turning-off can be prevented. Such knowledge may allow us to extend the life of a plant indefinitely.

DEATH

Death is the termination of senescence and therefore the termination of plant development. This intriguing subject is still a mystery. Indeed, it is a natural developmental phenomenon and all cells must eventually divide or die. But what is death and how can it be defined?

It is not too easy to distinguish between a living cell and one that has just died. In both cases, their chemical composition are the same, to the best of our knowledge. Even after death, some enzymatic activities continue, particularly hydrolytic activities. This is especially evident when one considers the function of homogenates outside the living cell. For instance, when a homogenate of mitochondria is prepared, the cell is killed, but the mitochondria still continue to function for sometime thereafter.

Intracellular Movements

Perhaps one difference between dead and living cells, is their intracellular movements. Diffusion and Brownian movement occur in both, but Brownian movement is very much more extensive in dead cells. On the other hand, cytoplasmic streaming does not occur in dead cells. Of course, it is not always evident in living cells either, so if cytoplasmic streaming cannot be observed in a cell, this does not mean that the cell is dead. But, if such streaming is observed, the cell cannot be dead.

Membrane Permeability

Another difference between living and dead cells is in the magnitude of the permeability of their membranes. Living cell membranes are differentially permeable and offer resistance to the passage of large molecules, such as dyes, through them. On the other hand, dead cells have membranes that are very permeable to such dyes. Therefore, the rate at which dyes enter the cell from the environment is a good test of whether the cell is living or dead. Along with this increase of permeability, the electrical resistance of the cell membrane is also destroyed when the cell dies.

Respiration

It should be noted, that when the cell dies, its respiration ceases, as does photosynthesis and most other synthetic reactions. This seems to contradict the evidence that homogenates of mitochondria remain functional for several hours after being removed from the cell, and isolated chloroplasts can undergo photosynthesis if the right conditions prevail.

Ultraviolet Absorption

Living and dead cells also differ in the extent to which they absorb ultraviolet radiation. This is useful for determining the extent of injury to a cell suspension, since the mortality rate can be found easily by observing the change in ultraviolet absorption by the population after treatment. However, some caution must be used here, since absorption of ultraviolet radiation of the shorter wavelengths injures, and may even kill, plant cells.

Chromosomal Aberrations

Since it has been reported that senescent cells have higher mutation rates and more chromosomal aberrations, perhaps these could account, in part, for irreversible alterations in metabolic regulation which could lead to the death of the cell.

Irreversible Changes

Another difference between living and dead cells is that what changes do occur in the dead cell, are irreversible. Cytoplasmic streaming cannot be restarted, nor can membrane permeability be decreased. In many respects, death seems to involve a cessation of the availability of metabolic energy.

Chapter Twenty

The Relevance of Plant Physiology

In this day and age a subject is of little interest to students unless its relevance to contemporary problems facing mankind can be shown. Perhaps this is as it should be. In spite of the fact that man is not the only biological entity in this world, he is the most important one, as far as man is concerned. No doubt, plants were not put on this earth for the benefit of man, but this does not prevent him from using them for that purpose. To reveal the relevance of plant physiology is the aim of this, the final chapter in this book. Perhaps it would have been better to have this chapter first, but to do so would require that too many statements would have to be made based upon facts unknown to the student.

PHYSIOLOGY OF CROP PRODUCTION

A botanist is a person who uses scientific methods to study plants. An agriculturist is one who uses such methods to study the production of plants and animals of value to mankind. Is plant physiology a botanical or an agricultural subject? Certainly it is a botanical subject, since it is concerned with the study of plants. To many, it is also an agricultural subject, since the results of the research of plant physiologists can often be used to improve crop production. In fact, some plant physiologists study plants with that aim in mind. In any respect, the findings of plant physiologists often lead to a better knowledge of the plant and to better ways to produce the plant for man's benefit.

What a plant is and what it will become depends upon the genetic make-up of the plant and upon its environment. Both act only as they alter the physiology of the plant, so that what a plant develops into depends upon the physiological activities which occur within it. Plant production can be improved either by producing better plants through selective breeding, by improving the environment, or by using chemicals to alter the physiology of the plant. Many attempts are being made by plant breeders to alter the genetic makeup of crop plants so that a better potential can be achieved, and new varieties are regularly appearing to attest to their success. But, even when the best genetic makeup of a plant exists, optimum development of this plant still requires an optimum environment, and the natural environment is not the optimum environment. Therefore, either the type of plant must be selected which grows best in the prevailing environment, or the environment must be altered to give the best growth for the plant which grows there. In the past, the former was practiced. This is the most desirable, in many respects, of the two alternatives, because to alter the environment requires not only an expenditure of energy, which may not be feasible due to limited energy reserves, but also requires facilities and economic aspects which may not be practical. Therefore, the culture of plants always does, and perhaps always will, involve producing the best plant possible with the facilities which are available. However, with an understanding of plant physiology, it is usually possible to better utilize what is available for environmental modification to improve plant development.

THE PLANT'S ENVIRONMENT

Once the seed has been planted, nothing can be done to alter the potential of the plant. Perhaps the time will come when such alterations will be possible, but that time is a long way off. However, the plant's chemical and physical environment can be altered at any time. In the past, such alterations have been limited to times when the plant has shown symptoms of environmental deficiencies. In the future, more attempts to disclose such deficiencies before they alter yield, or even before they occur, will be made. This can be done, as is routinely being done in the sugarcane fields of Hawaii, where periodic routine analyses are made of sheath moisture content, leaf nitrogen content, minimum and maximum temperatures, rate of leaf emergence, green weight of sheaths, age of the plant, total sugars, soil moisture, and light intensity to discover approaching deficiencies in plant development. With such analyses, impending deficiencies can be detected and prevented before they occur. Needless to say, the physiology of the plant must be well known before such practices can be meaningful. As such knowledge expands, so will such practices expand to other crops, and such analyses may become an integral part of farming in the future.

The environment to which a plant is exposed is made up of many factors, interacting in many ways, as illustrated in Fig. 20.1. In fact, it is so dynamic and so complex that the statement has been made that the plant's environment can never be understood. Perhaps the reasoning behind this statement is that even if all of the environmental factors affecting a plant were known, they would change before they could be defined. Be that as it may, not all environmental factors affect the plant with equal intensity. In fact, many of these factors have such small effects that they can be considered as insignificant. Therefore, a study of the plant's environment can be reduced to a study of those factors which significantly alter desirable characteristics of the plant, and these are fewer in number and quite understandable.

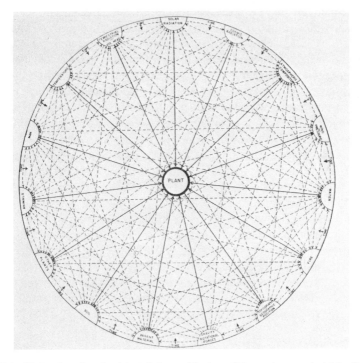

Fig. 20.1   Illustration showing interrelations of factors of the environment which affect the plant. (From Billings, H. C. *Quart. Rev. Biol.* 1952. 27(3).)

What are these desirable plant characteristics? The answer varies with the purpose for which the plant is produced. In some cases, the most valuable characteristic is the rate of growth. In other cases, it is the amount of fruit produced, or the quality of the floral structures, or the size or shape of the plant, or the quality and quantity of the chemicals produced by the plant.

Certainly, the climate controls plant development, and consequently desirable plant characteristics. In fact, it is often observed that due to climatic variations from year to year, year to year variations in field studies are greater than those variations due to controlled treatments. Climatic differences on two sides of a hill can have very differential effects on plant development, both with cultivated crops, as well as with native vegetation, as shown in Chapter 4.

PRIMARY FACTORS

In the field, the plant's development is controlled primarily by light, temperature, water, mineral nutrients, gases, and plant diseases. Therefore, these can be considered as the *primary factors of the environment* which control plant development, although artificial treatments are becoming more and more important and will, perhaps, someday have to be included among these primary factors. In the past, much of agricultural practice has been concerned with the control of these primary factors to increase plant yield. Now, such control appears to be approaching its upper limit, with only minor increases in yield being reported when these environmental conditions are improved. In some cases, below-optimum yields are due to our inability to impress greater control on the environment, and in other cases, they may be due to a lack of our knowledge of the physiology of the plant.

Crop yield is greatly determined by the relative rates of photosynthesis and respiration, the latter including both dark respiration and photorespiration. This is so because yield is reflected largely in the dry matter produced, and the dry matter accumulation is the sum of the dry matter produced through photosynthesis minus the dry matter destroyed through respiration. Therefore, the higher the photosynthesis to respiration ratio, the greater the crop yield.

Leaf Area Index

Productivity is often correlated with two determinations, namely, chlorophyll content and *leaf area index*. Chlorophyll content of waters has been found to be a good measure of the amount of yield produced by plankton in oceans, lakes, etc. However, this criterion does not hold for terrestrial plants, since the chlorophyll content per leaf is in excess of that needed for photosynthesis. Therefore, the term leaf area index is used. Leaf area index (LAI) is determined by measuring the land area and the area of the total leaf surface of the plants growing on this land. The latter value divided by the former gives the LAI, or

$$LAI = \frac{total\ leaf\ area}{total\ land\ area}$$

It has been found that maximum yield by terrestrial plants is often associated with a LAI of about three or greater, which means that total leaf area should be about three times that of the total land area, or if a plant is growing in a pot with a total soil surface of 1 sq ft, the total leaf area of the plant growing in this pot should be about 3 sq ft. This indicates that some of the shaded leaves are also important in photosynthesis, which might be expected as long as the incident light intensity exceeded the compensation point.

Engineering Plants

The term, "engineering plants for maximum production," is often heard today. This refers to developing the type of plant that will give the maximum yield per hectare. Surprising as it may seem, this does not necessarily mean that such a plant will give maximum yield per plant. In fact, such is not often the case. For example, in some of the graminaceous crops, it has been found much more desirable to have a plant with erect leaves than one with horizontal leaves. Such a plant captures less sunlight and therefore has a lower rate of photosynthesis and therefore gives a lower yield per plant but so many more of these plants can be grown on each hectare of land that the yield per hectare is greater.

It must also be remembered, that leaves in the canopy, leaves that are in the shade, have a very high photosynthetic efficiency, so they may contribute a great deal to plant yield even though they are shaded. Also, the relative amount of light to which these shade leaves are exposed will be greater on cloudy days than on bright days, due to the reflection of light from the clouds.

Since yield is so closely related to photosynthesis, efforts are being made to reduce the amount of sunlight that falls on bare ground, or to increase the LAI, since such energy is wasted. Numerous studies have been made to determine the pattern of planting crops to utilize more of this sunlight. For instance, planting in north-south rows has been shown to give best yield since more sunlight is captured and used in photosynthesis. Also, the spacing of the rows and the number of plants per row are important criteria, but these values differ greatly with the type of plant grown and the local soil and climate conditions. Yes, yield is greatly controlled by the extent of photosynthesis, and maximum yield can only be obtained by conditions which give maximum photosynthesis per unit area of land.

A perplexing problem associated with land use for maximum yield is associated with differences in plant size with age. Seedlings do not occupy as much area as these plants will at maturity, so much sunlight is wasted during the early stages of plant development. More use should be made of seedlings as food, instead of waiting until they reach maturity.

Efficiency of Photosynthesis

Perhaps the greatest contribution a plant physiologist could make to reducing the world's food shortage would be to find some way to increase the efficiency of photosynthesis. Photosynthetic efficiency is only a few percent, based on the amount of energy available from sunlight. If this efficiency could be only doubled, this should mean that twice as much food could be produced as is produced now and should give the potential to nearly double the world's food supply. This it could do by perhaps allowing twice the number of harvests per growing season and/or by increasing the amount of yield per harvest. No doubt this possibility will receive much attention in the future.

Metabolic Sink

As was pointed out in Chapter 6, one way in which the rate of photosynthesis can be controlled is by the extent and activity of metabolic sinks. A *metabolic sink* is any cell of the plant that is using the products of photosynthesis, namely, sucrose, either for storage, respiration, conversion to storage carbohydrates, or for other syntheses. The greater the capacity of the sinks in the plant, the more rapid the rate of photosynthesis in the leaves. Although the mechanism by which sink capacity can control the rate of photosynthesis is unknown, as more is learned of this mechanism, the greater will be the possibility to control sink activity to increase the rate of photosynthesis. This aspect of plant physiology should be very important in the future, and it is entirely within the realm of possibility that greater increases in photosynthetic rates will be obtained by manipulating sucrose distribution within the plant than by changing the plant's environment.

Temperature

As shown in Chapters 6 and 7, the rate of photosynthesis in each plant is also controlled by temperature, carbon dioxide concentration, leaf area, leaf angle, position of the leaf, amount of light reflected by the leaf, amount of light transmitted through the leaf, brightness of the sun and sky, position of the sun with respect to the horizon, and the efficiency of the photosynthetic mechanism. To improve any one of these factors could mean an increase in yield, and it is the aim of cultural methods to bring this about.

Water

As indicated in previous chapters, the best growth of the plant occurs when the water content of the plant is high, at least when it is high enough so that plasmolysis is avoided. On the other hand, too much water is detrimental either because of poor root aeration, or because of a high evapotranspiration loss which represents wasted water, a waste most societies can ill afford today. Therefore, optimum plant development requires that the plant be supplied with a sufficient amount of water and supplied with this at the time when the water is needed.

Since water deficits can, and often do, occur in plants during the growing season, it is necessary for maximum yield to get the water into the plant when it is needed and in the amounts needed. Even in humid areas with abundant rainfall, there are periods when soil moisture is deficient for maximum yield, so supplemental irrigation must become a more widespread practice. Only by the use of irrigation can the water supply be kept optimum. Also, the plant varies in the extent of yield reduction with the stage of its development at the time water becomes deficient. Water deficiency has little effect just prior to anthesis, but following anthesis, the yield of grain or fruits is greatly reduced by even moderate water deficits. However, too little is known about the stages of development that are most sensitive to water deficits.

Perhaps the first effect of water deficits on plants is to reduce cell enlargement, since such deficits reduce turgor pressure, which is needed for cell enlargement, as shown in Chapter 17. Such reduction results in wilted plants and, if imposed too long, will result in stunted growth and reduced leaf surface. This reduced leaf surface, plus the closing of stomates associated with moderate to severe water deficits, reduces the rate of photosynthesis and therefore the dry weight of the plant. Continued water deficits will subsequently alter most physiological activities of the plant with related consequences. Thus, a knowledge of the physiology of the plant illustrates more clearly the need for the maintenance of adequate water in the crop plant.

Transpiration rate usually appears to be greater than is necessary for optimum plant development. This then represents wasted water. Research is needed to determine the optimum transpiration rate and how to obtain and maintain it. The use of antitranspirants may be the answer, although to date, such studies have not been very fruitful. What is really needed is an antitranspirant chemical which can be sprayed on the foliage to reduce water loss, but which will not reduce carbon dioxide uptake and will not otherwise damage the leaf cells.

It cannot be too strongly emphasized that as important as any one factor may be in environmental control of plant development, all primary factors are greatly interrelated. For instance, mineral deficiencies may appear when the plant is exposed to high light intensity or has a good supply of available water, but will disappear when the plant is transferred to the shade or its supply of available water is decreased. Therefore, in the future, better control of plant development will be brought about after establishing a group of indices by which this development can be predicted—indices which take into account all of the primary factors involved—and routinely cataloging the physiology of the plant during its development.

Mineral Nutrients

Studies of the physiology of the plant with respect to mineral nutrients have greatly increased crop yields in the past. Perhaps, most of the large, crop improvements caused by elucidation of the roles of mineral nutrients are past, and future improvements will be small, and concerned primarily with improving fertilizer techniques. Diagnostic tests, such as chemical analyses of tissues and of soil solutions, will continue to be important and their use will expand. More work needs to be done to study the physiology of the toxicity caused by minerals, since these will become more widespread in the future. Also, it will be necessary to learn what the optimum concentration of each of the various mineral nutrients is in each part of the plant at each stage of development so that the supply can be altered periodically to allow optimum plant development.

Studies of the loss of mineral nutrients from plants are yielding good information as to the cause and extent of such losses, losses which must be compensated for in the future for optimum plant development.

Competition Among Plants

Competition among plants is a reality. This competition exists both among plants of different taxa and among those of the same taxon. This competition may be in the form of shading, competition for available soil water or mineral nutrients, or in the form of chemicals which are given off by one plant and which reduces the development of another. For instance, many species of plants are known to give off volatile chemicals which move through the atmosphere to neighboring plants and reduce their growth or other aspects of their development. Many species also produce water-soluble chemicals in their leaves or fruits which are leached out and get into the soil where they may inhibit the germination of seeds, thereby reducing competition. Yes, many kinds of competition are evident among plants, and all must be understood and considered when actions are taken to increase plant development.

PLANT PROPAGATION

An intriguing and important area of plant physiology is that associated with the physiology of propagation. Propagation refers to the increase in the number of plants, and can be either sexual or asexual. Sexual propagation is that associated with the production and germination of seeds and results in offspring which vary in genotype. Asexual (vegetative) propagation does not involve meiosis, but only mitosis, and each offspring is essentially like its parent. Such asexual propagation is carried out by runners, suckers, layering, the separation of bulbs or corms, by division of rhizomes, tubers, etc., by cuttings which form roots or buds, and therefore new plants, by grafting scions and rootstocks together to form a new plant, or by budding, or ultimately by isolating a single cell and inducing it to form a new plant. These methods were discussed in past chapters but each does involve a change in metabolic regulation which ultimately results in the formation of a new plant. How these changes are brought about is a challenge to physiologists.

PLANT DISEASES

Plant diseases take a tremendous toll each year. They do so by altering the physiology of the plant in such a way that its development is either delayed or the plant is destroyed. The mechanisms by which such diseases cause these physiological disturbances are varied but worth considering.

Diseases are either pathogenic or non-pathogenic. The non-pathogenic diseases are caused by adverse environmental conditions, and will not be considered here. The pathogenic diseases are caused by other living organisms called *pathogens*.

The presence of a microorganism on a plant is not evidence that the plant is diseased. Numerous such organisms live in contact with the plant under the best of conditions, but cause no injury to it. Perhaps only less than half of one percent of these microorganisms are actually pathogens. However, this small percentage is very important economically.

Entrance Into the Plant

To injure the plant, the pathogen usually must enter the plant. Such entrance is the exception rather than the rule, as plants are well protected against the invasion of pathogens. Such protection is afforded by the epidermal cells and the cuticle and wax layers covering them. Also, some plant chemicals, such as the alkaloids, have an undesirable or even a toxic effect on pathogens, which may prevent injury. However, some pathogens do get into the plant and produce characteristic disease symptoms.

Host Resistance Within the Cell

Once the pathogen gets into the cell, it faces a different environment. If this environment is not conducive to its further development, it will go no further and the disease will disappear or at least will be restricted in its distribution. If the pathogen is not a virus, it must find an appropriate nutrient medium within the cell—a medium that will furnish all of its needs. On the other hand, it must not find substances which will be injurious to it. Often, when a pathogen invades a host, the host produces a *phytoalexin*, a chemical which is toxic to the pathogen. The production of such a chemical can render the host immune to further invasions of similar pathogens, a characteristic which is similar to immunity to diseases in man. These phytoalexins are often phenolic compounds, but some are protein. They are produced only when the host is growing or has recently reached maturity. Their production in senescent hosts or tissues is very much reduced.

If the pathogen finds the environment suitable within the host, it multiplies, by cell division, and moves from one area to another. Sometimes such transport is limited to within a given tissue or organ and at other times, it is very extensive moving through the phloem at a rate similar to that of the movement of carbohydrates and other organic compounds.

Symptoms of Disease

The symptoms caused by the pathogen within the host are varied. Sometimes the pathogen stimulates the production site of ethylene, which results in epinasty, a common disease symptom. Sometimes it causes wilting, by blocking the vessel elements of the xylem. Sometimes it causes abscission, by causing the leaf to become senescent. This abscission is, in some respects, a protective measure, as far as the host is concerned. Why?

Most plant pathogens are fungi. These usually cause an increase in the permeability of the host plant's cells, and an associated increase in the transpiration rate.

In addition to the above symptoms, certain changes take place which alter biochemical activities of the cell. In most infections, the rate of respiration originally increases, but may later decrease. Also, changes in the respiratory pathway occur with glycolysis becoming less evident. The rate of photosynthesis declines and certain metabolites accumulate in the diseased areas. In general, most diseases eventually alter all physiological activities, but some are altered sooner than others or to a greater extent.

To complicate the matter, invasion of the host by a pathogen is often followed, shortly thereafter, by invasion by other organisms which may or may not alter the host's physiological activities.

The question might be asked as to the mechanism by which the pathogen alters the metabolism of the host to cause these disease symptoms. Virus diseases disturb the host's nucleic acid metabolism with the amino acids and nucleic acids of the host being used for the production of more viruses rather than for the host's normal metabolism. Other pathogens cause diseases either by producing phytotoxins (toxic chemicals), utilizing the host's metabolites or cell components, or by crowding out certain structural phenomena of the host. Some also synthesize and secrete enzymes which digest plant cell wall polysaccharides or other cell structures. Thus, they cause senescence within the cell and often the symptoms of senescence including alterations in the hormone balance of the plant.

ABSCISSION

Abscission may be desirable or non-desirable. It is not desirable in the Fall of the year if one has to rake up the leaves from the lawn. It is usually not desirable when flowers fall from fruit trees without forming fruit. It is not desirable when the leaves abscise from poinsettia plants before the Christmas holidays come around. It is not desirable when fruit falls to the ground before it can be harvested. In these cases, a study of abscission has revealed how plants can be treated to delay the abscission process. What treatments would be recommended to delay abscission? Certainly spraying the structure with auxins would help and this practice is used by agriculturists.

Value of Abscission

Abscission may be desirable for the thinning of fruit trees to remove some of the flowers before they set fruit to avoid too many fruits being produced on a given tree and therefore a decrease in the size and quality of the fruit. Abscission is desirable by the armed forces to clear a field of fire in the jungles or to uncover enemy forces hidden beneath foliage. It is also desirable for the cotton grower since, if abscission does not occur prior to harvest, the foliage will mix with, and reduce the quality of, the cotton fibers. Abscission of the lower branches is also important on lumber trees to reduce the size of knots in the finished product. In some of the applications listed above, it is desirable and feasible to stimulate abscission. This is done by the application of chemicals to the plants — chemicals which greatly stimulate abscission, either directly by causing a premature formation of the separation layer, or indirectly by injuring the leaves, etc. so that such injury will cause the part to abscise normally.

Sometimes abscission will not occur at the time desired, so it is necessary to thin the plant manually. For instance, many plants are pruned to either remove unsightly branches or to remove those lower branches that are a liability to the plant, rather than an asset. The lower branches, either due to senescence or shading of the leaves, have a higher respiration rate than photosynthetic rate, so they use more sugars than they produce and therefore are not of value to the plant. Pruning such structures will often result in better development of the remaining branches. Sometimes manual pruning is necessary to remove diseased branches.

HERBICIDES

The statement has been made that the best weed killer is a good hoe, a pair of gloves, and someone to use both. This does have the potential for the most satisfactory eradication, but most people do not wish to spend the time or energy necessary for such a task. Therefore, extensive searches have been made for chemicals that will do the job more easily. Such chemicals have been named *herbicides*.

Ideal Herbicide

An ideal herbicide will have all or most of the following characteristics: First, it will be effective in small amounts and economical to apply. Second, it will be able to be applied as a spray directly to the plant, since most herbicides are destroyed in the soil or leached out of the soil. Third, it will not remain as a residue for long periods of time to prevent desirable plants from being grown in the area shortly after its use. Fourth, it will be selective in that it will kill the weeds but will do no harm to the desirable species. Fifth, it will be readily absorbed and translocated to the roots. Many chemicals have been screened for their herbicidal abilities and none fulfill all of the above requirements, but some fill most of them. By far the most often used is 2,4-*D* (2,4-dichlorophenoxyacetic acid) which is an auxin but not a hormone. It is much more injurious to broad leaf plants than to grass plants so it is often used to remove broadleaf plants from lawns and pastures.

Although the herbicide 2,4-*D* does not occur naturally, it has been manufactured in many forms. It can be obtained as an ester, an acid, or as some kind of salt. The esters and salts are often more readily absorbed than the acid.

To be of any value as an herbicide, the 2,4-*D* must get into the plant, usually through the leaf. This it does quite readily but, the rate of entrance is influenced by many factors, such as the form of the 2,4-*D*, as mentioned above, the acidity of the solution, the presence or absence of detergents in the spray, the relative humidity of the air, the prevailing temperature at the time of application, and the length of time the 2,4-*D* remains on the plant before it is washed off by rain, etc. This spray is best applied when the temperature is about 20°C and when the sun is bright. All of these characteristics associated with absorption can be related to studies of membrane permeability and uptake through the cuticle and stomates as studied in subsequent chapters. In fact, by the use of knowledge gained earlier, it should have been possible to guess what these requirements for absorption would be.

In the case of perennial plants, uptake through the leaves is not sufficient. The herbicide must then be translocated to the roots, to kill the roots before the leaf and phloem cells are killed. Otherwise, root cells will differentiate and form new buds which will develop into new plants. This translocation occurs through the phloem, and its direction and rate of translocation are similar to those for other organic compounds, such as sugars, as previously studied.

Plant's Sensitivity

One reason the plant differs in its sensitivity to herbicides with stage of development is related to the rate and direction of herbicide movement through the phloem. As seen in Chapter 15, translocation is proceeded by, or accompanies, photosynthesis, or hydrolysis and movement of reserve foods. It occurs out of old, senescent leaves, but not out of young leaves, flowers, or fruits that are importing foods. It occurs from the lower leaves to the roots and from the upper leaves to growing shoots or fruits, and it is slowed down by high or low temperatures. This is true of herbicide translocation also, with the rate of translocation being slower in resistant species or in the same individual at the stages of its development when it is most sensitive to injury by the herbicide.

Within the plant, the herbicide may still not be lethal or even injurious. Many things can happen to neutralize it. Some of the herbicide molecules can become adsorbed on the cytoplasmic colloids, some can be accumulated in the vacuoles where it may not be injurious, some will be degraded within the plant before it can be injurious, and this rate of degradation varies from plant to plant, or some may be lost through the roots, etc. Any of these factors associated with adsorption, translocation, or destruction or neutralization of the herbicide can cause differences among individual plants or species in their resistance to injury by the herbicide.

Not only are these variations common, but even within one individual plant, they can change from time to time. For instance, susceptibility can vary from one stage of development to another, as seen in Chapter 19. In this case, as is commonly observed, the herbicide does most damage just prior to maturity and floral induction. Another factor to consider here is the relationship between injury and growth rate. The more optimum the growth rate, the more injurious the chemical. It is not facetious to say that wild morning glory (bindweed) can be killed more easily and more efficiently if these weeds are irrigated prior to spraying so that their growth will be more rapid when the herbicide hits them. Indeed, the herbicide is usually most destructive to meristematic cells. Why? The herbicide 2,4-*D* kills the plant by altering the physiology of the meristematic cells in such a way, or in such ways, as to cause senescence and eventual death. Beyond this, it is not possible to say with certainty what specific physiological activity within the cell is first altered to cause death. Interestingly enough, it seems to injure plants in the same way as high concentrations of IAA.

The discussion above, outlining the physiology of herbicidal action, although it applied to 2,4-*D* could also be applied to some other herbicides, with small differences. Uptake and translocation phenomena are similar, as are environmental and plant characteristics which lead to most injury. However, the specific action of the herbicide at the cell level will vary. For instance, the grass herbicide, dalapon, interferes with the formation of pantothenic acid within the cell, or the triazines, such as simazine and atrizine interfere with photosynthesis, as do the phenylureas and the acylanilides. Other herbicides have other effects, but in all cases, they act to kill the plant by altering its metabolic regulation.

Non-tillage Farming

One interesting possibility for the use of herbicides in the future is related to their use for non-tillage farming. With this practice, no plowing or cultivation of any type is carried out. Herbicides are applied to the field to kill the weeds and then the crop seeds are sown among the dead weeds with as little disturbance of the soil as possible. In some locations, this practice gives better yields than conventional methods.

GROWTH INHIBITORS AND RETARDANTS

Perhaps closely associated with these herbicides, but not included with them is a group of compounds which do not destroy the plant, but do inhibit its growth. These include the growth inhibitor, maleic hydrazide, which is used commercially to reduce the growth of hedges in backyards, so that these hedges will not have to be trimmed as often as they usually are.

Maleic hydrazide is a growth inhibitor. There are also chemicals known that do not inhibit growth completely, but retard it. These may interfere with cell division or with cell enlargement or both. Some act by interfering with GA action. Anyway, they bring about greater frost resistance, a reduction in the amount of vegetative growth, and a surprising stimulation of flowering. Not only are more flowers formed, but they appear earlier if the plants have been treated with a growth retardant. This is true of flowers on fruit trees, on certain other trees with a long growth period each season, and on many herbaceous plants. Most of these retardants are very water-soluble so they can either be applied as an aqueous spray, or poured directly on the soil in which the plant is anchored. Examples of such retardants include *Amo-1618* (1-(2,4-dichlorobenzyl)-1-methyl-2-3-pyridyl, pyrrolidinium chloride), *CCC* ((2-chloroethyl) trimethylammonium chloride), and *phosfon* (2,4-dichlorobenzyltributylammonium chloride). Although these chemicals do not occur naturally in plants, substances which behave similarly are endogenous. These *growth retardants* have other very interesting potential uses in agriculture. For instance, phosphon is often used to make potted flower plants shorter, thus giving more flower and less shoot. They are, however, often very specific, affecting just one or a few species and not others.

ALLERGIES

Plant physiologists are also involved in the problem of *allergies*. About one out of every twenty persons in the United States suffers from some type of allergy each year. Such allergies are usually caused by *allergens* produced in plants, such as those of the ragweed pollen or those found in sagebrush pollen. However, other allergens also may be present in other plant structures which result in allergies being contacted by some people when they eat certain foods derived from plants. Such allergens are often proteins or other polymers, and the study of the nature, occurrence, and synthesis of these may lead to methods of controlling such allergies as hay fever. Indeed, such studies will have to reveal the nature of the allergen and factors which alter its production within the plant. Much needs to be learned in this field.

Yes, a knowledge of plant physiology is important for the solution of many contemporary problems and for the progress and survival of mankind in the future. Now that you have been introduced to the subject, you are better prepared to interpret and perhaps even to become involved in furthering this progress.

It should be pointed out that the principles learned in this book are general and many exceptions do occur. In fact, as indicated in Chapter 1, each plant species is different, and future advances are going to be made by specialists who work with one, or perhaps a few, plant species to learn all they can about the physiology of that species. The need for such specialists is great and will increase in the future.

Index